Determinants
of
Spatial Organization

DETERMINANTS OF SPATIAL ORGANIZATION

The Thirty-Seventh Symposium of
The Society for Developmental Biology

Madison, Wisconsin, June 14-16, 1978

EXECUTIVE COMMITTEE:

1977-1978

IRWIN R. KONIGSBERG, University of Virginia, *President*
IAN M. SUSSEX, Yale University, *Past-President*
NORMAN K. WESSELLS, Stanford University, *President-Designate*
WINIFRED W. DOANE, Arizona State University, *Secretary*
MARIE DI BERARDINO, Medical College of Pennsylvania, *Treasurer*
GERALD M. KIDDER, University of Western Ontario, *Member-at-Large*

1978-1979

NORMAN K. WESSELLS, Stanford University, *President*
IRWIN R. KONIGSBERG, University of Virginia, *Past-President*
URSULA K. ABBOTT, University of California, *President-Designate*
WINIFRED W. DOANE, Arizona State University, *Secretary*
JOHN G. SCANDALIOS, North Carolina State University, *Treasurer*
GERALD M. KIDDER, University of Western Ontario, *Member-at-Large*

Business Manager
CLAUDIA FORET
P. O. Box 43
Eliot, Maine 03903

Determinants of Spatial Organization

Stephen Subtelny, Editor

Department of Biology
Rice University
Houston, Texas

Irwin R. Konigsberg, Co-Editor

Biology Department
University of Virginia
Charlottesville, Virginia

ACADEMIC PRESS   New York   San Francisco   London   1979
A Subsidiary of Harcourt Brace Jovanovich, Publishers

Copyright © 1979, by Academic Press, Inc.
ALL RIGHTS RESERVED.
NO PART OF THIS PUBLICATION MAY BE REPRODUCED OR
TRANSMITTED IN ANY FORM OR BY ANY MEANS, ELECTRONIC
OR MECHANICAL, INCLUDING PHOTOCOPY, RECORDING, OR ANY
INFORMATION STORAGE AND RETRIEVAL SYSTEM, WITHOUT
PERMISSION IN WRITING FROM THE PUBLISHER.

ACADEMIC PRESS, INC.
111 Fifth Avenue, New York, New York 10003

United Kingdom Edition published by
ACADEMIC PRESS, INC. (LONDON) LTD.
24/28 Oval Road, London NW1 7DX

LIBRARY OF CONGRESS CATALOG CARD NUMBER : 78—23508

ISBN 0—12—612983—5

PRINTED IN THE UNITED STATES OF AMERICA
79 80 81 82   9 8 7 6 5 4 3 2 1

Contents

CONTRIBUTORS AND PRESIDING CHAIRPERSONS vii
PREFACE.. ix
ACKNOWLEDGMENTS xxi

I. Cytoplasmic Localization in Early Development

The Ultrastructure and Role of the Polar Lobe in Development
of Molluscs 3
    M. R. Dohmen and N. H. Verdonk

Cytoplasmic Determinants of Tissue Differentiation in the
Ascidian Egg 29
    J. R. Whittaker

The Multiple Roles which Cell Division Can Play in the Localization
of Developmental Potential 53
    Gary Freeman

The Control of the Polar Deposition of a Sulfated Polysaccharide
in *Fucus* Zygotes 77
    Ralph S. Quatrano, Susan H. Brawley, and William E. Hogsett

Analysis of a Morphogenetic Determinant in an Insect Embryo
(*Smittia Spec., Chironomidae, Diptera*) 97
    Klaus Kalthoff

Germ Plasm and Pole Cells of *Drosophila* 127
    A. P. Mahowald, C. D. Allis, K. M. Karrer, E. M. Underwood, and
    G. L. Waring

II. Maternal Effect Mutants of Development

Temperature Sensitive Maternal Effect Mutants of Early Development in *Caenorhabditis elegans* 149
    David Hirsh

A Specific Case of Genetic Control of Early Development: the *o* Maternal Effect Mutation of the Mexican Axolotl 167
  Ann Janice Brothers

Maternal Effect Mutations that Alter the Spatial Coordinates of the Embryo of *Drosophila melanogaster* 185
  Christiane Nüsslein-Volhard

III. Pattern Formation in Developing Systems

An Analysis of Cell-Surface Patterning in *Tetrahymena* 215
  Joseph Frankel

Intercellular Interactions and Pattern Formation in Filamentous Cyanobacteria 247
  C. Peter Wolk

Development of Hydra Lacking Interstitial and Nerve Cells ("Epithelial Hydra") 267
  Richard D. Campbell

Pattern Formation, Growth Control and Cell Interactions in *Drosophila* Imaginal Discs 295
  Peter J. Bryant

Pattern Formation and Compartments in the Tarsus of *Drosophila* 317
  P. A. Lawrence and G. Morata

Subject Index 325

Contributors and Presiding Chairpersons

Numbers in parentheses indicate the pages on which the authors' contributions begin.

Session I *Chairperson:* Irwin R. Konigsberg, Department of Biology, University of Virginia, Charlottesville, Virginia

M. R. Dohmen and N. H. Verdonk, Zoological Laboratory, University of Utrecht, Utrecht, The Netherlands (3)

J. R. Whittaker, The Wistar Institute of Anatomy and Biology, Philadelphia, Pennsylvania (29)

Gary Freeman, Department of Zoology, University of Texas, Austin, Texas (53)

Session II *Chairperson:* Elizabeth Hay, Department of Anatomy, Harvard Medical School, Boston, Massachusetts

Ralph Quatrano, Susan H. Brawley* and William E. Hogsett, Department of Botany and Plant Pathology, Oregon State University, Corvallis, Oregon, and *Department of Botany, University of California, Berkeley, California (77)

Klaus Kalthoff, Department of Zoology, University of Texas, Austin, Texas (97)

Session III *Chairperson:* L. Dennis Smith, Department of Biological Sciences, Purdue University, West Lafayette, Indiana

A. P. Mahowald, C. D. Allis, K. M. Karrer, E. M. Underwood, and G. L. Waring, Department of Biology, Indiana University, Bloomington, Indiana (127)

David Hirsh, Department of Molecular, Cellular and Developmental Biology, University of Colorado, Boulder, Colorado (149)

Ann Janice Brothers, Department of Zoology, University of California, Berkeley, California (167)

Session IV *Chairperson:* David Nanney, Zoology Department, University of Illinois, Urbana, Illinois

Christiane Nüsslein-Volhard, European Molecular Biology Laboratory, Heidelberg, Germany (185)

Joseph Frankel, Department of Zoology, University of Iowa, Iowa City, Iowa (215)

Session V *Chairperson:* David Sonneborn, Department of Zoology, University of Wisconsin, Madison, Wisconsin

C. Peter Wolk, MSU-DOE Plant Research Laboratory, Michigan State University, East Lansing, Michigan (247)

Richard D. Campbell, Department of Development and Cell Biology, University of California, Irvine, California (267)

Peter J. Bryant, Center for Pathobiology, University of California, Irvine, California (295)

Session VI *Chairperson:* Irwin R. Konigsberg, Department of Biology, University of Virginia, Charlottesville, Virginia

P. A. Lawrence and G. Morata*, MRC Laboratory of Molecular Biology, Cambridge, England, and *Centro de Biologia Molecular, Universidad Autonoma de Madrid, Madrid, Spain (317)

Preface

Developmental phenomena, whether they occur during the genesis of a new individual or are involved in the maintenance and repair of the adult form, are characterized by a progressive increase in complexity that is expressed at all levels of biological organization. Not only do we observe the emergence of a diversity of cell types, each exhibiting a unique spectrum of macromolecules that restricts the specialized function of that cell type, but these differentiated cell types are precisely localized within the developing embryo, bud, or regenerating part.

The mechanisms involved in the generation of this high degree of spatial organization have continued to intrigue investigators since the emergence of the discipline of developmental biology. In examining these phenomena during the formative period of this science, several major observations were made suggesting the operation of mechanisms, unique to early development, that regulate subsequent gene expression in the various subgroups of the expanding cell population. The first of these was the establishment of the fact that developing zygotes of a wide variety of organisms contain morphogenetic determinants localized to discrete regions of the egg cytoplasm and that the blastomeres formed in these areas give rise to specific differentiated cell types.

These early-segregating cells were also shown to exhibit both gradient properties within a single morphogenetic area (for example, the amphibian gray crescent area) or interacting gradients between the blastomere lineages, which arise from each of two spatially separated cytoplasmic areas (such as in the developing sea urchin). Similarly, at later stages of development and during the regeneration of ablated structures, the existence of field or gradient properties also suggests that some system of cell communication represents a second-order mechanism for establishing spatial organization in developing systems.

It was my purpose in organizing this symposium to bring together a diverse group of investigators who are analyzing these problems from different vantage points, employing a variety of experimental systems in innovative ways. The development of a program that would adequately treat the topic within the constraints of time involved many hard choices as well as a number of seemingly trivial chores. One has to decide, for example, who goes first, who follows, and how to keep the unavoidable breaks for lunch and other physiological needs from interrupting the continuity. The scheduling became simpler when at some point during these deliberations, I became convinced that, indeed, the research interests of these symposium speakers were closely interrelated. I imagine other organizers of symposia have clutched at the same straw under similar circumstances. As Clifford Grobstein observed, however, in his preface to the twenty-first symposium of this Society, speakers frequently do not share the organizer's view of the unifying theme of a symposium. If such a division of opinion occurred in Madison, it was not evident, and I have every reason to hope that this volume, which reports the proceedings, reflects a synthesis of interests of investigators who are (1) probing the mechanism of localization and the nature of morphogenetic determinants in the developing zygote, (2) employing maternal-effect mutants to study the roles of cytoplasmic determinants and the expression of gradient properties in early development, and (3) using genetic, micromanipulative, and biochemical tools to study pattern formation in simple and more complex forms.

During the first session, which dealt with cytoplasmic localization of determinants, Gary Freeman introduced his presentation by projecting the frontispieces of two volumes: one, *The Cell in Development and Heredity,* published by Edmond B. Wilson in 1928, and the other, the second edition of Eric Davidson's *Gene Activity in Early Development* published in 1977. Freeman's appraisal was that very little had been added between the two publication dates. The five papers presented during the first day of the symposium (including Freeman's) indicate that the appraisal, although a good opening gambit, should be taken with a grain of jovial salt.

Within the past decade, at least, our knowledge of the cytoplasmic localization of developmental information has been extended by the application of new approaches to this old problem. Dohmen and Verdonk, for example, have studied the fine structure of cytoplasmic specializations of the polar lobe of molluscan embryos and observed a unique aggregate of vesicular structures in Bithynia. By the more classical centrifugation and deletion experiments this "vegetal body" appears to carry the lobe-associated determinants. Similarly, scanning electron microscopy reveals striking differences in the surface architecture in the prospective polar lobe region and the possible role of these surface specializations is discussed.

With a similar goal in mind, Mahowald characterized the fine structure of the polar granules of *Drosophila* eggs. Having an unequivocal marker for the germ cell determinants, he has recently been successful in first obtaining a granule-enriched fraction and then identifying a unique polar granule protein in 2-D gels. Following the synthesis of this protein during öogenesis, he is currently testing a scheme of the continuity of the polar granule in the germ cell lineage, which he postulated earlier on the basis of fine-structure studies. Polar granule morphology was also used earlier to confirm cytoplasmic transfer of germ plasm (with Illmensee). Such studies provided the first demonstration that the transfer of a specific cytoplasmic determinant to an "ectopic" region of the egg directs the development of the blastomeres that form in that region into the lineage specified by that determinant. Similarly, in studies of Ascidian development, Whittaker has now shown, by altering the cleavage pattern rather than by microinjection, that determinants of muscle cell proteins will alter the fate of blastomeres normally destined to form ectoderm.

Freeman has also employed techniques to delay cleavage and to alter the cleavage pattern, in both *Cerebratulus* and *Mnemiopsis*, to pose significant questions by-passed in the classical period. Using inhibitors of karyokinesis he has been able to uncouple develomental time from cleavage and to determine the respective roles played by cell division and cleavage planes in the localization of cytoplasmic determinants. These studies indicate that, contrary to the classical view, that determinants in the zygotes in some species are not definitively localized during the postfertilization cytoplasmic streaming but become progressively segregated during cleavage processes linked to the cell division cycle. One might anticipate that knowing how these determinants are translocated might provide, as well, an approach to their identity.

Quatrano's research on the polarity of rhizoid formation in *Fucus* has, in fact, followed such a trail and might provide a paradigm for the localization of specific macromolecules in embryonic anlagen. Starting with a consideration of how polar gradients are established and fixed, Quatrano and his colleagues moved on to the identification and characterization of rhizoid specific polysaccharides and of the localization of such molecules to preformed sites in the zygote.

The data presented by Kalthoff at this symposium also trace a line of research leading from initial phenomenological observations to the establishment of a body of evidence that suggests that a localized determinant of the cephalic region of the embryo of the choronimid *Smittia* may be one of a limited number of species of RNA. It was observed initially that UV microbeam irradiation of a precise region of the anterior pole at early stages leads to the induction, in high frequency, of bicaudal embryos. Action spectra suggested the presence of both a protein and nucleic acid moiety in the target area. The photoreversibility of the morphological effect again suggests the involvement of nucleic acid, which is further supported by Kalthoff's observation of the formation and decay of pyrimidine dimers in RNA after UV irradiation followed by exposure to visible light. Since either RNAase or UV irradiation applied to the same site results in double abdomen formation, the simplest hypothesis, Kalthoff suggests, is that both inactivate a single type of cytoplasmic determinant.

The strategy of the "new look" at cytoplasmic localization that seems to be emerging is first to define the phenomenon in more precise terms. This redefinition frequently involves a clearer characterization of the cell properties specified by a given determinant such as Quatrano's rhizoid-specific polysaccharides, Freeman's use of light emission by the photocyte, or Whittaker's exploitation of cell-type-specific histochemically demonstrable enzymes. Other studies have focused on a fine-structural identification of the cytoplasmic inclusion with which the determinant is associated (the "ventral body" of Dohmen and Verdonk and Mahowald's polar granule) or on the experimental lability of the determinant property. How these "handles" are used varies considerably from Mahowald's use of ultrastructural criteria to obtain granule-enriched fractions and examine subunit composition to Quatrano's analysis of the role of sulfation in the localization of fucoidin. What is most important, however, is that the more precise characterizations lead to more readily resolved questions, and if there is any doubt that such seemingly prosaic beginnings can lead to highly significant findings I suggest that the reader carefully think through Whittaker's paper, not only where he has been, but where he is going and how he intends to get there.

Since it is clear that cytoplasmic determinants must be synthesized and stored during öogenesis, maternal-effect mutants of developmental processes offer promising tools to investigate the time of synthesis, nature, and mechanism of expression of cytoplasmic determinants. In this class of mutants the developmental defect is an expression of the maternal genome and not that of the zygote. Whether the defect represents the absence of a particular gene product, the production of an altered gene product, or distortions of egg organization, the primary event must occur prior to fertilization.

This unique type of gene expression was first recognized in the pair of alleles that control dextral and sinistral cleavage in the egg (and subsequent coiling of the shell) in the fresh water snail *Lynnea stagnalis*. First described by Boycott and Diver (1923–1938), the mode of inheritance of these traits was subsequently analyzed by Sturtevant in 1923. The most extensive *developmental* study of a mutant of this type has been performed in the 0 mutant (ova deficient) of the Mexican axolotl. The mutant phenotype observed in the progeny of homozygous females is arrest of development at the gastrula stage. Brothers has reviewed the evidence that gastrular arrest is due to a deficiency in the synthesis of 0^+ factor normally synthesized during the germinal vessicle stages and released into the cytoplasm at the 1st meiotic division. The defect can be corrected by injecting either germinal vessicle nucleoplasm or egg cytoplasm from eggs of wild-type females into fertilized, uncleaved eggs of homozygous mutant females. Rescue of a blastula nucleus of an embryo fated for gastrula arrest can be effected by transplantation into an enucleated egg containing 0^+ substance provided the nuclear transfer is performed before the stage (late blastula) at which normal nuclei are activated by the factor. Conversely, late blastula nuclei of normal embryos, having been activated by 0^+ factor, support normal development when transplanted into enucleated eggs of homozygous mutant females. Activation in such nuclei is stable and heritable through at least 30 mitotic divisions in clonal serial transplants in enucleated eggs of homozygous mutant females. The basis of the maternal effect in the 0 mutant is the inability to synthesize and store a soluble gene product that at late blastula activates, in stable fashion, one or more gene functions required to carry the embryo through normal gastrulation and neurulation.

In theory, a variety of maternal-effect mutations should occur affecting not only cleavage and early morphogenetic and inductive processes but the establishment of embryonic symmetry and polarity as well. One mutation of the latter type, *bicaudal*, was described in *Drosophila* by Bull in 1966. Unfortunately the low penetrance and expressivity of the mutant gene precluded more extensive analysis. The most extreme expression of this mutant phenotype, however, was remarkably similar to the double abdomen embryos of *Smittia*, which Kalthoff has since experimentally produced.

Recently Nüsslein-Volhard has been able to increase the frequency of mutant expression by constructing hemizygous mutant females in which the single point *bic* mutation is balanced against a homolog carrying a deletion in the bicaudal region. One would assume, therefore, that *bicaudal* is a hypomorphic mutation producing a smaller amount of the normal gene product.

The mutant phenotype spectrum varies continuously from an embryo lacking only a head to completely symmetrical double abdomens with a distribution frequency, which suggests that the mutation shifts the pattern between the two more stable extremes. A number of observations suggest that the mutation affects the pattern of segments in both anterior and posterior halves of the embryo, thereby suggesting that the normal allele of *bicaudal* controls the establishment of a matrix of positional values. In the extreme *bicaudal* phenotype the anterior and posterior abdomens are mirror images of one another, gastrulation occurring at both ends of the embryo. Pole cells occur, however, only at the posterior tip, where we know (see Mahowald, this volume) the germ cell determinants are localized. The model presented by Nüsslein-Volhard to explain the wild-type pattern as well as the spectrum of mutant phenotypes assumes a gradient of a single morphogen, which increases continuously from anterior to posterior pole in the wild type but which in symmetrical mutants exhibits maxima at both poles. Assuming that the cellular response to morphogen concentration is specified by discrete threshold values that elicit either head or thorax or abdominal structures, each mutant phenotype can be described by the minimum morphogen level reached in the biphasic mutant gradient and by where, along the axis, this low point is located. Nüsslein-Volhard discusses, in addition to *bicaudal*, a newly isolated maternal-effect mutant (*dl*) of the dorso-ventral pattern, which also behaves as a mutation of the pattern-specifying mechanism. The primary value of models of this sort, of course, is that they provide a conceptual framework upon which a complex body of data can be arranged and summarized. If the model suggests additional experiments as well, its worth is considerably augmented. On the other hand, such models are often so seductive that there is some concern that the postulate might attain more significance than the actual data permit.

Although *bicaudal* and *dorsal* behave like mutations of the gradient-forming system, Stern's (1968) investigations of sex comb mutations in *Drosophila* indicate that all but one of these are mutations in the ability of cells to "read" the positional cue. One might expect, therefore, cell-autonomous mutants in which the response to the normal gradient is altered. Mutations that alter the thresholds of response to the postulated morphogen, for example, might shift the proportions of normal body segments.

Maternal-effect mutants of development, such as ova deficient in the Mexican axolotl and *bicaudal* in *Drosophila*, were fortuitous discoveries made by astute investigators, thoroughly familiar with their organism. In each case the mutant reached a developmental stage sufficiently advanced to enable the mutant phenotype to be characterized and its developmental significance established.

By selecting for temperature-sensitive mutants and screening these for strict maternal-effect mutants the element of chance is reduced and the probability of detecting mutations of early development enhanced.

Temperature-sensitive mutants not only provide a convenient method of determining the time of mutant gene expression but also facilitate carrying the mutant stock.

David Hirsch has applied such technology to the soil nematode *Caenorabditis elegans* and, using a three-staged screen, has distinguished 11 of 24 zygote defective mutants as being strict maternal-effect mutants. The phenotype of all of these mutants can be changed by temperature shifts at various times during öogenesis or early embryogenesis or both. Predicted patterns of temperature sensitivity can be constructed based on whether the temperature sensitivity is a measure of thermolability or temperature sensitivity of the synthesis of that protein and whether it is translated from maternal mRNA in the oocyte, the unfertilized egg, or after fertilization. Each of his maternal-effect mutants can be matched to one or the other of the predicted patterns, providing, hopefully, a useful model for future tests.

The most striking feature of these zygote-defective mutants (the five maternal effect and two male rescue mutants shown) is the extreme, mutant specific, abnormality of first cleavage. If these cleavage patterns do not reflect displacements of localized determinants, they should result in the abnormal segregation of normally localized determinants. Unfortunately, at this writing a marker for only one lineage (the gut) has been described. In view of the significance of these questions it is difficult to imagine, however, that this deficiency will long go uncorrected.

Interest has been renewed recently in those situations in developing zygotes, organ primordia, and regenerates, in which morphogenetic expression appears to exhibit gradient or field properties. This interest was stimulated initially by attempts to fit all such observations into a single, unifying concept. Briefly, the concept postulates that spatial ordering is the resultant of two component processes: the assessment by each cell of its location in a cellular matrix (sensing a graded or pulsed common signal), and then each cell, constrained by its genome and previous developmental history, differentiates in accordance with its position in the field. This restatement of earlier explanations of the development of pattern suggests a number of corollaries currently being tested in a variety of organisms.

Although we owe the first formal model of positional information to Wolpert (1968), the concept of a bipartite system of assessment of position and appropriate response was explicitly invoked earlier in the analysis of pattern in two widely divergent organisms. Although couched in other terms, the concept was employed by Stern (summarized in 1968) in his analysis of bristle pattern formation in *Drosophila* and is also deeply entrenched in the literature of the maintenance and inheritance of the cortical pattern in ciliates (see Frankel's introduction, this volume.)

The relevance and importance of an understanding of the generation of the cortical pattern in ciliates to the central question of spatial ordering in developmental systems has long been recognized. Frankel's presentation at this symposium is, in a sense, a continuation of the discourse initiated by Vance Tartar at the Society's third symposium at Dartmouth and continued at the fourteenth symposium at Amherst. This scholarly review deals principally with the more recent analyses of intracellular patterning in Tetrahymena, to which this speaker has made substantial contributions. These studies suggest that two different mechanisms, a short-range ("nearest-neighbor") and a long-range ("gradient-field") mechanism, act in concert to determine the overall cortical pattern. Throughout Frankel's analysis he relates, where appropriate, pattern specification in the ciliates to analogs in multicellular forms and expresses a final thoughtful and not untimely reservation over the heuristic value of evoking gradients of chemical "morphogens" as the basis of morphogenetic fields.

One source of Frankel's disaffection with the concept of the diffusion of chemical morphogens or activators in pattern formation stems from Campbell's studies of the development of "epithelial hydra," which were also presented at this symposium.

Campbell's work on the developmental properties of nerve-free hydra suggests that the role of nerve cell secretions and purified "head factor" in the control of hydra regeneration (see Schaller, 1978) needs reevaluation. Campbell has demonstrated that nerve cells can be virtually eliminated from hydra exposed to two cycles of treatment with colchicine. Hydra so treated are viable, bud, and allow the establishment of clones of animals permitting the selection of clones that are completely free of the interstitial cell precursors of nerve cells and nematocytes. Such clones of "epithelial hydra," consisting of only ectodermal and endodermal epithelial cells, exhibit some peculiarities associated with their paralytic condition but are otherwise normal in morphology. The possibility that epithelial cells assume compensatory neurosecretory functions has been examined and deemed unlikely.

The developmental behavior of "epithelial hydra" is no different from normal animals with respect to (1) the cycle of epithelial cell replacement, (2) the ability to regenerate basal disk or tentacles, (3) the capacity of hypostomal tissue to induce a secondary axis when grafted in the gastric region of either normal or "epithelial hydra," and (4) the development of inductive capacity in subhypostomal tissue following hydranth removal. Most importantly, segments of the gastric region obtained from normal and "epithelial" hydra exhibit the same polarity of regeneration and the same kinetics of reversal of polarity.

Campbell concludes that the predominant control over morphogenetic patterns in complete hydra is exerted by the epithelial cells. Neurosecretory factors, he suggests, may govern primarily interstitial cell determination or, if they exert a strong morphogenetic effect, it is overridden by stronger influences of the epithelial cells. In conclusion he poses the interesting possibility that patterning is controlled in hydra by mechanical rather than chemical mediation.

In the filamentous cyanobacteria, a system less familiar, I suspect, to most developmental biologists, three specialized cell types can be distinguished: a vegetative, photosynthetic cell, a heterocyst or nitrogen-fixing cell, and the dormant spore cell. These cell types are arranged within the linear filament in a specific order. Depending upon the species, the morphologically distinguishable heterocysts are either spaced at semiregular intervals within the filament of vegetative cells or restricted to the termini of each filament. Spores, when they form, develop from vegetative cells contiguous with the heterocysts.

In his presentation Wolk described a multifaceted approach to determine the rules of cell ordering, the biochemical basis of heterocyst and vegetative cell differentiation, and the consequent cooperative metabolic coupling of these two cell types. In addition, he evaluated the evidence that heterocyst differentiation inhibits similar differentiative expression in nearby vegetative cells.

Interdependency of the two specialized cell types in its simplest terms consists of the transfer of a photosynthetic product to the heterocyst, which is in turn oxidized via the pentose-PO_4 shunt, providing a source of electrons for the reduction of nitrogen to ammonium ions. Nitrogen in this form is then transported back into the vegetative cell as glutamine. Although the adaptive value of controlling the ratio of cell types and their spacing is clear, the precise mechanisms involved are less clear.

In the presence of NH_4^+, heterocyst differentiation is suppressed. When such filaments are then deprived of a source of fixed nitrogen, spaced protoheterocysts arise and must inhibit adjacent vegetative cells. Decreases in levels of cell components characteristic of vegetative cells occur during this transition, and the synthesis of heterocyst proteins is inhibited in *all* cells. If, as such observations indicate, all cells can initiate heterocyst differentiation, there must be some mechanism restricting its realization to a minority population. Wolk considers alternative mechanisms and how these might be resolved.

Drosophila imaginal discs have proven to be convenient material for studies of the generation of pattern since the interval between determination and expression is sufficiently long to permit the analysis of altered pattern after experimental manipulation of the geometrically simple organ primordia. Bryant has reviewed studies of the regulation of pattern (duplication or regeneration) of disc fragments in terms of the polar coordinate model (or "clock" model) of French, Bryant, and Bryant that is applicable to a number of other regenerating systems as well (see, for example, S. Bryant, 1978). This model predicts that, depending on the plane of the cut and the manner in which the fragments heal, a particular disc fragment will duplicate itself, regenerate the entire disc, or differentiate without first increasing in size. For example, fragments from opposite ends of the wing disc duplicate when cultured separately, but when mixed with each other, each fragment regenerates, contributing parts of the wing pattern not contained in either fragment and the entire pattern in some cases. Using genetic bristle color markers, it can be demonstrated that genital disc fragments from each of the mutants (which *in situ* form two morphologically distinct components of the genital apparatus) when mixed together form a chimera in which cells from both disc areas form specific genital structures normally formed exclusively by one or the other. Such observations support the postulate that regeneration occurs by the intercalation of cells of missing positional values between the cells of different fate that form the edges of the minced, excised areas of disc tissue. Employing this same strategy to combinations of area from different discs (viz., haltere and leg), Bryant and his colleagues have obtained evidence supporting one of Wolpert's corrolaries (1969) that in all of the different fields of an organism positional information may be specified by a universal mechanism employing properties common to all cells.

Peter Lawrence's presentation at this symposium, as he points out, summarizes only his more recent contributions to the compartment concept of the formation of integumentary patterns in insect development. This seemed appropriate since at the previous symposium of this Society the compartment hypothesis was reviewed in detail (Morata and Lawrence, 1978) in the context of the application of clonal analysis to developmental problems. Since the present symposium focuses on the determination of spatial ordering *per se*, it would have been seriously incomplete without the inclusion of this work.

Lawrence, although he has confined himself to his more recent observations, skillfully uses this information to illustrate the basic principles of compartmentalization and the response of cells of different genetic constitution to the boundaries of the compartment. In his opening remarks, Lawrence commented that "this has been a very good symposium—the right proportion of data and theory—and delivered in the right order." I should like to turn this same compliment back to Peter Lawrence, although I would prefer the word "excellent."

Irwin R. Konigsberg

REFERENCES

Boycott, A. E. and C. Diver (1923). *Proc. Roy. Soc. (London) Ser. B Biol.* **95,** 207.
Bryant, S. V. (1978). *36th Symp. Soc. Devel. Biol.,* p. 63.
Davidson, E. (1977). "Gene Activity in Early Development." Academic Press, N.Y.
Morata, G. and P. A. Lawrence (1978). *36th Symp. Soc. Devel. Biol.,* p. 45.
Schaller, H. C. (1978). *35th Symp. Soc. Devel. Biol.,* p. 231.
Stern, C. (1968). "Genetic Mosaics and Other Essays." Harvard Univ. Press.
Sturtevant, A. H. (1923). *Science* **58,** 269.
Wilson, E. B. (1928). "The Cell in Development and Heredity." The Macmillan Co.
Wolpert, L. (1968). *In* "Towards a Theoretical Biology" (C. H. Waddington, ed.), Vol. 1, p. 125. Edinburgh Univ. Press.
Wolpert, L. (1969). *J. Theoret. Biol.* **25,** 1.

Acknowledgments

This symposium was held on the campus of the University of Wisconsin at Madison. We would like to express our sincere appreciation of all of the help and support that we received from our colleagues at the University and to Claudia Foret who maintained liaison with the local committee.

We acknowledge our indebtedness to the chairman of the local committee, Bruce H. Lipton, and record the fact that the Society at its business meeting unanimously passed a vote of acclamation for his efforts.

The Society deeply appreciates the financial support of the Developmental Biology Program of the National Science Foundation, which made it possible to bring this outstanding group of scientists to speak at this symposium.

The logo on both the program and this volume is after T. Gustafson and many others and was designed by Irwin R. Konigsberg and executed by Lucy Taylor.

Finally, both of the editors acknowledge their indebtedness and thanks to all of the speakers and session chairpersons whose cooperation and performance at the symposium (and later) were admirable.

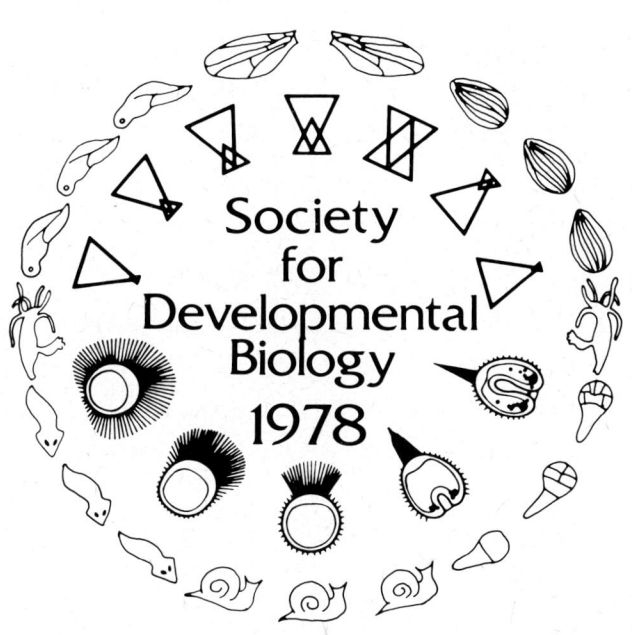

I. Cytoplasmic Localization in Early Development

The Ultrastructure and Role of the Polar Lobe in Development of Molluscs.

M. R. Dohmen and N. H. Verdonk
Zoological Laboratory
University of Utrecht
Padualaan 8, Utrecht, The Netherlands

I.	Introduction	3
II.	The Cytoplasm of Polar Lobes	6
	A. Special Structures in Small Polar Lobes	6
	B. Experimental Evidence for the Morphogenetic Significance of the Vegetal Body in *Bithynia*	10
	C. The Ultrastructure of Large Polar Lobes	11
	D. Centrifugation Experiments on Large Polar Lobes	15
III.	The Relation between the Cortex of the Polar Lobe and Cytoplasmic Localizations	16
IV.	The Possible Role of RNA as a Morphogenetic Factor in Polar Lobes	22
V.	Summary	25
	References	26

I. INTRODUCTION

In theories on the influence of cytoplasmic determinants on development, the polar lobe formed by eggs of many annelids and molluscs has played an important role. This polar lobe is a transient protrusion at the vegetal pole of the egg, in which part of the cytoplasm is set apart during cleavage (cf. Figs. 1, 2). At the end of the first cleavage it fuses with one of the blastomeres (CD) of the 2-cell stage. At second cleavage a lobe is formed again, which flows into the D-blastomere at the 4-cell stage. In this way the vegetal region of the egg is shunted to the D-quadrant, which will become the dorsal quadrant of the future embryo.

As was first shown by Crampton (1896), the polar lobe, which in favorable cases is connected only by a thin strand with the egg, can easily be removed without immediate damage to the egg, which continues cleaving. The consequences of a removal of the lobe for development are however, quite dramatic. In lobeless embryos of gastropods adult structures such as shell, foot, operculum, statocysts, eyes, tentacles, and

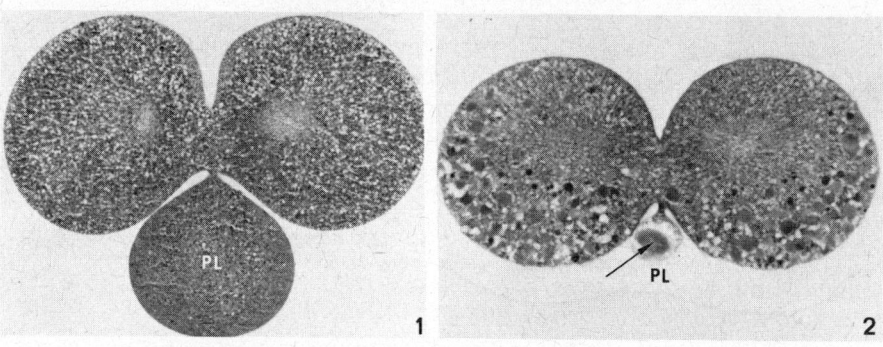

Fig. 1. *Dentalium*. Section of an egg at first cleavage, the so-called trefoil stage. The polar lobe (PL) contains about 1/3 of the egg cytoplasm. Haematoxylin-eosin staining.

Fig. 2. *Bithynia*. Section of an egg at first cleavage, showing the small polar lobe (PL) with the vegetal body (arrow). The polar lobe contains less than 1% of the egg cytoplasm. Haematoxylin-eosin staining.

heart are absent, as was shown by Clement (1952) for *Ilyanassa*, which has a large polar lobe (about 1/3 of the egg volume), and by Cather and Verdonk (1974) for *Bithynia*, which has a small polar lobe (less than 1% of the egg volume). Part of these structures, such as heart and intestine, are directly lobe-dependent, as they originate from the D-quadrant, which receives the lobe material; others, such as eyes and tentacles, are not formed from the D-quadrant, but their appearance is dependent on an interaction with the D-quadrant, which according to Cather (1971) acts as a primary organizer in molluscan development.

Whereas removal of the lobe results in absence of the lobe-dependent structures, equalization of first cleavage by a treatment with cytochalasin B in *Dentalium* results in a duplication of lobe-dependent structures (Guerrier *et al.*, 1978).

While there is ample information available on the influence of the polar lobe on organ formation, far less is known about the influence of the lobe on early development. Clement (1952) showed that in normal development of *Ilyanassa* the cleavage rhythm in the D-quadrant is different from the other quadrants. The mesentoblast 4d appears 3 hours ahead of the cells 4a-4c, and the division in the cell 1d and its descendants lags behind the corresponding cells of the other quadrants. In *Dentalium* van Dongen and Geilenkirchen (1974, 1975), following the cell lineage of normal and lobeless embryos up to the trochophore stage, showed that in normal development the cleavages in the D-quadrant succeed each other at a faster rate than in the A, B, and C quadrants. In lobeless embryos, both of *Ilyanassa* and *Dentalium*, the cell divisions in the quadrants are synchronized and all quadrants follow exactly the same time schedule. This suggests that the polar lobe contains factors that control the initiation of cell cyclic processes in the cell lines to which the lobe is segregated during cleavage.

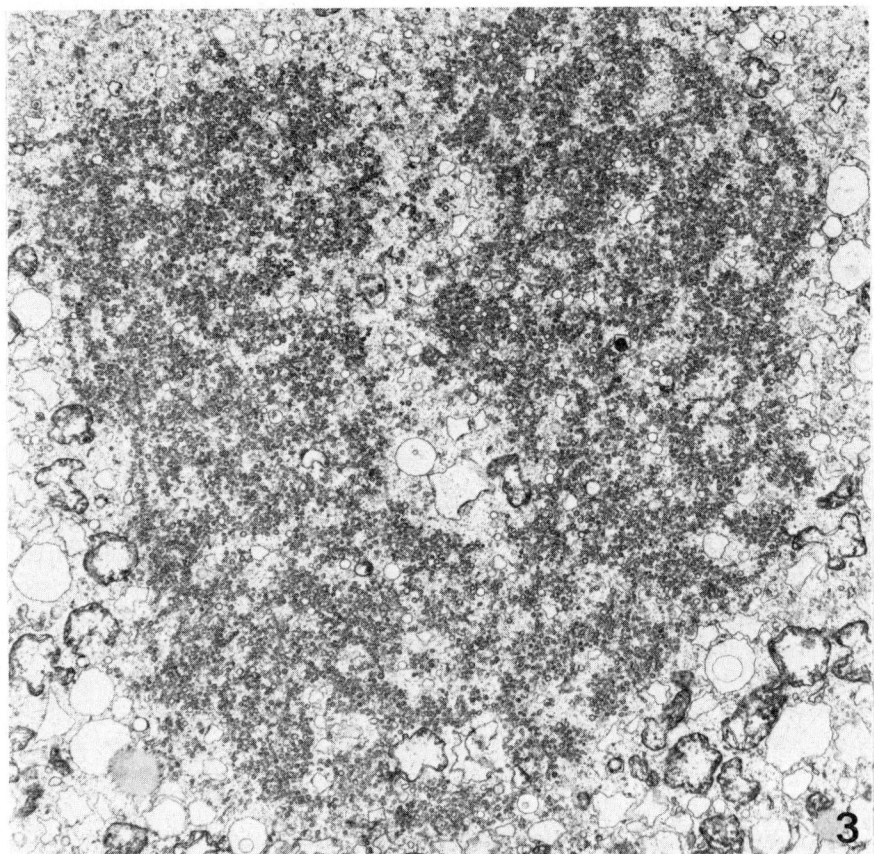

Fig. 3. *Bithynia*. Electron micrograph of the vegetal body in the first polar lobe. x 10500.

In normal development not only the time schedule of the divisions in the D-quadrant is different, but also the pattern of the divisions. In *Ilyanassa* some cells (e.g. 1d, 1d^1, and 1d^{12}) are smaller, others (e.g. 2d^{11}) are larger than the corresponding cells in the other quadrants (Clement, 1952). Similar phenomena were observed in *Dentalium* by van Dongen and Geilenkirchen (1974, 1975). These differences also disappear after removal of the lobe. Consequently the polar lobe contains factors that control the position and the orientation of the cleavage planes in the cells of the D-quadrant. In order to elucidate the role of the polar lobe in development, the composition of the lobe in various species of molluscs has been studied, both in normal and centrifuged eggs.

II. THE CYTOPLASM OF POLAR LOBES

A. *Special Structures in Small Polar Lobes*

The nature and the localization of the morphogenetic factors in polar lobes have been studied by centrifugation, microsurgery, biochemical, cytochemical, and electron microscopical methods, etc. Most of these techniques have been applied both to eggs with large polar lobes (e.g. *Ilyanassa, Dentalium, Mytilus*) and with small lobes (e.g. *Bithynia, Crepidula, Buccinum*). In some respects the small polar lobes have yielded the most promising results. In these lobes specific structures have been found which are supposed to contain the morphogenetic factors, and strong evidence supporting this view is accumulating. The egg of *Bithynia tentaculata* has been most thoroughly studied so far. In the polar lobe of this species a conspicuous structure has been described: the vegetal body (Dohmen and Verdonk, 1974) (Figs. 2, 3). This body consists of a large number of small electron dense vesicles which probably contain RNA.

A study of the origin of the vegetal body has shown that this structure is localized already at the prospective vegetal pole at an early stage during oogenesis. The earliest observation of a nascent vegetal body was made at the beginning of vitellogenesis, when this body consists of a small cluster of electron dense vesicles, about 3 μm in diameter, located at a distance of about 3 μm from the plasma membrane. During vitellogenesis it grows by the addition of vesicles budding off from the endoplasmic reticulum. At the same time the body moves closer to the plasma membrane at the future vegetal pole. In the full-grown oocyte the vegetal body is a flat disc, about 4 μm thick, and located at a distance of about 0.5 to 1.0 μm from the plasma membrane. Concomitantly with the formation of the polar lobe it assumes a cup-shape and this shape is

STRUCTURE AND ROLE OF THE POLAR LOBE IN MOLLUSCS

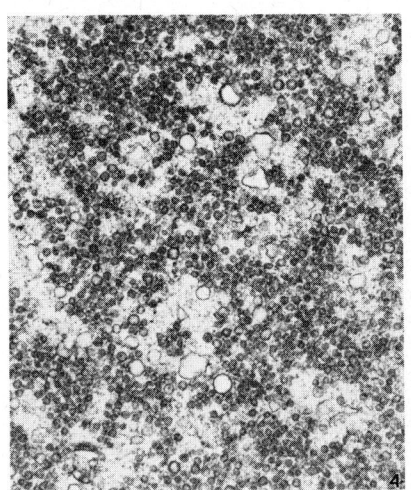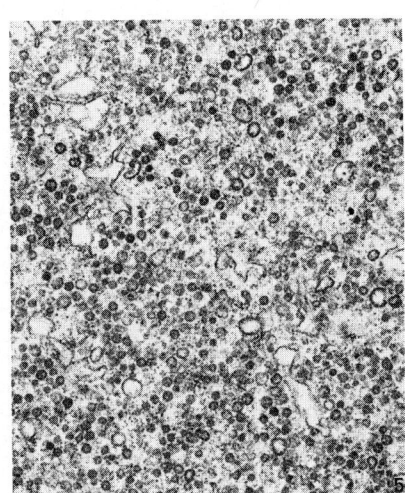

Fig. 4. *Bithynia*. Detail of the vegetal body in the first polar lobe. The vesicles are not homogeneously distributed; they form a network of clusters with vesicle-free areas in between. x 16000.

Fig. 5. *Bithynia*. Detail of the vegetal body in the second polar lobe. Clustering of the vesicles is much less than in earlier stages (cf. Fig. 4). x 16000.

retained after the resorption of the lobe into the CD-cell.

Just before the second cleavage the vegetal body cannot be detected anymore in histological sections. In electron microscopical preparations, however, a large aggregate of vesicles is still seen to be present at the vegetative side of the CD-cell. The individual vesicles are apparently unchanged, but the cup-shape of the whole aggregate is lost. Whereas in earlier stages the vesicles are arranged in dense clusters with vesicle-free areas in between (Figs. 3,4), this pattern practically disappears just before the second cleavage (Fig. 5). These structural changes may be due to the loss of some binding substance, which holds the vesicles together. The disappearance of such a substance may also explain the loss of stainability with haematoxylin-eosin in histological sections at this stage. It would also fit with the results from blastomere-deletion experiments (Verdonk and Cather, 1973; Cather et al., 1976), which can be explained by assuming that the contents of the vegetal body are distributed to the C- and the D-cell, as these cells have about equal developmental potential. It has not yet been possible to ascertain whether there is indeed a segregation of part of the vegetal body into the C-cell. At the second cleavage a large mass of vesicles is segregated into the second polar lobe (Fig. 6) and then shunted into the D-cell. During the 4-cell stage these vesicles disappear nearly completely and their fate cannot be followed anymore by electron microscopy.

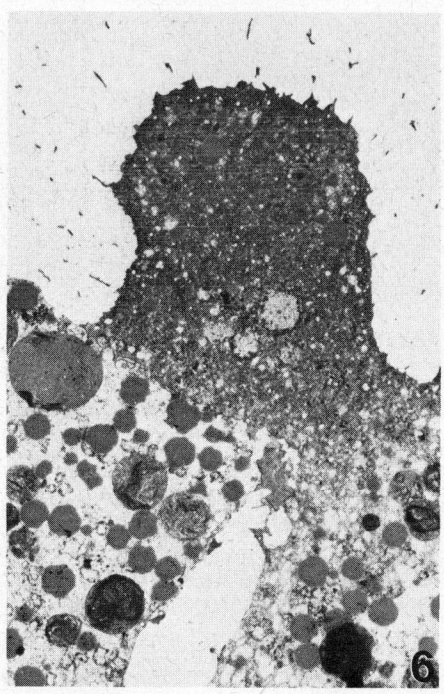

Fig. 6. *Bithynia*. Second polar lobe, filled with dense cytoplasm mainly consisting of vesicles of the vegetal body. Long villi emanate from the surface of the lobe. x 2600.

Conspicuous aggregates of vesicles have also been found in the polar lobe of *Crepidula fornicata* (Dohmen and Lok, 1975) and *Buccinum undatum*. In these species the aggregates do not assume a cup-shape and the vesicles differ in size, shape, and probably also in contents from those of *Bithynia*. In *Crepidula* the vesicles are elongated and occur in a few separated aggregates. In *Buccinum* two types of vesicles are intermingled in the aggregate: small ones (diameter ca. 0.1 μm) and large ones (diameter ca. 0.5 μm) (Fig. 7). We should be cautious, however, in concluding from these examples that vesicular aggregates are a universal morphogenetic component of small polar lobes. The polar lobe of the egg of *Littorina obtusata* has been searched without results so far. This may be due to the hypothetical occurrence of very small vesicular aggregates which may easily escape detection. In the polar lobe of *Littorina saxatilis* we did indeed find such a minimal aggregate (Fig. 8) of elongated electron dense vesicles, but proof of the morphogenetic significance of these small aggregates is hard to give. Ultrastructural studies of small polar lobes in other molluscan species have not been made, as far as we know.

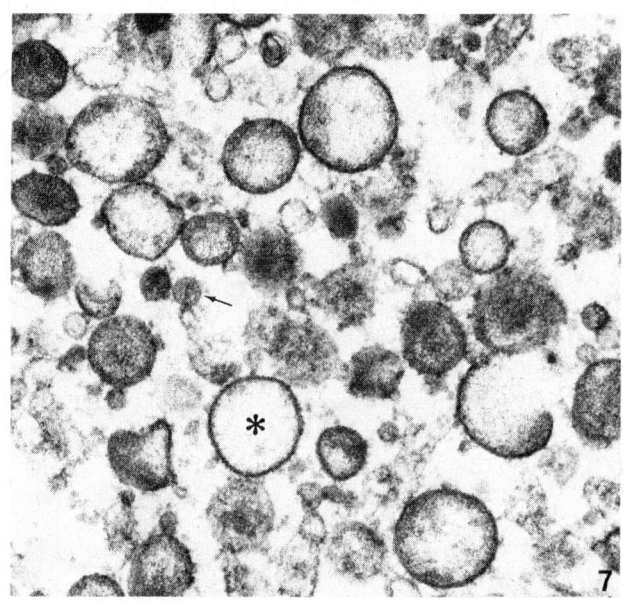

Fig. 7. *Buccinum*. Detail of the vesicular aggregate in the polar lobe. Two types of vesicles are present, small ones (arrow) and large ones (*). x 40000.

Fig. 8. *Littorina saxatilis*. Small aggregate of elongated vesicles with electron dense contents in the polar lobe. x 50200.

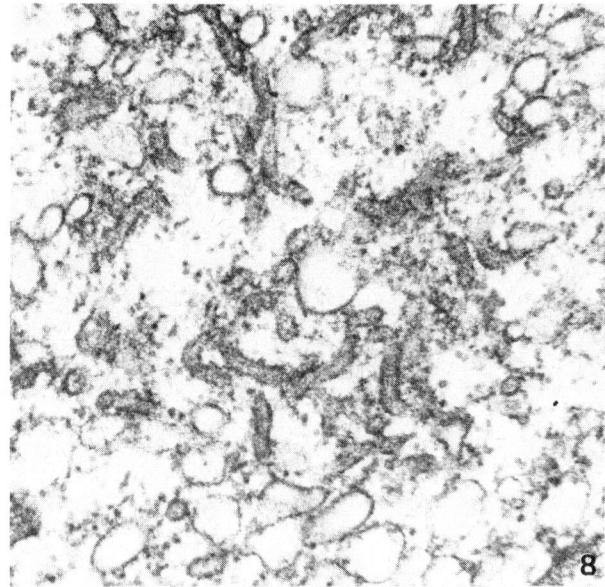

B. *Experimental Evidence for the Morphogenetic Significance of the Vegetal Body in Bithynia.*

In order to study the influence of the special cytoplasm, localized in the so-called vegetal body of the first polar lobe of *Bithynia*, one may try to disperse or displace the constituents by centrifuging eggs before first cleavage. The vegetal body appears to be bound rather strongly to the cortex, so that it can be removed only by a relatively strong centrifugal force (about 1400 g), which separates part of the eggs into two halves. When the vegetal body is removed from the vegetal pole of the egg it is never dispersed but always displaced as a whole (cf. Fig. 9). Seventy eggs, centrifuged about one hour before first cleavage, were fixed at the moment of first cleavage and studied in sections. In 30 eggs the vegetal body was found outside the polar lobe in the cytoplasm of one of the two blastomeres (cf. Fig. 10).

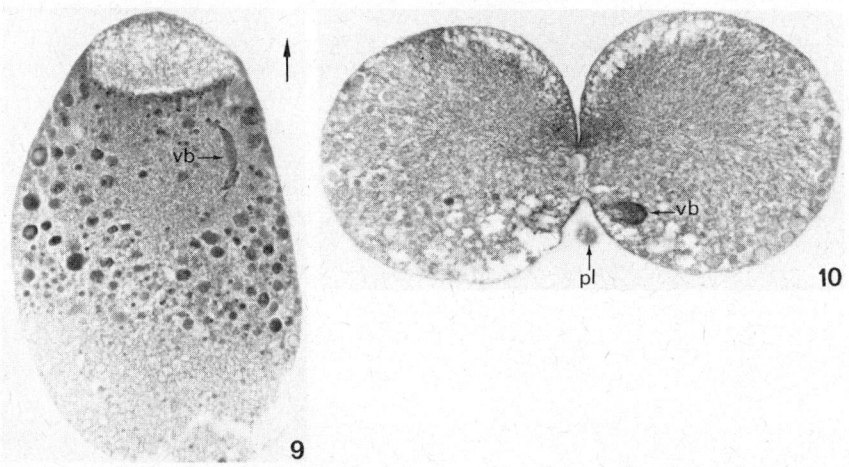

Fig. 9. *Bithynia.* Centrifuged egg, showing the vegetal body (vb) displaced as a whole. The direction of the centrifugal force is indicated by the arrow.

Fig. 10. *Bithynia.* Egg centrifuged one hour before first cleavage and fixed at first cleavage. The vegetal body (vb) is now absent from the polar lobe and located in one of the blastomeres. pl. grazing section through the polar lobe.

As yet we have no means to establish the presence or absence of the vegetal body in the polar lobe of a living egg. We know, however, from experiments of Cather and Verdonk (1974) that removal of the first lobe in normal, uncentrifuged, eggs always results in an embryo that fails to form adult structures such as eyes, tentacles, foot, shell, etc. (cf. Fig. 11).

After removal of the polar lobe in centrifuged eggs, apart from this lobeless type of embryos, also completely normal embryos were obtained (cf. Fig. 12). After separation of the two blastomeres at the 2-cell stage, the CD-blastomere, which receives the polar lobe, forms adult structures, which are absent in the AB-embryos (Verdonk and Cather, 1973). After removal of the polar lobe and subsequent separation of the blastomeres, both blastomeres fail to form adult structures (Cather and Verdonk, 1974). When this experiment is done with centrifuged eggs, one of the blastomeres may form adult structures. These data indicate that in *Bithynia* the morphogenetic determinants are localized in the vegetal body. They perfectly agree with the observation that the vegetal body is not dispersed but displaced as a whole by centrifugation.

C. *The Ultrastructure of Large Polar Lobes*

In large polar lobes nothing resembling the special cytoplasms found in small polar lobes has been detected. In ultrastructural studies of the egg of *Mytilus* (Reverberi and Mancuso, 1961; Humphreys, 1964) nothing

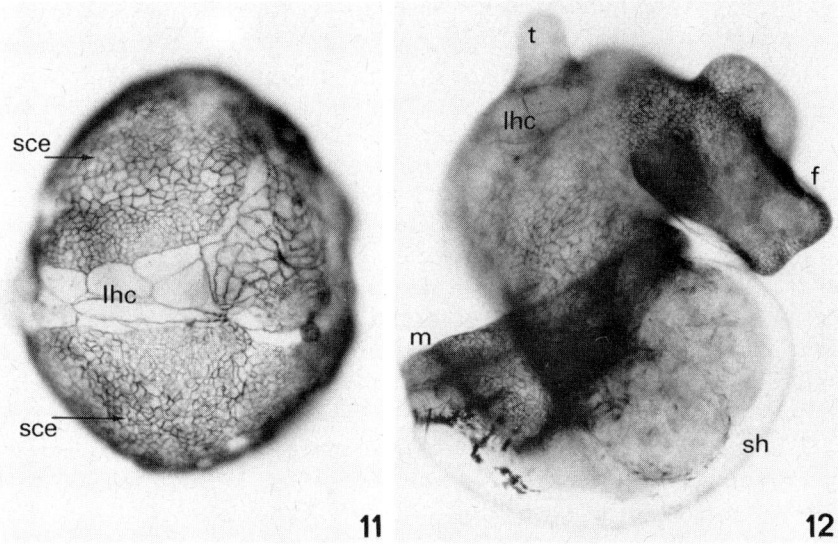

Fig. 11. *Bithynia*. Embryo showing severely defective development after removal of the first polar lobe without prior centrifugation. sce. small-celled ectoderm; lhc. larval head cells.

Fig. 12. *Bithynia*. Embryo showing normal development after removal of the first polar lobe. The egg was centrifuged before the deletion of the lobe. lhc. larval head cells; t. tentacle; f. foot; m. mantle; sh. shell.

special was found in the polar lobe. A cytochemical study (Pucci, 1961) also showed a uniform distribution of the investigated components over the polar lobe and the blastomeres.

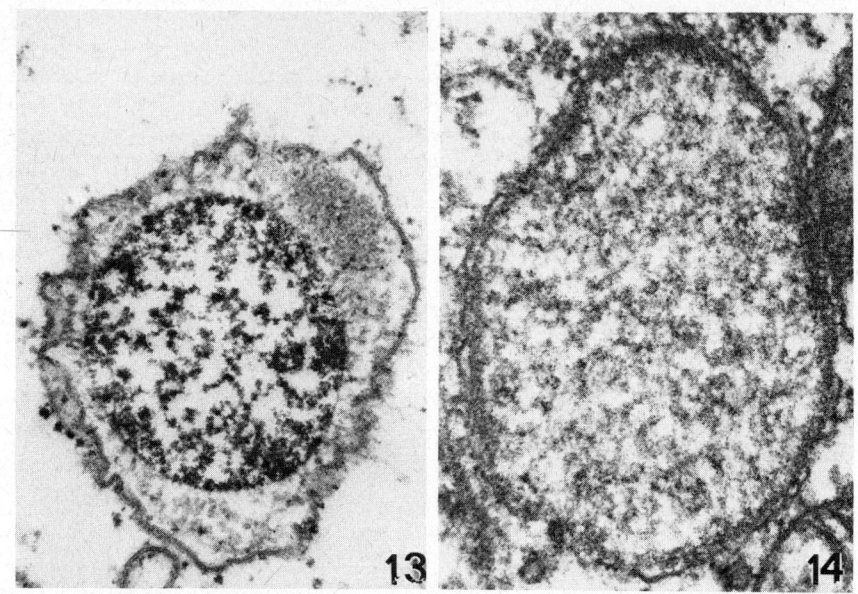

Fig. 13. *Nassarius reticulatus.* Double-membrane vesicle. The dense contents of the central vacuole have been preserved by fixation in a mixture of glutaraldehyde and osmium tetroxide. x 75000.
Fig. 14. *Dentalium.* Multisheet vesicle. x 50900.

The cytoplasm in the egg of *Ilyanassa obsoleta* shows a pronounced segregation: lipid droplets and mitochondria in the animal hemisphere, and yolk in the vegetal hemisphere (Clement and Lehmann, 1956). When the polar lobe forms, a large part of the yolk is included in it. Two kinds of special structures have been observed between the yolk granules: double-membrane vesicles (d.m.v.) and "ribosome clusters". Double-membrane vesicles were first described by Crowell (1964). They are also present in the polar lobe of *Nassarius reticulatus* (Schmekel and Fioroni, 1975), and in this species they show the same structure: an outer unit-membrane, an inner amorphous "membrane" enclosing a vacuole, and in between these membranes an electron dense body (Figs. 13, 15). Occasionally the central vacuole is seen to contain a fibrillar material (Crowell, 1964; Schmekel and Fioroni, 1975) (Fig. 15). When a fixative consisting of a mixture of glutaraldehyde and osmium tetroxide was used, we observed consistently that the central vacuoles of all d.m.v. were filled with a dense substance (Fig. 13) whose composition could not

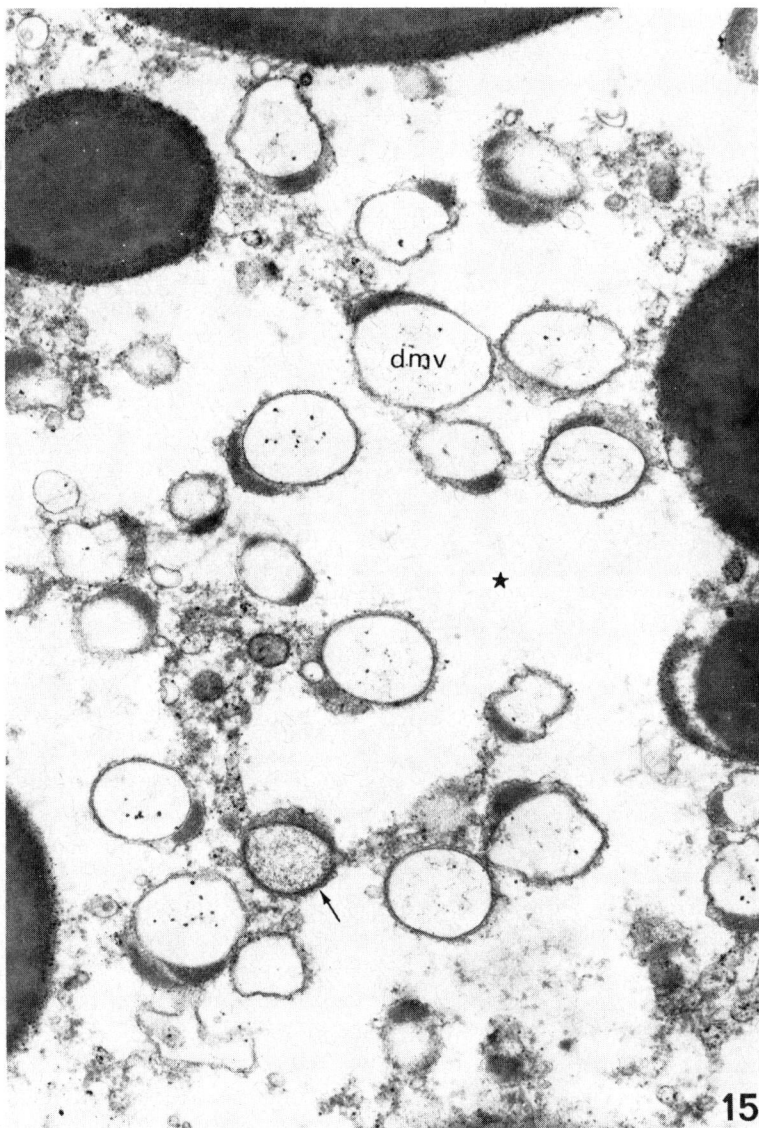

Fig. 15. *Nassarius reticulatus*. Detail of the vegetative cytoplasm of the uncleaved egg, showing a number of double-membrane vesicles (dmv). In one of these the central vacuole is filled with fibrillar material (arrow). The large empty spaces (*) are in fact filled with glycogen, but this is masked by treatment of the section with methanolic uranyl acetate. Fixation: glutaraldehyde, followed by osmium tetroxide. x 18000.

yet be determined. Gérin (1972), who described their origin from perinuclear bodies, mainly consisting of protein (Gérin, 1971), suggests that the central vacuole is caused by a gradual breakdown of the perinuclear bodies. A transformation of the substance of the perinuclear bodies into the electron dense material we find inside the vacuole of the d.m.v. cannot be excluded, but a reinvestigation of the origin of these organelles seems to be necessary. Ultrastructural data on a possible precursor of the d.m.v. in the oocyte, termed polymerosome, have been reported by McCann-Collier (1977).

Pucci-Minafra et al. (1969) described 700 Å-particles, composed of 120 Å subunits, present in large patches at the periphery of the egg and also occupying almost all the space between the yolk granules. The particles are sensitive to both amylase and ribonuclease, and from this it was concluded that the particles are clusters of ribosomes embedded in an amylase-sensitive matrix. Geuskens and de Jonghe d'Ardoye (1971) conclude, however, that these particles are α-glycogen rosettes. Similar structures are present in the egg of *Nassarius reticulatus*. In this species the particles are stained by the periodic acid-thiocarbohydrazide-silver proteinate method (Thiery, 1967), which is specific for polysaccharides. They also show the masking effect after methanolic uranyl acetate staining, which is specific for glycogen (Bhatnagar and Leeson, 1975). These reactions indicate that in *Nassarius reticulatus* the particles are glycogen. Application of the same cytochemical methods to eggs of *Ilyanassa obsoleta* should establish whether there is reason to re-examine the effect of RNase-digestion.

The cytoplasm of the polar lobe of *Dentalium* cannot be distinguished as clearly from the blastomeres as in *Nassarius* (Fig. 1). Reverberi (1970) reported that the polar lobe contains many mitochondria and multisheet vesicles (m.s.v.), few pigment and yolk granules, and no cortical granules. He attributes a morphogenetic role to the DNA in small yolk granules and in mitochondria (Reverberi, 1970, 1972), and also to the multisheet vesicles, which he considers to be similar to the double-membrane vesicles in *Nassarius*. Our investigations of *Dentalium* show that the m.s.v. are quite different from the double-membrane vesicles as regards structure as well as localization. The m.s.v. consist of an outer unit-membrane, an inner amorphous layer closely applied to the outer membrane, and a central space filled with a substance of variable density (Fig. 14). The compartment between the outer membrane and the inner layer, which in double-membrane vesicles contains a dense body and other material, is absent in m.s.v. Reverberi's (1970) report that m.s.v. occur exclusively in the polar lobe could not be confirmed. We found that m.s.v. are very numerous in the cortical layer stretching from the constriction of the polar lobe to the animal pole. In the polar lobe the m.s.v. do not occur specifically in the cortex, but they are scattered throughout the cytoplasm. The m.s.v. in the cortex of the blastomeres are called immature cortical granules by Reverberi (1970), but our preliminary cytochemical investigations, using the carbohydrate-specific staining methods of Thiery (1967) and Rambourg (1967), suggest that they are identical to the m.s.v. in the polar lobe, as they react identically. The inner amorphous layer of the m.s.v., both in the cortex of the blastomeres and in the polar lobe, is heavily contrasted by these staining methods.

D. *Centrifugation Experiments on Large Polar Lobes*

After centrifugation of *Dentalium* eggs the cytoplasmic constituents are stratified in various configurations. Although the polar lobe receives different cytoplasms, centrifugation does not change the development of intact or lobeless embryos (Verdonk, 1968). In *Ilyanassa* the yolk is mainly situated in the vegetal region of the egg, which is incorporated in the polar lobe. Taking advantage of the distribution of the yolk, Clement (1968) centrifuged eggs of *Ilyanassa* either in animal-vegetal direction or in reverse. After re-centrifugation nucleated animal or vegetal halves, free of yolk, were obtained. The animal halves never formed a polar lobe and lobe-dependent structures, whereas vegetal halves, from which the yolk was removed, formed a polar lobe of appropriate size. About half of these vegetal fragments differentiated lobe-dependent structures. From these experiments it is evident that the displaceable components of the polar lobe cytoplasm do not contain the lobe-specific morphogenetic determinants.

In order to find out which organelles cannot be displaced, the eggs of *Dentalium* and *Nassarius reticulatus* were studied with the electron microscope after centrifugation. In *Nassarius* eggs, centrifuged in reverse, the polar lobe is filled with lipid droplets. The normal cytoplasm, consisting mainly of yolk, double-membrane vesicles, and glycogen, seems to be displaced as a whole to the animal pole. The components of this polar lobe cytoplasm are not stratified in separate layers by a force of about 600 g, as applied by Clement (1968). These observations are a strong argument against a morphogenetic role of yolk granules, double-membrane vesicles, and glycogen. We did not find a non-displaceable component in the polar lobe. In centrifuged eggs of *Dentalium* we found that cortical granules and multisheet vesicles, present in the cortex of the blastomeres, are not displaced by a force of about 400 g, as used by Verdonk (1968). In the polar lobe these organelles do not occur specifically in the cortex but are scattered throughout the cytoplasm. Hence they are easily displaced and stratify in a thin layer between the yolk granules and the hyaline zone. Whatever the reason for their presence in the polar lobe, their being displaced by centrifugation indicates that they probably do not determine lobe-dependent structures. This also applies to the mitochondria and small yolk granules. As in *Nassarius*, we did not find any non-displaceable organelle in the polar lobe of *Dentalium*.

The question, therefore, arises whether the morphogenetic determinants are situated in the cytoplasm or, as supposed already by Morgan (1933, 1935), in the surface layer of the vegetal hemisphere. This view received strong support by recent unpublished experiments of van den Biggelaar, who with a small pipette sucked the cytoplasm out of the polar lobe of *Dentalium* at the trefoil stage. The cytoplasm of the lobe is then gradually replaced by cytoplasm flowing in from the rest of the egg. Although nearly the whole content of the lobe was removed, all eggs, which continued cleaving, developed into normal embryos. Van den Biggelaar also removed cytoplasm of the second polar lobe of *Dentalium* and injected this cytoplasm into the B-blastomere of another egg at the 4-cell stage. This attempt to give to a B-blastomere, which does not receive a part of the polar lobe in normal development, the quality of a D-blastomere failed. No duplication of dorsal structures was found, whereas 79% of these embryos showed a normal development.

The above data are best understood on the assumption that in eggs with large polar lobes, such as in *Dentalium* or *Ilyanassa*, the morphogenetic determinants are not situated in the cytoplasm. They are either bound to or localized in the plasma membrane of the vegetal hemisphere of the egg.

III. THE RELATION BETWEEN THE CORTEX OF THE POLAR LOBE AND CYTOPLASMIC LOCALIZATIONS

The processes which lead to the typical localization of morphogenetic determinants in eggs are still largely obscure. Several mechanisms are probably involved. In ctenophores, for example, the morphogenetic factors are initially distributed in a uniform manner. They become localized in different regions of blastomeres as a consequence of the cytoplasmic movements accompanying cleavage (Freeman, 1976, 1977). Another mechanism is found in molluscan polar lobe-forming eggs, where the lobe-specific factors are strictly localized at the vegetal pole at an early stage, in *Bithynia* already during oogenesis.

Many phenomena related to cytoplasmic localization can be explained by assuming that local areas of the cortex of the egg may acquire special properties, resulting in an attraction of morphogenetic determinants or other cytoplasmic substances. Such a mechanism can explain the origin of cytoplasmic localizations, their resistance to displacement, or their return to the original site after having been displaced by centrifugation (see Kühn, 1965; Raven, 1970). The best direct evidence presently available is the existence of local surface differentiations which correlate with cortical or cytoplasmic localizations. Polar lobe-forming eggs provide some striking examples of these correlations.

In *Dentalium* a tuft of bacteria typically attaches to the vegetal pole area. The bacteria remain at this site during development and they are ultimately found in the follicle of growing oocytes (Geilenkirchen *et al.*, 1971; van Dongen, 1977). The bacteria do not penetrate into the egg, but they are attached in pits in the surface of the egg (Fig. 16). Their presence in a limited area suggests that the vegetal pole surface contains specific

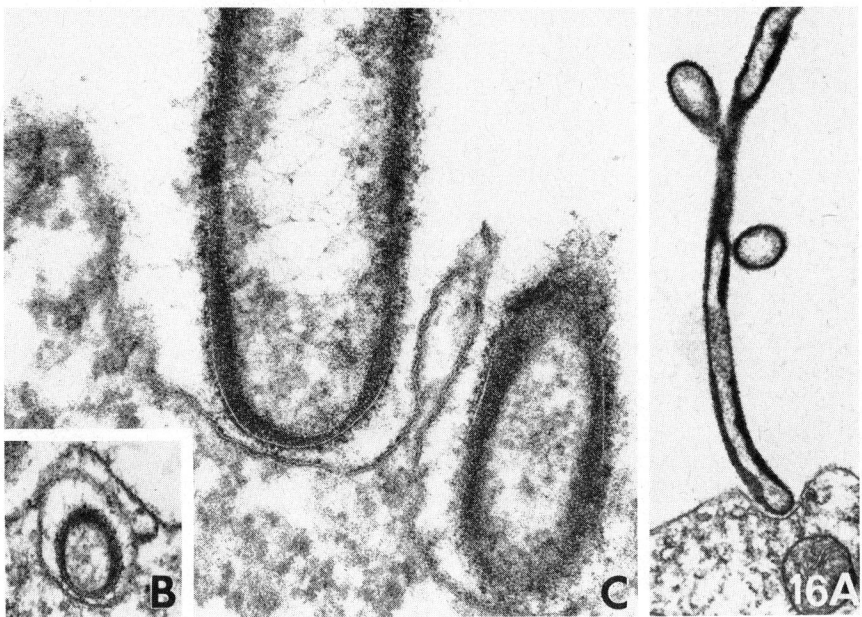

Fig. 16. *Dentalium*. Bacterium attached in surface pit at the vegetal pole (Fig. A). Fig. B shows a cross-section through a bacterium at its base in the pit. Fig. C shows that the plasma membrane of the bacterium is not in direct contact with the plasma membrane of the egg. Fig. B x 40500, Fig. C x 73000; Fig. A x 18000.

substances, not present elsewhere, that allow the bacteria to attach only at this site. Another possibility is that the vegetal pole lacks inhibiting substances which prevent the attachment of the bacteria elsewhere. On the cytoplasmic side of the cortex of the vegetal pole Reverberi (1970) has observed another local differentiation in *Dentalium*; the absence of cortical granules and cortically bound multisheet vesicles.

In *Nassarius reticulatus* the vegetal pole area is characterized by the absence of the uniform type of microvilli that covers the rest of the egg. Instead other surface structures develop during maturation and cleavage (Dohmen and van der Mey, 1977) (Fig. 17).

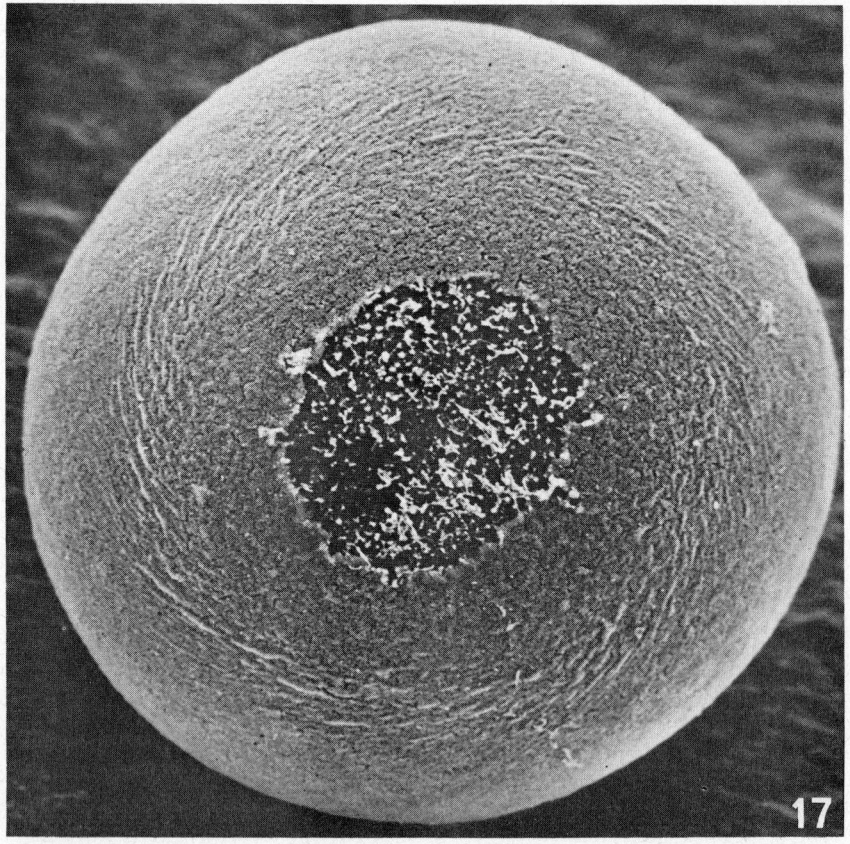

Fig. 17. *Nassarius reticulatus*. Scanning electron micrograph of an uncleaved egg at first meiotic division, viewed on the vegetal pole. The vegetal area is characterized by the absence of the uniform type of microvilli that covers the rest of the egg. Instead, other surface structures develop during maturation and cleavage. In this preparation a few large villi can be discerned in this area. x 650.

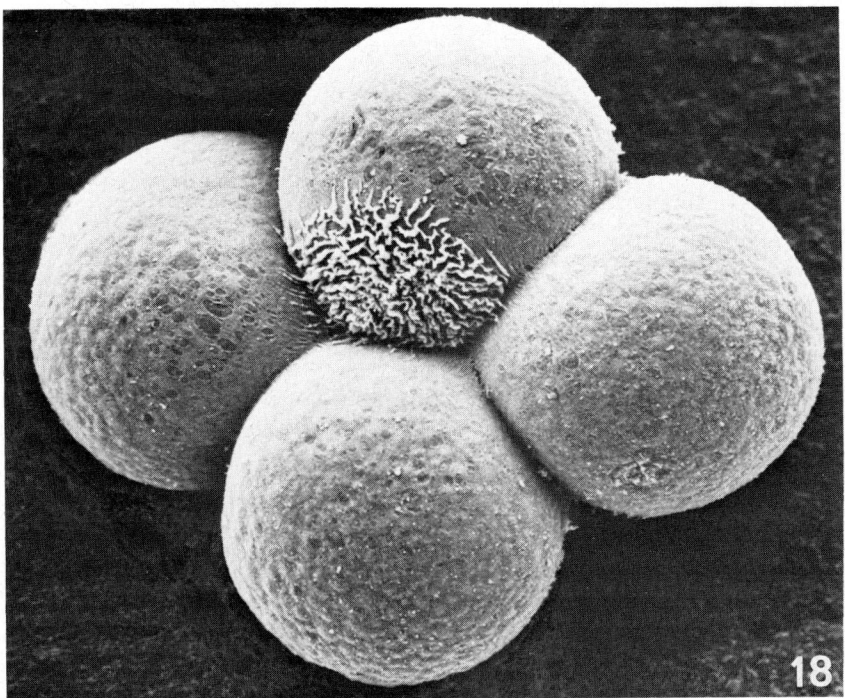

Fig. 18. *Crepidula fornicata*. Four-cell stage, vegetal side. Surface folds are present on the D-cell, in the area where the second polar lobe has been resorbed. x 400.

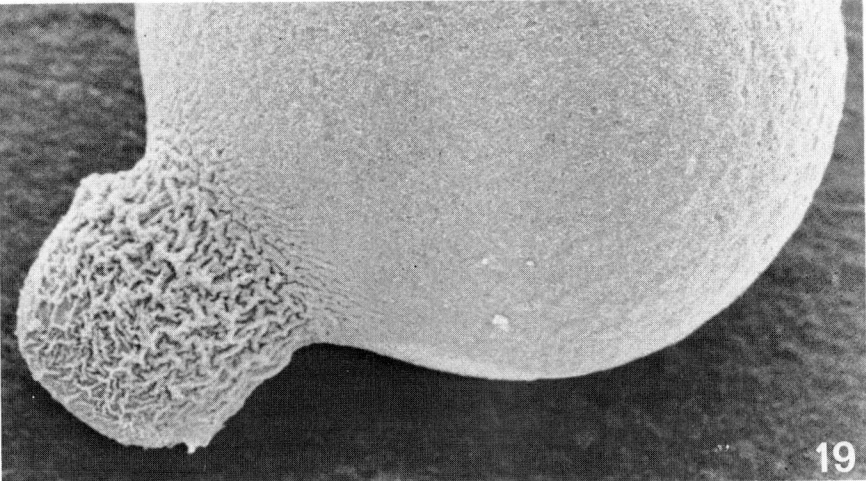

Fig. 19. *Buccinum undatum*. Polar lobe of an uncleaved egg. The surface ridges are restricted to the lobe area; they are small in the center of the area and large in the periphery. x 350.

In *Crepidula fornicata* and *Buccinum undatum* the whole polar lobe shows a special surface architecture (Dohmen and van der Mey, 1977). In *Crepidula* the surface of this area is thrown into folds, which are present already before the appearance of the polar lobe and persist on the D-quadrant after the resorption of the lobe (Fig. 18). In *Buccinum* the surface of the polar lobe differentiates into protrusions bearing each a large number of microvilli. In this way surface ridges are formed, small ones in the center of the area and large ones in the periphery (Fig. 19). On the cytoplasmic side of the cortex a dense fibrillar layer, about 0.15 μm thick, is always present in the polar lobe.

In *Bithynia* the first and second polar lobes are covered with long slender villi, which are occasionally well preserved in fixed preparations (Fig. 6). They can also be observed in living eggs with Nomarski optics. In eggs of *Littorina obtusata*, *Littorina saxatilis*, and *Nucella lapillus* we also observed a special surface architecture on the polar lobe, consisting of longer and more elaborate villi than on the rest of the egg.

Apart from studying the surface architecture of polar lobes, we also investigated whether this area may be characterized at a molecular level by the presence of specific substances, e.g. glycoproteins or glycolipids. The carbohydrate part of these molecules can be conveniently studied by means of lectins, and in this way regional surface differentiations have been detected in several eggs, e.g. in *Ascidia malaca* (Ortolani et al., 1977). Up till now we did not find any special localization of carbohydrates on the surface of a polar lobe, but in *Nassarius reticulatus* we observed that an area which shows a strong binding of concanavalin A at the surface correlates with a zone of lipid droplets in the cytoplasm. This correlation is maintained from the uncleaved egg until late cleavage stages. Its significance is not yet clear and is not relevant to the present discussion of polar lobes, but its existence means that we should expect similar correlations to occur on polar lobes as well. Further study of polar lobes by means of lectins and other surface probes may provide interesting data.

The presence of a special surface architecture on polar lobes obviously correlates with the presence of morphogenetic localizations in the cortex or the cytoplasm. The crucial question to be answered is which kind of relation exists between the two phenomena. At present we can only speculate about the function of the surface structures, as experimental data are not yet available. They may serve to bring about or maintain the localization of morphogenetic factors, or they may be necessary for the activation of the morphogens or for some other aspect of their functioning. Another possibility is that they are only secondarily related to the morphogenetic factors in the polar lobe. They may, for instance, be responsible for the appearance of a polar lobe at the right place, or they may be a reaction of the cell surface on processes going on in the vegetal pole plasm without having a primary significance for development.

A common characteristic of all the above described surface differentiations is the increased surface area. This may indicate an increased transport through the membrane, e.g. of ions, resulting in an electrical current. Electrical currents have been thoroughly studied in eggs of brown algae. In the *Pelvetia* egg, for instance, Nuccitelli (1978) has shown that ooplasmic segregation and secretion are accompanied by a membrane-generated current. This author suggests that the currrent may control segregation by locally activating a mechanical force such as a microfilament array or by generating an electrical field to electrophorese cytoplasmic vesicles.

The egg of *Bithynia* seems to be a suitable object for studying the relation between surface structures and cytoplasmic localizations because of the presence of the easily observable vegetal body. This body is localized already at the prospective vegetal pole before the appearance of the long villi that are observed on the polar lobes. It is not likely, therefore, that in this species the surface differentiations bring about the localization of the body. We may try to elucidate the mechanism which binds the vegetal body to the cortex by morphological methods, e.g. electron microscopy, as well as by experimental methods, e.g. destroying the bond by means of agents such as cytochalasin B or local anesthetics. The discovery of a specific agent, capable of detaching the vegetal body from the cortex, would be an important step towards understanding the mechanism of localization, apart from being a useful tool in studying other aspects of the role of morphogens.

A preliminary study of the cortical layer, termed the attachment zone, that separates the vegetal body from the plasma membrane has not revealed any special structures. Microtubules and microfilaments are virtually absent. This agrees with the observations of Zalokar (1974), who found that in eggs of the ascidian *Phallusia* ooplasmic segregation is not inhibited by colchicine and cytochalasin B. Peaucellier *et al.* (1974), studying the effect of cytochalasin B on the development of *Sabellaria*, conclude, however, that developmental information may be detached from the cortex by this drug and thus be lost. Also Arnold and Williams-Arnold (1974, 1976) found that, in *Loligo* eggs, treatment of a prospective organogenic area with cytochalasin B leads to specific defects.

In order to establish whether surface differentiations on polar lobes play a role in the activation of morphogenetic factors, we must first of all know when this activity starts. In *Ilyanassa* and *Dentalium* polar lobe deletion experiments, resulting in an alteration of the time schedule as well as the pattern of the cleavages, suggest that at least part of the activity of the morphogens starts during cleavage (Clement, 1952; van Dongen and Geilenkirchen, 1975). This view is supported by studies of protein synthesis in the progeny of isolated blastomeres (Donohoo and Kafatos, 1973) and RNA synthesis in lobeless embryos (Collier, 1977). The analytical methods used are probably not sufficiently sensitive to detect the earliest signs of morphogenetic activity, so we cannot draw conclusions on the eventual relation between this activity and surface structures. In *Bithynia* such a relation seems unlikely, as the centrifugation experiments described above show that the vegetal body may be displaced one hour before first cleavage and still be capable of inducing normal development in lobeless eggs.

IV. THE POSSIBLE ROLE OF RNA AS A MORPHOGENETIC FACTOR IN POLAR LOBES

At the present moment the accumulated evidence on the mode of action of polar lobe determinants does not provide a coherent picture, and the nature of these determinants remains fully unknown. The most obvious hypotheses are that the determinants act by 1) inducing precocious transcription of genes in a cell lineage-specific manner, or by 2) translational regulation of prelocalized RNA templates.

Evidence has been presented that deletion of the polar lobe reduces both transcription (Davidson *et al.*, 1965; Koser and Collier, 1976; Collier, 1977) and translation (Abd-el-Wahab and Pantelouris, 1957; Collier, 1961) in the lobeless embryo. The diminution in RNA synthesis after removal of the polar lobe suggests that the polar lobe regulates the differentiation of the embryo by controlling differential gene transcription. There is no proof, however, that the genes which are not transcribed in lobeless embryos are primarily responsible for the development of lobe-dependent structures in normal embryos. On the contrary, there is evidence in favour of a translational control mechanism. Guerrier (1971) has demonstrated that in the annelid *Sabellaria* lobe-dependent structures can develop in embryos in which transcription is inhibited by actinomycin D. In *Ilyanassa* deletion of the polar lobe results in a modified pattern of protein synthesis (Freeman, 1972; Newrock and Raff, 1975). Isolated AB- and CD-progeny also produce different proteins (Donohoo and Kafatos, 1973). There is evidence that the differences in protein synthesis between normal and lobeless embryos are independent of concomitant embryonic transcripton (Raff *et al.*, 1976). These data suggest that the polar lobe exerts its control by direct regulation of the translation of preformed and probably prelocalized messengers coding for cell lineage-specific proteins.

The existence of mRNA in the polar lobe has been demonstrated in *Ilyanassa* (Clement and Tyler, 1967; Geuskens, 1969; Geuskens and de Jonghe d'Ardoye, 1971), but it is not known whether this is polar lobe-specific mRNA responsible for the development of lobe-dependent structures. In the polar lobe of *Bithynia* the presence of a large amount of RNA in the vegetal body has been demonstrated cytochemically by methylgreen-pyronin staining, with controls reacting negatively after digestion by RNase (Dohmen and Verdonk, 1974). This is confirmed by the results of fluorescence microscopy of eggs incubated with the nucleic acid-specific stains acridine orange and bisbenzimid fluorochrom Hoechst 33258 (Riedel-De Haën AG, Seelze-Hannover, GFR). The last one, applied at pH 2 and pH 7, allows to distinguish between DNA and RNA (Hilwig and Gropp, 1975) (Fig. 20). The polar lobe of the oyster *Gryphaea angulata* has also been reported to contain a large amount of RNA (Pasteels and Mulnard, 1957), and there are indications that this may also be the case in the polar lobe of *Buccinum*.

At present we do not know which kind of RNA is accumulated in the RNA-rich polar lobes. Ribosomes do not occur in sufficient concentrations to account for the positive RNA reactions in the polar

Fig. 20. *Bithynia*. Fluorescence micrograph of first cleavage stage, treated with the fluorochrom Hoechst 33258 at pH 2. The bright fluorescence of the vegetal body in the polar lobe indicates the presence of a large amount of RNA.

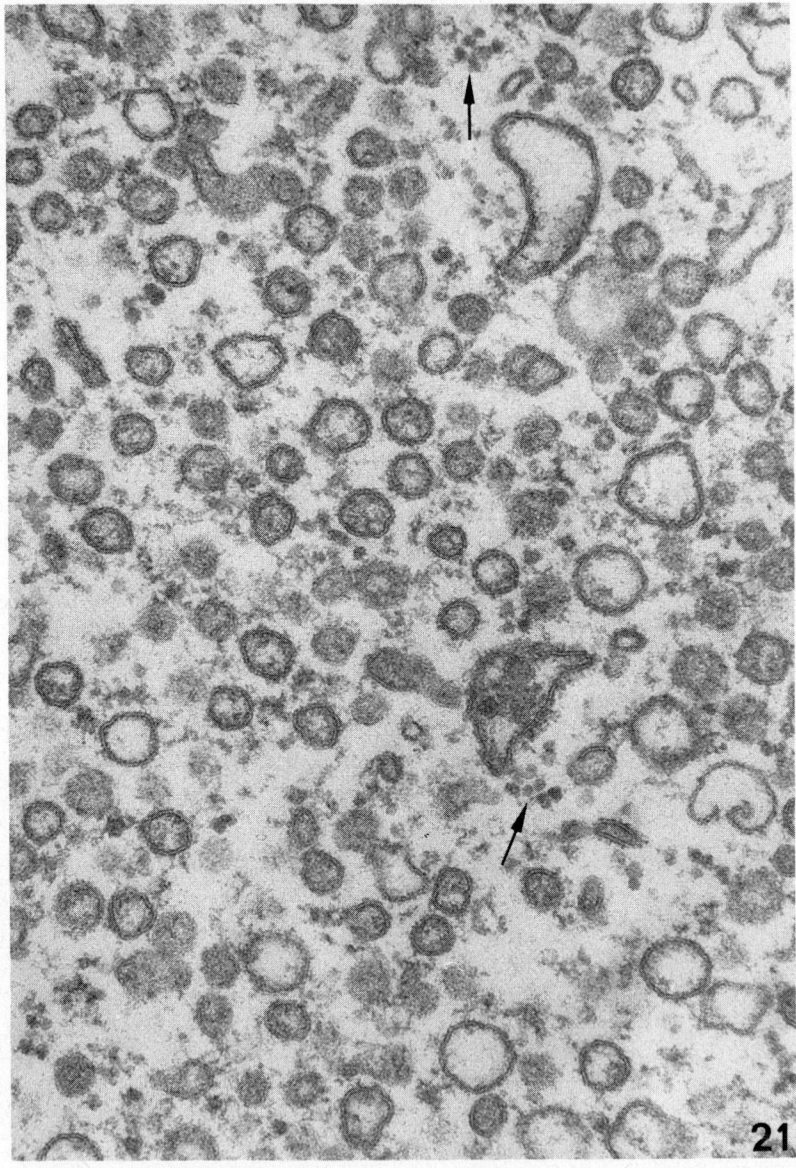

Fig. 21. *Bithynia*. Detail of the vegetal body in the first polar lobe. A few ribosomes and polysomes (arrows) are present in between the vesicles. × 105000.

lobe of *Bithynia* (Fig. 21). Experiments are presently being carried out to find out if there is a concentration of poly(A)-containing mRNA in this lobe by means of *in situ* hybridization of radioactively labelled poly(U) with the poly(A)-group.

The possible role of RNA as a morphogenetic determinant is a problem which is not encountered in polar lobes only. The germ plasm in the eggs of insects and amphibians also contains RNA (Blackler, 1958; Mahowald, 1971), but its role, if any, in determining germ cells is not yet known. Morphogenetic RNA is not necessarily messenger RNA. From chick embryonic heart Deshpande *et al.* (1977) isolated a low molecular weight RNA species, containing poly(A) but not translatable, which is capable of inducing heart differentiation in early chick blastoderm. These authors also discuss other reports of RNA species that seem to be involved in the regulation of translation or transcription. Apparently the possibility exists that RNA in polar lobes can exert a morphogenetic influence without necessarily being messenger RNA. This influence may be the regulation of translation as well as transcription. Both processes probably play a role in the origin of lobe-dependent structures.

Prospects for future research are greatly enhanced by the growing evidence that the special cytoplasms in small polar lobes, especially the vegetal body of *Bithynia*, contain the lobe-specific morphogenetic factors. This offers the opportunity to experiment on a clearly defined and visible structure, including its biochemical isolation and the subsequent purification and assay of morphogenetic constituents: RNA or other substances.

V. SUMMARY

From the available morphological and experimental evidence it is concluded that the localization of morphogenetic determinants is different in large and small polar lobes. Small polar lobes generally seem to contain special plasms, consisting of vesicular aggregates. Combined deletion- and centrifugation-experiments show that in *Bithynia* the morphogenetic determinants are indeed localized in a large vesicular aggregate termed the vegetal body. In large polar lobes special plasms have not been detected. From the failure of efforts to displace the morphogenetic determinants by centrifugation or to remove them by sucking out the polar lobe cytoplasm it is concluded that in large polar lobes the determinants are localized in the cortex.

The nature of the specific factors in polar lobes remains fully unknown. The possible role of RNA, whether mRNA or another RNA species, as a morphogenetic determinant is discussed, in view of the presence of a large amount of RNA in the polar lobe of *Bithynia* and probably in other species as well.

Several examples of local surface differentiations on polar lobes are described. Their presence correlates with the cortical or cytoplasmic localizations of morphogenetic factors. Possible functions of the surface structures are discussed.

REFERENCES

Abd-el-Wahab, A. and Pantelouris, E.M. (1957). *Exp. Cell Res.* **13**, 78-82.
Arnold, J.M. and Williams-Arnold, L.D. (1974). *J. Embryol. Exp. Morphol.* **31**, 1-25.
Arnold, J.M. and Williams-Arnold, L.D. (1976). *Amer. Zool.* **16**, 421-446.
Bhatnagar, R. and Leeson, T.S. (1975). *Stain Technol.* **50**, 213-217.
Blackler, A.W. (1958). *J. Embryol. Exp. Morphol.* **6**, 491-503.
Cather, J.N. (1971). *Adv. Morphogen.* **9**, 67-125.
Cather, J.N. and Verdonk, N.H. (1974). *J. Embryol. Exp. Morphol.* **31**, 415-422.
Cather, J.N., Verdonk, N.H. and Dohmen, M.R. (1976). *Amer. Zool.* **16**, 455-468.
Clement, A.C. (1952). *J. Exp. Zool.* **121**, 593-626.
Clement, A.C. (1968). *Develop. Biol.* **17**, 165-186.
Clement, A.C. and Lehmann, F.E. (1956). *Naturwissenschaften* **43**, 478-479.
Clement, A.C. and Tyler, A. (1967). *Science* **158**, 1457-1458.
Collier, J.R. (1961). *Acta Embryol. Morphol. Exp.* **4**, 70-76.
Collier, J.R. (1977). *Exp. Cell Res.* **106**, 390-394.
Crampton, H.E. (1896). *Arch. Entwicklungsmech.* **3**, 1-19.
Crowell, J. (1964). *Acta Embryol. Morphol. Exp.* **7**, 225-234.
Davidson, E.H., Haslett, G.W., Finney, R.J., Allfrey, V.G., and Mirsky, A.E. (1965). *Proc. Nat. Acad. Sci. U.S.* **54**, 696-704.
Deshpande, A.K., Jakowlew, S.B., Arnold, H.H., Crawford, P.A. and Siddiqui, M.A.Q. (1977). *J. Biol. Chem.* **252**, 6521-6527.
Dohmen, M.R. and Lok, D. (1975). *J. Embryol. Exp. Morphol.* **34**, 419-428.
Dohmen, M.R. and van der Mey, J.C.A. (1977). *Develop. Biol.* **61**, 104-113.
Dohmen, M.R. and Verdonk, N.H. (1974). *J. Embryol. Exp. Morphol.* **31**, 423-433.
van Dongen, C.A.M. (1977). *Proc. K. Ned. Akad. Wet. Ser. C* **80**, 372-376.
van Dongen, C.A.M. and Geilenkirchen, W.L.M. (1974). *Proc. K. Ned. Akad. Wet. Ser. C.* **77**, 57-100.
van Dongen, C.A.M. and Geilenkirchen, W.L.M. (1975). *Proc. K. Ned. Akad. Wet. Ser. C.* **78**, 358-375.
Donohoo, P. and Kafatos, F.C. (1973). *Develop. Biol.* **32**, 224-229.
Freeman, S.B. (1972). *J. Embryol. Exp. Morphol.* **26**, 339-349.
Freeman, G. (1976). *Develop. Biol.* **49**, 143-177.
Freeman, G. (1977). *J. Embryol. Exp. Morphol.* **42**, 237-260.
Geilenkirchen, W.L.M., Timmermans, L.P.M., van Dongen, C.A.M. and Arnolds, W.J.A. (1971). *Exp. Cell Res.* **67**, 477-479.

Gérin, Y. (1971). *J. Embryol. Exp. Morphol.* **25**, 423-438.
Gérin, Y. (1972). *J. Microscopie* **13**, 57-66.
Geuskens, M. (1969). *Exp. Cell Res.* **54**, 263-266.
Geuskens, M. and de Jonghe d'Ardoye, V. (1971). *Exp. Cell Res.* **67**, 61-72.
Guerrier, P. (1971). *Exp. Cell Res.* **67**, 215-218.
Guerrier, P., van den Biggelaar, J.A.M., van Dongen, C.A.M. and Verdonk, N.H. (1978). *Develop. Biol.* **63**, 233-242.
Hilwig, I. and Gropp, A. (1975). *Exp. Cell Res.* **91**, 457-460.
Humphreys, W.J. (1964). *J. Ultrastruct. Res.* **10**, 244-262.
Koser, R.B. and Collier, J.R. (1976). *Differentiation* **6**, 47-52.
Kühn, A. (1965). "Vorlesungen über Entwicklungsphysiologie". 2nd ed. Springer Verlag, Berlin.
Mahowald, A.P. (1971). *J. Exp. Zool.* **176**, 345-352.
McCann-Collier, M. (1977). *J. Morphol.* **153**, 119-127.
Morgan, T.H. (1933). *J. Exp. Zool.* **64**, 433-467.
Morgan, T.H. (1935). *Biol. Bull.* **68**, 268-279.
Newrock, K.M. and Raff, R.A. (1975). *Develop. Biol.* **42**, 242-261.
Nuccitelli, R. (1978). *Develop. Biol.* **62**, 13-33.
Ortolani, G., O'Dell, D.S., and Monroy, A. (1977). *Exp. Cell Res.* **106**, 402-404.
Pasteels, J.J. and Mulnard, J. (1957). *Arch. Biol. Liège* **68**, 115-163.
Peaucellier, G., Guerrier, P. and Bergerard, J. (1974). *J. Embryol. Exp. Morphol.* **31**, 61-74.
Pucci, I. (1961). *Acta Embryol. Morphol. Exp.* **4**, 96-101.
Pucci-Minafra, I., Minafra, S. and Collier, J.R. (1969). *Exp. Cell Res.* **57**, 167-178.
Raff, R.A., Newrock, K.M., Secrist, R.D. and Turner, F.R. (1976). *Amer. Zool.* **16**, 529-545.
Rambourg, A. (1967). *J. Histochem. Cytochem.* **15**, 409-412.
Raven, C.P. (1970). *Int. Rev. Cytol.* **28**, 1-44.
Reverberi, G. (1970). *Acta Embryol. Exp.* **12**, 31-43.
Reverberi, G. (1972). *Acta Embryol. Exp.* **14**, 135-166.
Reverberi, G. and Mancuso, V. (1961). *Acta Embryol. Morphol. Exp.* **4**, 102-121.
Schmekel, L. and Fioroni, P. (1975). *Cell Tiss. Res.* **159**, 503-522.
Thiery, T.P. (1967). *J. Microscopie* **6**, 987-1018.
Verdonk, N.H. (1968). *J. Embryol. Exp. Morphol.* **19**, 33-42.
Verdonk, N.H. and Cather, J.N. (1973). *J. Exp. Zool.* **186**, 47-61.
Zalokar, M. (1974). *Wilhelm Roux' Arch.* **175**, 243-248.

Cytoplasmic Determinants of Tissue Differentiation in the Ascidian Egg

J. R. Whittaker

The Wistar Institute of Anatomy and Biology
36th Street at Spruce
Philadelphia, Pennsylvania 19104
and
Marine Biological Laboratory
Woods Hole, Massachusetts 02543

I. Introduction 29
II. Evidence of Morphogenetic Determinants 32
   A. Visible Cytoplasmic Segregations 32
   B. Determinate Cleavage and Cell Lineage 33
   C. Restricted Developmental Potential 34
   D. Differentiation without Cleavage 37
   E. Differentiation Change by Altered Cytoplasmic
      Segregation 41
III. The Mitochondrial-Associated Muscle Determinant 42
IV. Maternal RNA as the Alkaline Phosphatase
    Determinant 43
V. Messenger RNA Synthesis during Development 46
VI. Possible Nature of Other Determinants 47
VII. Conclusions 49
    References 50

I. INTRODUCTION

Ascidians (Subphylum Tunicata or Urochordata; Class Ascidiacea) are sessile, filter feeding animals that have solved certain ecological problems of dispersal and site selection by evolving a rapidly developing and transitory larva with highly specialized tissues adapted to these ecological needs. This is the so-called tadpole larva, having a notochord, dorsal tubular nervous system (Fig. 1) and, in some cases, "gill slits". The larva is believed to be an invention of the tunicates and also the progenitor of the vertebrate body plan (Berrill, 1955). These two opinions have provoked some interesting challenges and disagreements, yet the obvious vertebrate connection has caused ascidians to remain objects of considerable interest to biologists for over a century.

The larva has a locomotory tail which contains muscle tissues and the notochord (Fig. 1A,C). There is also a nervous system which innervates the tail dorsally from a cerebral ganglion or brain in the head region. Two sensory structures within the brain are concerned with larval responses to light and gravity, each containing a large melanocyte as part of the structure (Fig. 1B). In addition, there are attachment organs (palps) on the head, and a hydrodynamic streamlining of the larva contributed by a thin, smooth, translucent test surrounding the larva; this test is flattened to a wide fin in the tail region.

Depending on the needs of particular species, some or all of the main features of the early postmetamorphic zooid may be precociously differentiated in the larva. This frequently includes an early development of branchial stigmata that are assumed to be possible precursors of the vetebrate gill slits. After metamorphosis, which involves attachment and tail resorption processes, the strictly larval tissues eventually cytolyze and become phagocytized.

Special interest in the embryology of the group lies in the very early specification and differentiation of cells associated with transient larval structures, a characteristic the ascidians also share with many other invertebrate groups. Fates of cells are blocked out in relation to an egg polarity during the initial cleavage stages of development, and cleavage stage blastomeres already have unusual autonomy in the expression of their determined fates. Investigation of these properties by the classical experimental embryologists resulted in the concept of developmental mosaicism: the egg has a mosaic pattern of localized organelles or substances, the morphogenetic determinants, that are responsible for establishing many of the larval tissue differentiations.

Ascidians provide one of the more extreme examples of such mosaicism, but the developmental mechanisms so implied occur commonly enough in other animal groups (Wilson, 1925; Davidson, 1976). Even vertebrates retain some properties of mosaic development, notably in amphibian eggs which have embryonic axis-determining factors in the gray crescent region (Brachet, 1977) and vegetally localized germ plasm determinants (Beams and Kessel, 1974; Eddy, 1975). It is very

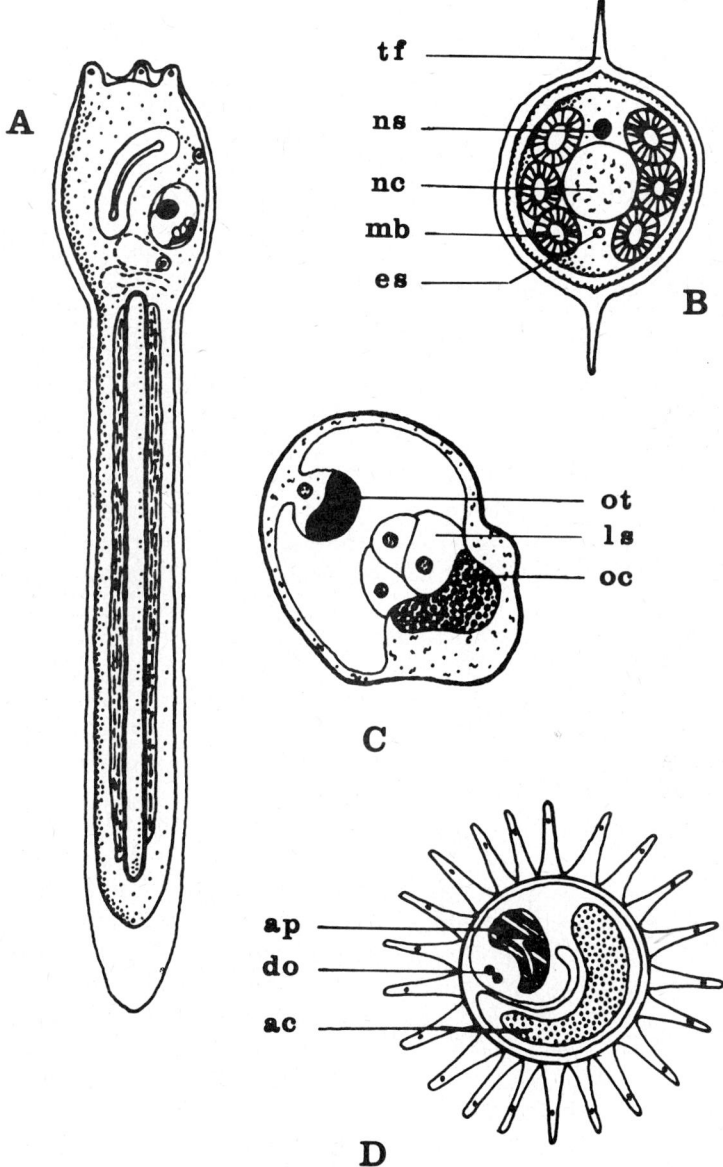

Fig.1. The *Ciona intestinalis* tadpole larva. A, structure of the larva. B, cross-section of tail. C, sensory vesicle of brain showing the two melanocytes. D, 12-hr embryo showing the location of three histospecific enzymes. Ac, acetylcholinesterase; ap, alkaline phosphatase; do, dopa oxidase; es, endodermal strand; ls, lens cells; mb, muscle band; nc, notochord; ns, nervous system; oc, ocellus melanocyte; ot, otolith melanocyte; tf, tail fin. After Berrill (1947, 1955).

likely that morphogenetic determinants represent an extreme and somewhat precocious use of processes that underly *all* patterns of cellular determination in animal development. Ordinarily, these processes are not seen so clearly, especially in vertebrate embryos, presumably because they function later in development and are obscured by the complexities of many intervening events.

This paper will reexamine the principal lines of evidence supporting the inference of a mosaic pattern of discrete morphogenetic determinants in the ascidian egg, and review recent work that pertains to uncovering the possible nature and identity of these determinants.

II. EVIDENCE OF MORPHOGENETIC DETERMINANTS

Five kinds of observations on ascidian embryos serve to establish a foundation for the theory that there are morphogenetic determinants localized in the egg cytoplasm. This evidence firmly supports the idea that such materials are segregated during early development into cells of particular tissue lineages and are in some way related to mechanisms by which different pathways of cell differentiation are selected (i.e. the process of cell determination). The evidence concerns: (i) visible cytoplasmic materials whose segregation can be correlated with particular tissue differentiations, (ii) demonstration of invariate cell lineages produced by a determinate cleavage pattern, (iii) a restricted developmental potential in isolated cells correlated with these cell lineages, (iv) a similarly restricted developmental potential in the autonomously developing cells of cleavage-arrested embryos, (v) determination changes caused by altered cytoplasmic segregation.

Recent advances in the analysis of this problem have been made possible by the histochemical identification of certain tissue-specific enzyme localizations during larval development (Fig 1D). Durante (1956) showed that muscle acetylcholinesterase (AChE) develops in the tail rudiment as early as the neurula stage. A tyrosinase (dopa oxidase) is found during early development of the brain melanocytes (Minganti, 1951; Whittaker, 1966, 1973a). An endodermal localization of alkaline phosphatase first occurs at late gastrulation and persists thereafter in the branchial and digestive tissue rudiments (Minganti, 1954).

A. *Visible Cytoplasmic Segregations*

E.G. Conklin (1905) found with *Styela partita* that material in a yellow pigmented crescent of the egg, which forms during the immediate postfertilization movements of cytoplasm, becomes localized during

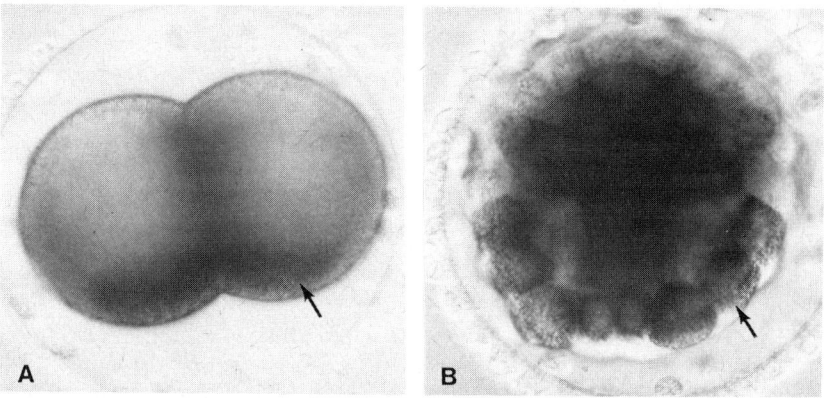

Fig. 2. Location of yellow crescent material in live embryos of *Styela plicata* by Nomarski interference phase microscopy. A, 2-cell stage. B, 64-cell stage. X 300.

progressive cleavages into the presumptive tail muscle cells of the developing larva. This is illustrated in Fig. 2 with photographs from a related species, *Styela plicata*. Only quite recently have we learned that this colored crescent of some species is caused by pigmented yolk granules adherent to mitochondria which are segregated into these cells (Berg and Humphreys, 1960); work of several investigators over the years had established that a large number of mitochondria are segregated into the presumptive muscle cells even of those species in which there are no easily observable crescents.

At the time of fertilization, cytoplasmic movements which are involved in localizing the posterior yellow crescent in *Styela* also produce an anterior gray crescent, a transparent hyaloplasm of the animal hemisphere, and the yolk-filled area of the vegetal hemisphere. In species which do not have pigmented crescents it is still possible to see some of these regional differences in living or otherwise in fixed and sectioned material (Conklin, 1911; Cohen and Berrill, 1936). Regional cytoplasmic differences involve obvious chemical or organellar variations. That such recognizable cytoplasm becomes visibly segregated into certain tissues has been historically a strong argument in favor of the existence of morphogenetic determinants.

B. *Determinate Cleavage and Cell Lineage*

Van Benedin and Julin (1884) were the first to relate early developmental stages of the ascidian egg to the later embryonic stages. They followed cleavage, cell by cell, as far as the 44-cell stage and established with considerable accuracy the relationship of these cells to the germ layers of the embryo. They determined the relations of the egg axes and early cleavage stages to axes of the gastrula and larva; theirs was the first demonstration in the history of embryology that the principal axes of the larva could be identified in the unsegmented egg. We now know that embryos, in general, have these mechanistically determinate cleavage patterns in relation to an essentially preformed axial organization of the egg. Yet, a century later, there is still no clear understanding of the structural basis and causes underlying such determinate cleavage sequences.

When Conklin (1905), who had previously studied cell lineage in molluscs, turned to *Styela*, the yellow crescent marking and other cytoplasmic differences of the egg permitted an unprecedented degree of accuracy to be obtained in charting the lineages because there were an already visible set of axial markings in the uncleaved egg. Conklin was able to follow lineage as far as the 218-cell stage and to learn what particular differentiated tissue types each of these cells gave rise to in the developing larva. Details of Conklin's lineage assignments have been confirmed and amended by the elegant cell marking experiments of Ortolani (1955, 1957, 1962). She tagged individual cells at each stage by sticking colored chalk particles to the cell membrane and followed these particles to their final destination in the larva. The lineage diagram shown in Fig. 3 is a composite summary of her results and Conklin's. It shows clearly the potential expression of the cells at each stage. There is, then, in ascidians a strictly determinate cleavage pattern that results in an invariant cell lineage by apportioning fixed regions of the cytoplasm into particular tissue lines.

Fate maps of this kind do not in themselves establish the certainty of a mosaic pattern to development, that is, a rigidly restricted potential expression of the cells. There remains the possibility that cells at each stage may have wider potentialities than those which they ordinarily express *in situ*. Indeed, such regulatory properties can be demonstrated in the embryos of some animal groups. It is the microsurgical isolation experiments that have shown ascidian lineage maps to be an accurate depiction of cell developmental potential at each stage.

C. *Restricted Developmental Potential*

A seldom remarked fact is that the history of modern experimental embryology began with the blastomere destruction experiments of Laurent Chabry (1887) using embryos of the ascidian *Ascidiella aspersa*. His work preceded the Roux experiments by one year and the equally famous Driesch experiments by five years. There have been many subsequent

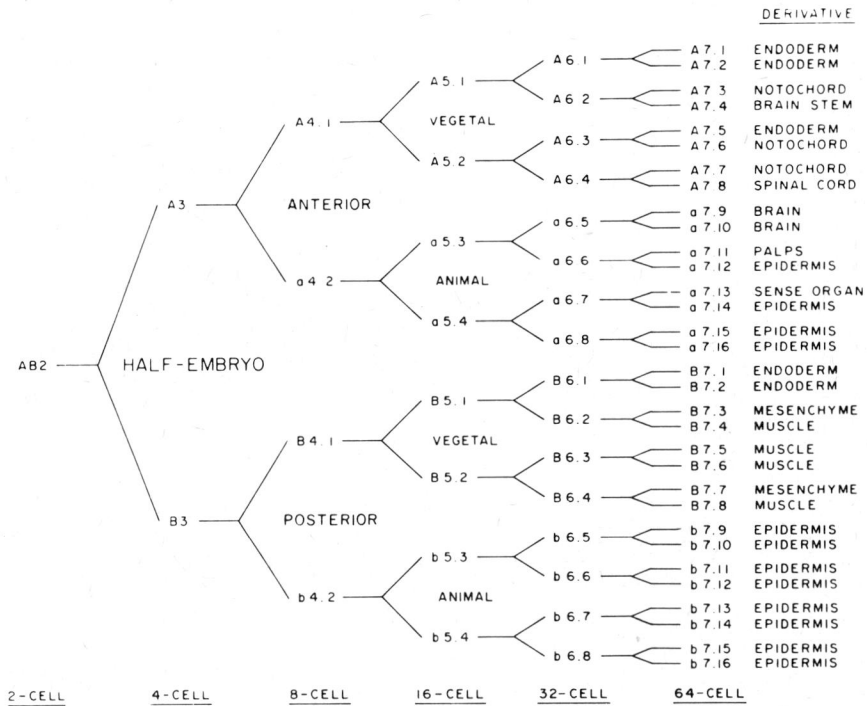

Fig. 3. Lineage fate map (bilateral half-embryo) of ascidian embryonic development. Constructed according to data of Conklin (1905), with corrections and additions by Ortolani (1955, 1957, 1962).

studies of developmental potential in surgically isolated early blastomeres of ascidians (reviewed by Reverberi, 1961), but Chabry's original conclusions have remained correct: larvae resulting from partial early embryos in which blastomeres at the 2- and 4-cell stages have been destroyed are defective in proportion to the amount of cell material removed. Later work, particularly that of Reverberi and Minganti (1946, 1947), has shown the potentiality of isolated early blastomeres never to be qualitatively more than what is predicted by the Conklin-Ortolani fate map.

An interesting recent confirmation of this restricted potential concerns the ability of muscle lineage blastomeres to give rise to tissues capable of producing the histospecific AChE of larval muscle (Whittaker et al., 1977). From lineage maps and the results of previous surgical isolations it is known that two cells at the 8-cell stage, the B 4.1 cell pair, carry the potential to produce larval muscle tissues. When these cells are removed and placed in isolation (Fig. 4A), they give rise to tissues which

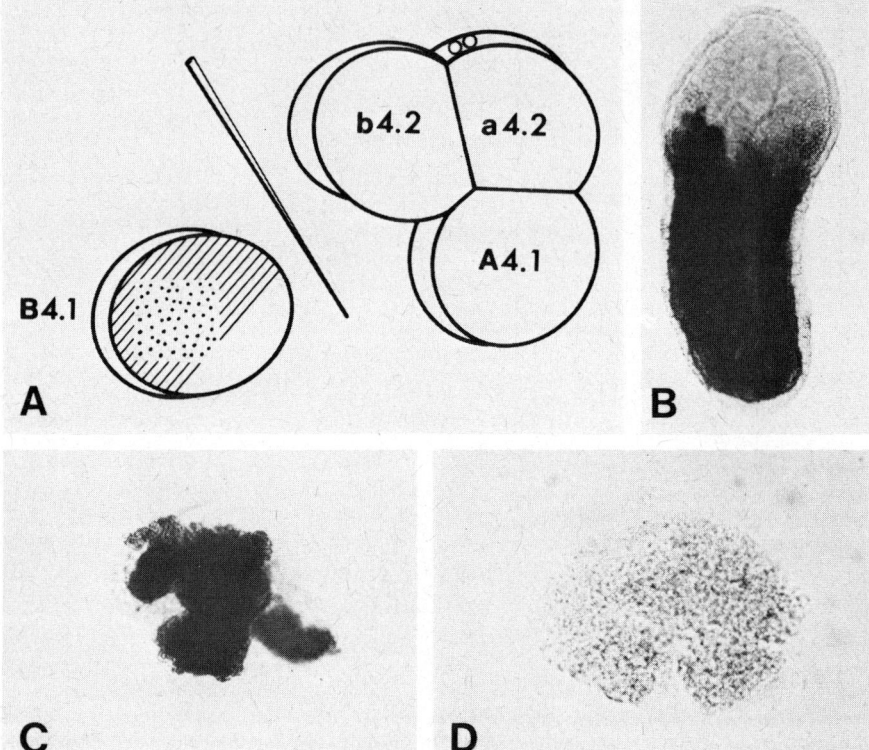

Fig. 4. Acetylcholinesterase development in progeny cells of muscle lineage blastomeres isolated at the 8-cell stage. A, diagram of the surgical isolation; B, enzyme localization in early tailbud control larva. C, enzyme localization in progeny cells of the isolated muscle lineage pair; same time as in B. D, embryo minus B4.1 cell pair; absence of reaction of "tailbud" stage. Photographs X 240. From Whittaker *et al.* (1977).

eventually produce muscle AChE (Fig. 4C). If, however, the cells are destroyed at the 8-cell stage, the remaining 6/8 of an embryo fails to produce any AChE-containing tissues (Fig. 4D). A similar series of experiments has verified that the 4 vegetal blastomeres at the 8-cell stage, which are lineage blastomeres for the endodermal tissues (Fig. 3), have the potential in isolation to give rise to tissues containing endodermally localized alkaline phosphatase (Whittaker *et al.*, 1979a).

Ascidian embryos are not completely mosaic. There is clear experimental evidence that cellular contact between neural ectoderm and either notochord or endodermal tissues is necessary for the normal development of brain and the differentiation of sensory structures containing the melanocytes (Reverberi *et al.*, 1960). Hence, isolated neural ectoderm cells never express their complete lineage potential, and in the absence of this inductive interaction between tissues there is no differentiation of melanin pigment in the melanocytes. Whittaker, Ortolani and Farinella-Ferruzza (unpublished data) have found, however, that the tyrosinase enzyme frequently develops in progeny cells of the isolated A 4.2 blastomere pair (Fig. 4A). Possibly inductive interactions are important in regulating aspects of higher structural organization, and differentiation at a certain biochemical level proceeds with relative autonomy.

D. *Differentiation without Cleavage*

Studies by F.R. Lillie (1902) on cleavage-arrested eggs of the annelid *Chaetopterus pergamentaceus* were a further indication of strong developmental autonomy in cells of mosaic embryos. When unsegmented *Chaetopterus* zygotes were treated with isotonic KCl, many of the eggs did not cleave and a few of these uncleaved eggs developed cilia and superficial characteristics of the annelid trochophore larva. This startling result was verified a number of times in the classic literature, but only recently has the idea that differentiation can occur independently of cleavage been used to probe further into the nature of mosaic development.

Ciona intestinalis zygotes which are cleavage-arrested at various stages with cytochalasin B prove to have the capacity to develop larval tail muscle AChE in some blastomeres of the cleavage-arrested embryos (Whittaker, 1971, 1973b). Cytochalasin B appears to inhibit cytokinesis by affecting assemby of microfilaments; the nuclei, however, continue to divide. If different early cleavage stages beginning with the 1-cell zygote are placed in cytochalasin B and then reacted histochemically for AChE activity at a time after normal production of the enzyme in development, activity is found localized in cells that match the pattern and (maximum) number of muscle lineage blastomeres shown by Conklin's study of yellow crescent segregation and indicated by the Conklin-Ortolani diagram (Fig. 3). These AChE-containing cells are depicted at various cleavage-arrested stages in Fig. 5.

Development of a localized endoderm-specific alkaline phosphatase behaves similarly in cytochalasin-arrested embryos (Whittaker, 1977). That is, enzyme develops in the cells according to the cell number and pattern of endodermal lineage. Up to the 4-cell stage, every cell has the potential to produce an intense alkaline phosphatase concentration (Fig. 6A,B) although all cells in each embryo do not necessarily do so. At the 8-cell stage, only the 4 vegetal quartet cells retain the ability to develop

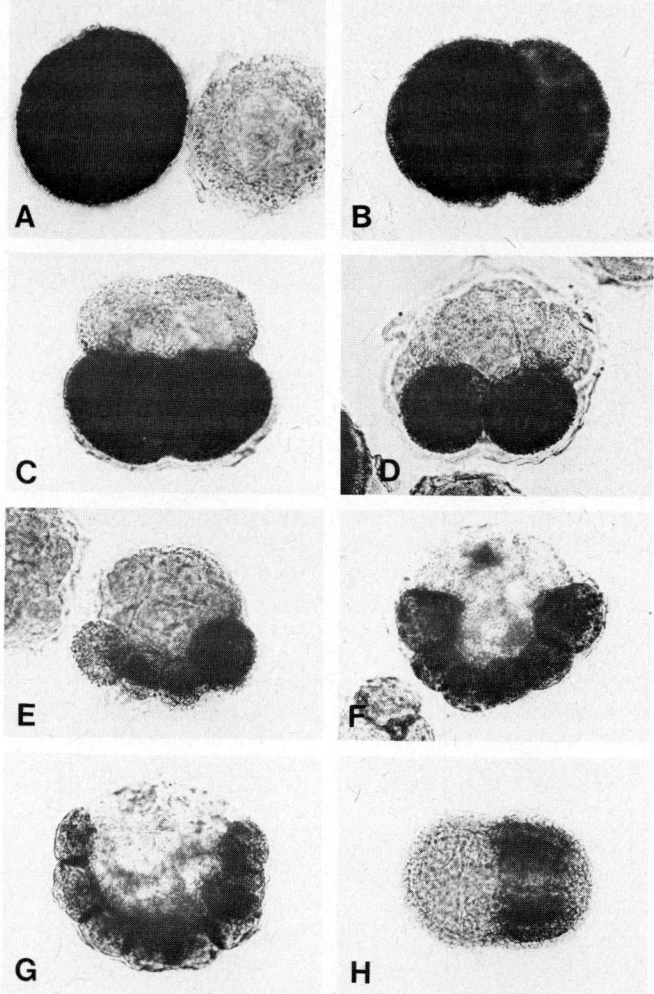

Fig. 5. Acetylcholinesterase development in cytochalasin B-arrested cleavage stage *Ciona* embryos. A, 1-cell. B, 2-cell. C, 4-cell. D, 8-cell. E, 16-cell. F, 32-cell. G, 64-cell. Histochemical reactions carried out 15-16 hr after fertilization (18°C). H, localization in 9 hr control embryo. X175. From Whittaker (1973b).

alkaline phosphatase (Fig. 6C), and there is a progressive restriction of possible expression that parallels the known endodermal lineage (Fig. 6D).

Tyrosinase (dopa oxidase) of the brain melanocytes also develops in cells of cleavage-arrested *Ciona* embryos (Whittaker, 1973b, 1976). *Ciona* larvae have two giant black melanocytes in the brain: one is a gravity

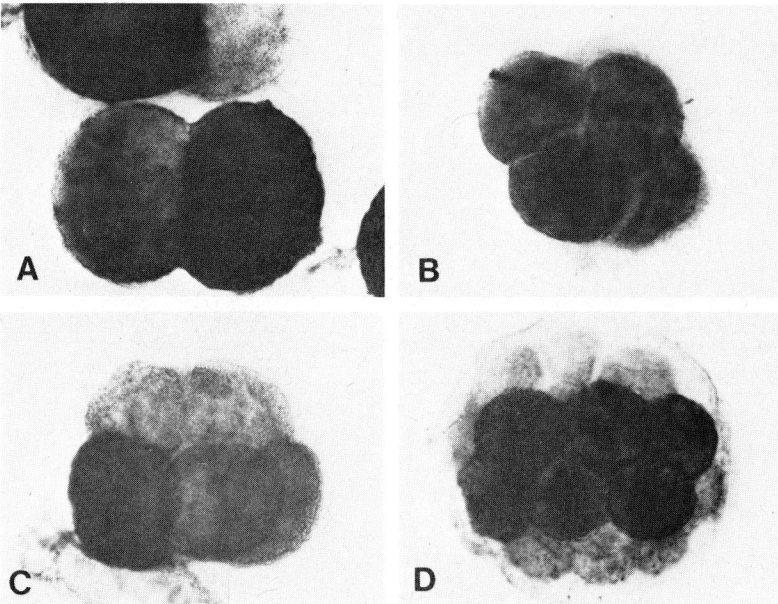

Fig. 6. Alkaline phosphatase development in cytochalasin B-arrested cleavage-stage *Ciona* embryos. A, 2-cell. B, 4-cell. C, 8-cell. D, 16-cell. Histochemical reactions carried out 14-16 hr after fertilization (18°C). X245. From Whittaker (1977).

receptor element, the otolith; the other, the ocellus melanocyte, is part of a primitive eye structure (Fig. 7A). Each melanocyte is the end product of a single lineage from one-half of the bilaterally symmetrical embryo, there being 9 cell divisions in each lineage from fertilized egg to terminal melanocyte. At each cleavage only one daughter cell inherits the full lineage potential. The last division of the lineage results in one large and one smaller cell; all 4 can be identified after histochemical staining for dopa oxidase (Fig. 7B). Only the two larger cells produce additional tyrosinase and go on to develop melanin pigment.

Cytochalasin-arrested embryos from the 8-cell stage onwards develop tyrosinase in pairs of lineage blastomeres, but at the earlier stages only a few embryos do this and both cells do not necessarily do so. At the 32-cell and 64-cell stages most of the embryos develop tyrosinase in one or both of the blastomeres. After the 5th cleavage (32-cell stage) increasing numbers of the embryos will eventually make melanin in the melanocyte lineage cells. This pattern is identical to the lineage pattern for sensory organs worked out by Ortolani (1962).

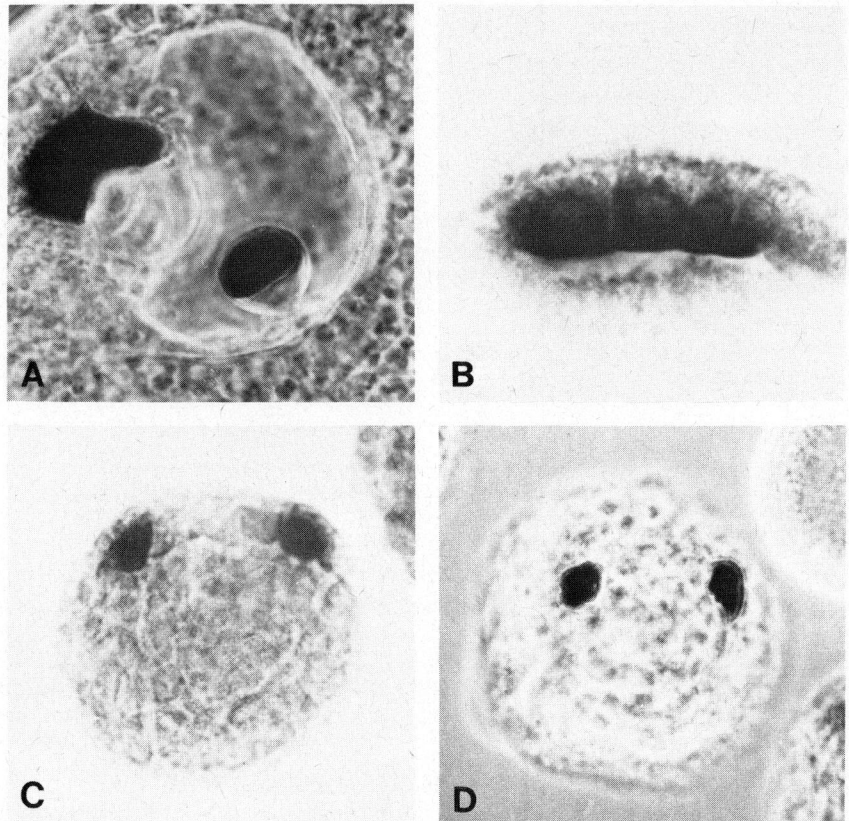

Fig. 7. Differentiation of the larval brain melanocytes of *Ciona*. A, brain vesicle showing ocellus (upper left) and otolith melanocytes. Live. X856. B, melanoblasts of the brain in 10 hr embryo stained for dopa oxidase activity; the two larger cells at the left become differentiated melanocytes. X950. C, embryo cleavage-arrested at 7 hr with cytochalasin B; reacted for dopa oxidase at 11 hr. X270. D, embryo cytochalasin B-treated at 7 hr and left to produce melanin until 22 hr. X270. Development times at 18°C. From Whittaker (1979).

Embryos which are cleavage-arrested just after the 7th cleavage of the melanocyte lineage (neural plate stage) almost all develop tyrosinase and melanin in both lineage cells (Fig. 7C,D). Quantitative measurements of tyrosinase activity and melanin synthesis in such embryos indicate that these embryos and normal terminal larvae produce the same amounts of tyrosinase and melanin (Whittaker, 1976). Melanoblasts which are, therefore, 4 times the final volume of the terminal melanocytes, and which contain 4 nuclei, produce the same quantity of differentiation end products as the melanocytes. In part, excessive dilution of the reasonably small amount of the enzyme produced undoubtedly accounts for the failure to find it histochemically in the earliest cleavage-arrested stages, and the failure of melanogenesis in these early stages.

Collectively, studies with these three enzymes indicate that morphogenetic determinants related to the development of single histospecific proteins (or isozyme groups) are localized in the zygote and segregated during cleavage into appropriate future tissue regions. In itself, this is a considerable advance in the analysis of complex developmental characters in mosaic embryos. It is possible for the first time to consider the polypeptide expressions of particular genes in relation to these supposed morphogenetic factors in the egg cytoplasm.

E. *Differentiation Change by Altered Cytoplasmic Segregation*

Perhaps the ultimate proof of the occurrence of morphogenetic determinants in the egg cytoplasm is a change of cellular fates caused by selective alterations in cytoplasmic relationships. Morgan (1910) attempted to alter the cytoplasmic segregation pattern of *Ciona* eggs by compressing the egg during third cleavage. As predicted by Hertwig's Rule, that the mitotic spindle will form with its long axis in the direction of the longest axis of the cell, the cleavage planes at third cleavage of a compressed egg are meridional rather than equatorial as in normal ascidian third cleavage. When such embryos are released from compression they are an 8-cell stage with cytoplasm of the muscle lineage distributed into 4 rather than 2 cells. Unfortunately, Morgan noted only that such embryos became grossly abnormal on further development.

I have now repeated this experiment using *Styela plicata* eggs, which have a very pronounced yellow-orange crescent (Whittaker, 1979). Here it is possible to see directly after the experiment that yellow crescent material occurs in 4 instead of 2 of the 8 blastomeres. These embryos become abnormal because they fail to gastrulate; it is also difficult to know whether the fate of any cells has changed. When, however, such 8-cell embryos are kept from further cleavage by treatment with cytochalasin B immediately after their release from compression, one finds some embryos eventually developing AChE in 3 and 4 cells instead of just the two cells found in cleavage-arrested normal 8-cell stages.

The cytoplasmic alteration caused by changing the cleavage planes does cause a change in the developmental fate of cells. It is interesting that this procedure also affects a shift in the nuclear cytoplasmic relationships: two of the 4 nuclei in cells now containing myoplasm are nuclei that would normally be found in ectodermal cells. These results imply that it would be possible to cause "muscle" cell transformations by microinjection of myoplasm into non-lineage early blastomeres.

The hypothesis of localized morphogenetic determinants as a cause of mosaic development phenomena presumes that autonomous and stable nuclear lineages in ascidian and other mosaic embryos are not responsible for the programming of cell fates (Wilson, 1896). Tung *et al.* (1977) transplanted nuclei from different early larval tissues into enucleate *Ciona* egg fragments and found that cell and tissue differentiations of the resulting partial embryos seem to be conditioned by the regional cytoplasmic composition rather than the source of the nucleus. The effect of changing the nuclear cytoplasmic relationship by compression also argues strongly against autonomous nuclear lineage as a basis for cell determination.

III. THE MITOCHONDRIAL-ASSOCIATED MUSCLE DETERMINANT

Conklin (1931) concluded from the results of his centrifugation experiments on unfertilized *Ciona* and *Styela* eggs that mitochondrial localizations were not the cause of muscle determination. Mitochondria may be driven out of the finely granular plasm in which they are found without preventing the formation of muscles during subsequent development. When the plasm itself is displaced, the larval muscles are also displaced. A decade later, Tung *et al.* (1941) noted that mitochondria centrifugally displaced to neural and ectodermal regions did not cause these cells to develop myofibrillae. Some centrifugation and other studies appear to contradict this conclusion (reviewed by Reverberi, 1961, 1971), and opinion has persisted that the "crescent" mitochondria may in some sense be (a) permissive, (b) selective, or even (c) instructive of muscle differentiation.

The difficulty with various experiments attempting to resolve this question is that techniques which alter the distribution of mitochondria otherwise cause considerable trauma to the embryo. Nature, however, has obligingly provided an example of disjunction between muscle differentiation and mitochondrial localization and segregation. Certain species of the family Molgulidae have secondarily evolved anural larvae by suppressing development of the tail and other features of the typical urodele tadpole larvae (Berrill, 1931). These species live on sand flats where they no longer have the problems of dispersal and site selection that the original tadpole larva was adaptively selected to overcome.

One of these species, *Molgula arenata*, persists in having a modest vestigial development of muscle AChE, although muscle and other urodele larval features do not develop past the early neurula stage. When mitochondrial distribution was examined using histochemical staining for succinic dehydrogenase, there was no indication of mitochondrial localization and segregation. At the 4-cell stage, enzyme was found equally distributed to the 4 blastomeres, and not preferentially distributed to the two muscle lineage blastomeres as seen with normal urodele embryos. *M. arenata* did not at any stage of development show mitochondria localized in the presumptive muscle cells.

This result clearly indicates that mitochondrial localization is not causally related to muscle determination since there is differentiation of histospecific AChE in its absence. One cannot rule out the possibility that the same mechanical processes for localizing and segregating the mitochondria may be used to segregate and localize the morphogenetic determinants. If so, they operate independently of one another.

IV. MATERNAL RNA AS THE ALKALINE PHOSPHATASE DETERMINANT

Some evidence now suggests that the morphogenetic determinant for endodermal alkaline phosphatase development is a maternal mRNA for the enzyme. Inhibition of RNA synthesis, using even very high concentrations of actinomycin D, appears to be unable to suppress a large part of the localized alkaline phosphatase development (Whittaker, 1977). At a concentration of 20 μg/ml actinomycin D, other enzymes (AChE and tyrosinase) have specific time periods at which their development becomes susceptible to inhibition. This is hr 5-6 for AChE and hr 6-7 for tyrosinase. Embryos exposed continuously to actinomycin D from 5 hr onwards develop neither enzyme; when treatment is started at 6 hr some AChE develops but no tyrosinase, and treatment from 7 hr does not prevent at least small amounts of both enzymes from developing. The implication of these time differences is that each represents the beginning of mRNA synthesis for an enzyme. Concentrations of actinomycin D as high as 120 μg/ml do not prevent at least some of the localized alkaline phosphatase from developing (Fig. 8).

The following experiment indicates the extent to which synthesis of alkaline phosphatase is conserved (Whittaker *et al.*, 1979b). *Ciona* eggs were first treated for 1 hr with 50 μg/ml actinomycin D, then fertilized and permitted to develop in that concentration of inhibitor. Embryos developing in this way continued to divide for at least 5-6 hr but formed only a multicellular blastula; they did not gastrulate. After 16 hr of development, these and control embryos were measured quantitatively

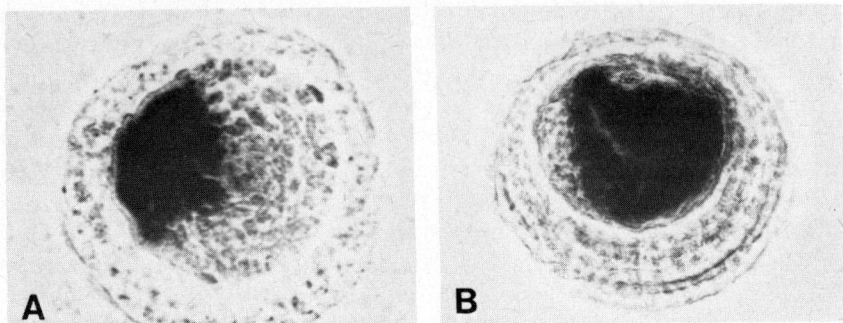

Fig. 8. Histochemical localization of alkaline phosphatase activity in *Ciona* embryo treated from fertilization with actinomycin D and reacted at 15 hr development (A). Reaction in 15 hr control embryo (B). X240. From Whittaker (1977).

for their levels of alkaline phosphatase. Actinomycin-treated embryos in which morphogenesis has been blocked completely have about half of the enzyme activity found in paired 16-hr control embryos (Table I). Embryos of the same groups displayed a very clearly localized concentration of alkaline phosphatase when reacted histochemically.

Other inhibitors of RNA synthesis, chromomycin A3, cordycepin and daunomycin, produce the same effects as actinomycin D. There is the same differential in their ability to prevent AChE and tyrosinase development, and they are unable to prevent development of localized alkaline phosphatase (Whittaker, 1977). It is hardly likely that each of these produces the same secondary effects as actinomycin D; cordycepin exerts its inhibitory effect by a different mechanism than the other three drugs. All are more toxic than actinomycin D.

Interestingly, there are two other localizations of alkaline phosphatase in the *Ciona* embryo. A few "notochordal" cells (4-6) at the tip of the tail begin to produce a strong alkaline phosphatase reaction beginning at 16 hr of development (Whittaker, 1977). In the hatched and mature larva there is a development of alkaline phosphatase in a small subnotochordal strand of endodermal tissue attached to the main endodermal mass and running along the proximal part of the tail. Enzyme begins to appear 8-12 hr after hatching in this "endodermal strand". Unlike alkaline phosphatase development in the main endodermal mass, enzyme development in each of these other structures has a particular actinomycin D sensitivity period. Failure of actinomycin D to prevent the earlier endodermal enzyme development cannot easily be ascribed to some unusual resistance of the alkaline phosphatase genome to actinomycin D.

TABLE I

Alkaline Phosphatase Activity in Actinomycin D-treated Ciona intestinalis Embryos

Series	0 hr (A)	alkaline phosphatase activity[a] (ΔOD min^{-1} mg protein^{-1} × 10^4)		% in actinomycin (C/B)
		16 hr (B)	16 hr actinomycin[b] (C)	
1	58	569	315	55
2	53	485	150	31
3	123	407	284	70
4	110	880	466	53
				(mean = 52%)

[a] measured according to Pfohl (1975).
[b] 50 µg/ml actinomycin D from 1 hr before fertilization until 16 hr afterwards (at 18°C).

In the course of observing that essentially all *Ciona* vegetal quartets isolated at the 8-cell stage had the ability to develop alkaline phosphatase in some of their progeny cells we found that isolated animal quartet embryos would produce localizations of alkaline phosphatase about half of the time (Whittaker *et al.*, 1979a). This result was very surprising since animal cells do not have an endodermal lineage component (Fig. 3), and intact cleavage-arrested embryos do not appear to develop alkaline phosphatase in non-endodermal lineage blastomeres (Whittaker, 1977). A possible explanation is that maternal alkaline phosphatase mRNA occurs in the egg in an animal-vegetal gradient, with translation not ordinarily being activated in animal cells.

Experiments with actinomycin D show an interesting distinction between the two expressions. The inhibitor prevented alkaline phosphatase development in the isolated animal half-embryos but did not prevent expression in vegetal half-embryos. This result does not disprove the gradient hypothesis, but is consistent with the idea that alkaline phosphatase localizations in animal embryos probably arise from new mRNA synthesis. It adds further weight to the contention that only gut and branchial alkaline phosphatase synthesis is resistant to the effects of actinomycin D.

The evidence is consistent with the idea that maternal mRNA for the endodermal alkaline phosphatase occurs in the unfertilized egg and is segregated during cleavage into the endodermal tissues. This is the first complete identification, albeit tentative, of a so-called cytoplasmic determinant that acts at a histospecific level. It is also the first indication, in any species, of the occurrence of a maternal mRNA related to expression of a histospecific polypeptide. Unfortunately, one must hold certain reservations about conclusions based solely on the effects of RNA synthesis inhibitors.

Although Mansueto-Bonaccorso (1971) has shown autoradiographically that 20 µg/ml actinomycin D prevents [^{3}H]uridine incorporation into cellular RNA of *Ciona* embryos, it is possible that low levels of alkaline phosphatase mRNA are synthesized even when cells are exposed to high drug concentrations. In an experiment similar to that described in Table I, we have found that 50 µg/ml actinomycin D inhibits only 86% of [^{3}H]uridine incorporation into whole embryo RNA during a labeling period from fertilization to 6 hr (Whittaker *et al.*, 1979b). Alkaline phosphatase synthesis ordinarily begins at 6 hr in normal embryos. We have not yet determined whether the 14% of remaining activity in total embryo RNA is cytoplasmic or if it has any of the properties of mRNA. Studies in progress may succeed in showing that RNA synthesized in the presence of actinomycin is not cytoplasmic mRNA, but for the moment this 14% must be considered a possible source of alkaline phosphatase mRNA. If so, it remains puzzling why there are such differential responses to actinomycin D.

Production of endodermal alkaline phosphatase in an *in vitro* protein-synthesizing system by translation of polyadenylated RNA isolated from the unfertilized egg would be a more satisfactory demonstration of a specific maternal mRNA. The major alkaline phosphatase component found in adult *Ciona* digestive tissues proves to be a single protein when butanol-extracted, vacuum dialyzed, and separated by electrophoresis on starch and acrylamide gels. Isoelectric focusing also indicates a single protein (Whittaker, Troianello and Tachovsky, unpublished data). This same protein appears in early embryonic development and seems to be the endodermal enzyme. We are in the process of making preparations of suitable purity for antibody production, and we hope to identify and measure quantities of translation product in the *in vitro* reticulocyte system using a specific antibody to this phosphatase.

V. MESSENGER RNA SYNTHESIS DURING DEVELOPMENT

On the basis of isotope incorporation studies by investigators in two other laboratories there is serious question about whether ascidian embryos actually synthesize any mRNA during larval development (Lambert, 1971; Puccia et al., 1976). For this and other reasons, the suggestion has persisted in the review literature that ascidian larval development may be programmed solely by preformed mRNA. Our own findings, however, indicate that mRNA *is* synthesized during larval development (Meedel and Whittaker, 1978).

Polyribosomes were isolated from *Ciona* embryos treated 2 hr with [³H]uridine during early gastrulation to neural plate stage (hr 5-7 after fertilization), and the labeled RNA species characterized by release with EDTA treatment, sucrose density sedimentation analysis, and oligo (dT)-cellulose chromatography. Since more than half of the polyribosome-associated labeled RNA was polyadenylated and all of it sedimented heterogenously, mRNA is apparently synthesized during the labeling period. Synthesis of heterogeneously sedimenting, polyadenylated RNA at various other stages of development from mid-cleavage to metamorphosis indicates that mRNA synthesis occurs throughout larval development. Our autoradiographic studies verify that *Ciona* embryos are synthesizing RNA from [³H]uridine; the incorporated activity is ribonuclease-digestible. Uridine incorporation was localized in the nucleus and cytoplasm of embryonic cells and not in the accessory cells.

While it seems likely from the studies of alkaline phosphatase and from work by Minganti (1959) on andromerogonic hybrids that some preformed maternal mRNAs are factors in the early development of ascidians, there is no reason to suppose that this is the exclusive basis of mosaic development and regulation of larval characteristics, or even that development of many larval features is controlled by histospecific maternal mRNA. Our molecular studies with *Ciona* indicate that mRNA is being synthesized during development. Results with actinomycin D inhibition support the view that most developmental processes in ascidians require new RNA synthesis (Whittaker, 1977).

Various developmental processes in *Ciona* embryos have discrete susceptibility periods to actinomycin D: gastrulation and other morphogenetic changes up to tail bud stage, AChE development in tail muscle, tyrosinase development, melanin granulogenesis, tail tip alkaline phosphatase development, hatching enzyme synthesis, siphon muscle AChE development, alkaline phosphatase synthesis in the endodermal strand, tail resorption. So far, only endodermal alkaline phosphatase development has been found resistant to the effects of actinomycin D.

VI. POSSIBLE NATURE OF OTHER DETERMINANTS

As shown diagrammatically in Fig. 9, AChE and tyrosinase first appear

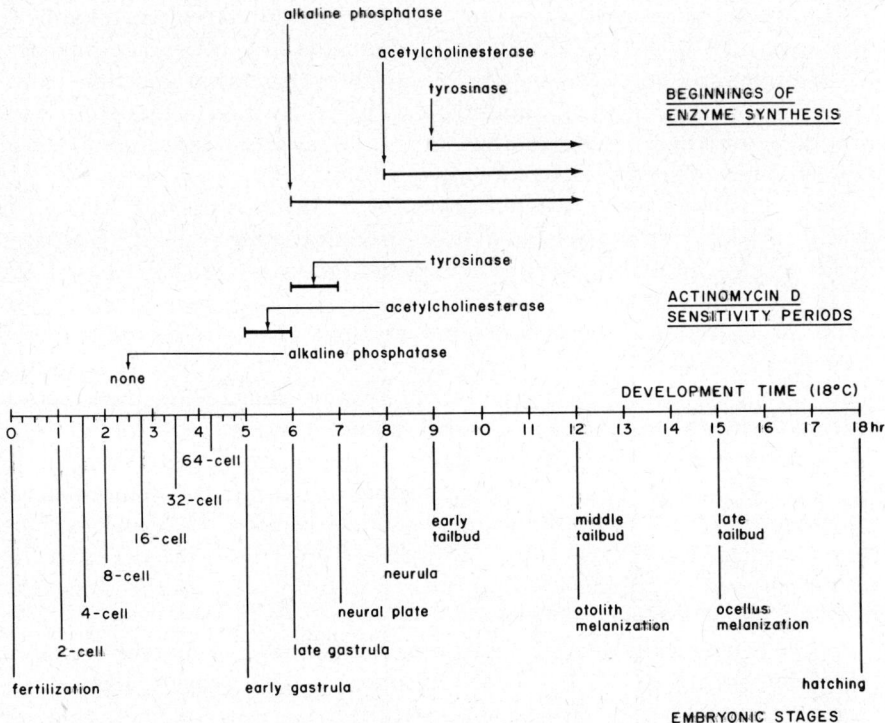

Fig. 9. Diagram relating the time of first synthesis of three enzymes and their actinomycin D sensitivity periods to the embryonic stages of *Ciona intestinalis*. From Whittaker (1977).

at different distinct times. Since there is a puromycin sensitivity period for each enzyme coincident with the times of first histochemical detection, the enzymes are probably synthesized *de novo* and are not themselves the cytoplasmic elements being segregated, even in a proenzyme form. Each enzyme has an actinomycin D sensitivity period, which suggests that preformed mRNA for the enzymes does not occur and is also unlikely to be the morphogenetic determinant. The segregated determinants related to these enzyme expressions are conceivably agents that activate the genome at an appropriate time to initiate transcription of mRNA. Morgan (1934) postulated that the cytoplasmic determinants so obvious in mosaic embryos were probably activating the genes responsible for larval characters. His suggestion is still one of the most likely explanations of how at least some of these factors may function (Davidson and Britten, 1971; Davidson, 1976).

As yet, there is no clue to the chemical nature of the AChE and tyrosinase determinants. Some cytoplasmic factors may be RNA other than the mRNA for particular differentiation end product polypeptides, and Ortolani and Marino (1973) report a provocative experimental result bearing on this possibility. When animal quartets, isolated from the 8-cell stage, were treated with total RNA prepared from *Ciona* ovaries a few of these partial embryos developed tail-like structures. Untreated animal quartets never produced "axial" structures, and animal quartet blastomeres do not contain the lineage potential for any of the structures normally associated with tail development.

What is needed for the next phase of this work is an assay system in which fractionated materials (nucleic acids and proteins) could be tested for their ability to induce the synthesis of histospecific enzymes. Microinjection of material into blastomeres, and later identification of enzyme products unrelated to the lineage potential of the blastomeres used, would probably be the most satisfactory method. As concluded from the results of microcompression experiments (section II-E), cytoplasmic components can artificially alter the differentiation pathways of cells. This finding strongly suggests that a microinjection approach would be feasible.

VII. CONCLUSIONS

Present evidence indicates that there are least three kinds of morphogenetic factors laid down in the ascidian egg: (i) enzymes and structural proteins as part of localized organelles, e.g. mitochondria, (ii) mRNA for the production of histospecific proteins, (iii) agents of unknown composition that have some relation to the activation of specific gene expression. The occurrence of these factors is the essence and basis of what is called mosaic development. An essential point is that such agents include a range of possible elements of the regulatory system, from factors limiting the selection of differentiation pathways to particular end products of a regulatory sequence. Probably any stable element of a regulatory sequence can be adaptively selected by the organism to be part of a preformed egg organization.

At one extreme, preformation of mitochondria in the oocyte is an adaptation to the needs of rapid development in oviparous embryos. Presumably the embryo is unable to synthesize sufficient mitochondria during a brief development time to meet the high energy requirements of tail muscle tissue, and must prepare these in advance of embryonic development. On the basis of enzyme activity measurements, it appears that approximately half of the mitochondria found in the larva are already present in the unfertilized egg (D'Anna and Metafora, 1965; D'Anna, 1966), and about half of these become segregated into the muscle blastomeres (Berg, 1956, 1957). This mitochondrial localization is not, however, instructional information for epigenetic regulation.

For understanding regulation of development the other factors are obviously more important because they are instructional. It is interesting that some proteins, such as endodermal alkaline phosphatase, may arise initially from a preformed maternal mRNA. One can easily imagine a cascade of events originating from the local metabolic effects of a single enzyme. Most interesting of all, however, are the putative gene regulatory factors. These are presumably those elusive regulators that have been sought, so far unsuccessfully, in other developmental systems. Ascidian mosaicism has at least the advantage of showing clear evidence that such cytoplasmic determinants exist.

ACKNOWLEDGEMENT

This work was supported by USPHS Research grant HD-09201 from the National Institute of Child Health and Human Development.

REFERENCES

Beams, H.W. and Kessel, R.G. (1974). *Int. Rev. Cytol.* **39**, 413-479.
Beneden, E. van and Julin C. (1884). *Arch. Biol.* **5**, 111-126.
Berg, W.E. (1956). *Biol. Bull.* **110**, 1-7.
Berg, W.E. (1957). *Biol. Bull.* **113**, 365-375.
Berg, W.E. and Humphreys, W.J. (1960). *Develop. Biol.* **2**, 42-60.
Berrill, N.J. (1931). *Phil. Trans. Roy. Soc. Lond.* B **219**, 281-346.
Berrill, N.J. (1947). *J. Mar. Biol. Assoc. U.K.* **27**, 245-251.
Berrill, N.J. (1955). "The Origin of Vertebrates." Oxford Univ. Press, London.
Brachet, J. (1977). *Curr. Top. Develop. Biol.* **11**, 133-186.
Chabry, J. (1887). *J. Anat. Physiol.* **23**, 167-319.
Cohen, A. and Berrill, N.J. (1936). *J. Exp. Zool.* **74**, 91-117.
Conklin, E.G. (1905). *J. Acad. Nat. Sci. Philadephia* **13**, 1-119.
Conklin, E.G. (1911). *J. Exp. Zool.* **10**, 393-407.
Conklin, E.G. (1931). *J. Exp. Zool.* **60**, 1-120.
D'Anna, T. (1966). *Boll. Zool.* **33**, 351-360.
D'Anna, T. and Metafora, S. (1965). *Acta Embryol. Morph. Exp.* **8**, 267-277.
Davidson, E.H. (1976). "Gene Activity in Early Development." 2nd Edition. Academic Press, New York.
Davidson, E.H. and Britten, R.J. (1971). *J. Theor. Biol.* **32**, 123-130.
Durante, M. (1956). *Experientia* **12**, 307-308.
Eddy, E.M. (1975). *Int. Rev. Cytol.* **43**, 229-280.
Lambert, C.C. (1971). *Exp. Cell Res.* **66**, 401-409.
Lillie, F.R. (1902). *Wilhelm Roux' Archiv.* **14**, 477-499.

Mansueto-Bonaccorso, C. (1971). *Lincei Rend. Sc. fis. mat. e nat.* **50,** 776-778.
Meedel, T.H. and Whittaker, J.R. (1978). *Develop. Biol.,* **66,** 410-421.
Minganti, A. (1951). *Pubbl. Staz. Zool. Napoli* **23,** 52-57.
Minganti, A. (1954). *Pubbl. Staz. Zool. Napoli* **25,** 9-17.
Minganti, A. (1959). *Acta Embryol. Morph. Exp.* **2,** 244-256.
Morgan, T.H. (1910). *Wilhelm Roux' Archiv.* **29,** 205-224.
Morgan, T.H. (1934). "Embryology and Genetics." Columbia Univ. Press, New York.
Ortolani, G. (1955). *Experientia* **11,** 445-446.
Ortolani, G. (1957). *Acta Embryol. Morph. Exp.* **1,** 33-36.
Ortolani, G. (1962). *Acta Embryol. Morph. Exp.* **5,** 189-198.
Ortolani, G. and Marino, L. (1973). *Acta Embryol.Morph. Exp. 1973,* 235-236.
Pfohl, R.J. (1975). *Develop. Biol.* **44,** 333-345.
Puccia, E., Mansueto-Bonaccorso, C., Farinella-Ferruzza, N. and Morello, R. (1976). *Acta Embryol. Exp. 1976,* **167-177.**
Reverberi, G. (1961). *Adv. Morphogen.* **1,** 55-101.
Reverberi, G. (1971). *In* **"Experimental Embryology of Marine and Fresh-water Invertebrates." (G. Reverberi, ed.), pp. 507-550. Elsevier North-Holland, New York.**
Reverberi, G. and Minganti, A. (1946). *Pubbl. Staz. Zool. Napoli* **21,** 199-252.
Reverberi, G. and Minganti, A. (1947). *Pubbl. Staz. Zool. Napoli* **21,** 1-35.
Reverberi, G., Ortolani, G. and Farinella-Ferruzza, N. (1960). *Acta Embryol. Morph. Exp.* **3, 296-336.**
Tung, T.C., Ku, S.H. and Tung, Y.F.Y. (1941). *Biol. Bull.* **80,** 153-168.
Tung, T.C., Wu, S.C., Yeh, Y.F., Li, K.S. and Hsu, M.C. (1977). *Scientia Sinica* **20,** 222-233.
Whittaker, J.R. (1966). *Develop. Biol.* **14,** 1-39.
Whittaker, J.R. (1971). *Biol. Bull.* **141,** 407-408.
Whittaker, J.R. (1973a). *Develop. Biol.* **30,** 441-454.
Whittaker, J.R. (1973b). *Proc. Nat. Acad. Sci. U.S.* **70,** 2096-2100.
Whittaker, J.R. (1976). *Biol. Bull.* **151,** 434.
Whittaker, J.R. (1977). *J. Exp. Zool.* **202,** 139-153.
Whittaker, J.R. (1979). in preparation.
Whittaker, J.R., Ortolani, G. and Farinella-Ferruzza, N. (1977). *Develop. Biol.* **55,** 196-200.
Whittaker, J.R., Ortolani, G., Farinella-Ferruzza, N. and Durante, M. (1979a). in preparation.
Whittaker, J.R., Troianello, L. and Meedel, T.H. (1979b). in preparation.
Wilson, E.B. (1896). *Wilhelm Roux' Archive.* **3,** 19-26.
Wilson, E.B. (1925). "The Cell in Development and Heredity." 3rd Edition. Macmillan Co., New York.

The Multiple Roles which Cell Division can Play in the Localization of Developmental Potential

Gary Freeman
Department of Zoology
University of Texas at Austin
Austin, Texas 78712

I. Introduction 53
II. The Ontogeny of Localization in Cleavage Stage Embryos 57
III. The Coupling of Localization and Cleavage 60
IV. The Plane of Cleavage and Determinant Localization 68
V. Concluding Remarks 74
References 75

"All morphogenetic events are based on the inhomogeneous distribution of material". (A. Kühn, 1971)

I. INTRODUCTION

The phrase, "localization of developmental potential", is a rubric for a set of ideas that has been used to explain how the cytoplasmic organization of the egg is related to the spatial pattern of cell differentiation in the developing embryo. Three ideas are usually associated with this rubric: 1) Cytoplasmic components present in the egg have the ability to bias the pathway of differentiation taken by the blastomeres that inherit these components. 2) During oogenesis and/or early embryogenesis these cytoplasmic components become localized at specific sites that are defined by the symmetry properties of the oocyte or early embryo. 3) As a consequence of cell division during early embryogenesis there is an unequal distribution of these cytoplasmic components, or determinants, in the daughter cells that form during a division or a set of divisions.

These ideas concerning the role of the cytoplasmic organization of eggs during early development originated at the turn of the century (Baxter, 1974). They still provide the paradigm in terms of which our attempts to understand cell specification during early development are framed (compare Wilson, 1925 with Davidson, 1976). While the localization of developmental potential is not the only mechanism which can lead to meaningful differences in the cytoplasmic composition of cells during early embryogenesis (Bonner, 1952; Hillman et al., 1972), it is probably the most important mechanism employed by the early embryo. The demonstration of localized determinants in the embryos of virtually every group of animals that has been examined for them supports this claim (Davidson, 1976).

In some cases there are correlations that associate the ability to specify a given pathway of differentiation with a set of cellular organelles, or a class of molecules in a given egg region. However, there is no case in which these determinants have been well characterized. The way these determinants function to bias differentiation in the cells that inherit them is not understood. It is generally thought that determinant function is translated either directly or indirectly into specific patterns of gene activity, and that these patterns of gene activity confer upon these cells and their descendants specialized functional properties. In some situations, organelles and macromolecules that play a major role in carrying out a given physiological activity may become concentrated in a specific blastomere by virtue of some prior localization process. This may confer a precocious state of differentiation on the cell or provide part of the phenotype in terms of which the cell is defined as differentiated that would not necessarily reflect a program of gene activity which is unique to that cell. Mitochondrial localization in certain blastomere lineages of many kinds of invertebrate embryos (Reverberi, 1971) and the initial steps in holdfast cell differentiation in the alga *Fucus* (Quatrano, 1972) may reflect this kind of situation.

In any discussion of the effect of a localized determinant in biasing a given pathway of cell differentiation, one has to specify at an operational level what is meant by the differentiation. Frequently this term is used for terminally differentiated cells like nerve or muscle cells which are specialized for carrying out specific activities within an organism that has passed through a given stage of development. Localized determinants frequently specify terminally differentiated cells in situations where embryogenesis occurs during a relatively short time interval or where these cells form precociously.

The term differentiation, can also refer to the different functional states a cell and its descendants pass through prior to their final differentiation as terminally specialized cells. Among animals in which embryonic induction plays a major role in development, the localizations of developmental potential that have been documented generally function to specify an inductive potential or the capacity to respond to an inductor. However, only a few cases have been studied in which localized determinants have been shown to bias the differentiation of a competence or inductive potential because it is difficult to devise tests to assay these differentiated states (Jacobson, 1966). Since inductive interactions play such an important role in embryogenesis in most animals, a large proportion of the localized determinants probably bias the kind of differentiation needed to make this mechanism work, especially in those cases in which inductive events occur during the first stages of development.

The ontogeny of a localization of developmental potential involves several steps. The determinant which will be localized must be synthesized. At some period during oogenesis and/or embryogenesis a promorphological scaffold will be established; this scaffold specifies where localization will occur. The determinant then moves to the appropriate site, either at the time the scaffold is established or at some later period. A promorphological scaffold is generally thought of as one or more axial coordinates with the property of polarity. In embryos with more than one axis of symmetry the axes appear to form sequentially. In several kinds of animal embryos, at least one axis of symmetry is established prior to fertilization. In many cases some kind of a localized signal from the environment such as a site of fertilization, plays a role in establishing where an axis of symmetry will be set up. However these signals are not necessary for axis formation (e.g. Ancel and Vintemberger, 1948 for amphibians). The physical bases for the axes of symmetry in oocytes and early embryos have not been well defined. Jaffe, (1969) has argued that electrical currents flowing through cells provide the basis for their symmetry properties. A number of investigators have invoked a cell cytoskeleton composed of microfilaments and microtubules as a structural basis for cell symmetry (Raff, 1977). Frequently it is not clear how these physical models for an axis of symmetry function to specify the positional information needed in order to localize determinants in a given region of an oocyte or early embryo.

Since the nineteen twenties, investigators have recognized that determinants can become localized at different stages of development. The localization of a determinant can occur during oogenesis, it can occur during oocyte maturation, it can occur in the uncleaved egg as a consequence of fertilization, and it can occur in blastomeres during cleavage stages of development. Different determinants in the same egg frequently become localized at different stages of development; a set of determinants which are initially localized at one site may move to another site as development proceeds.

In many kinds of maturing oocytes and eggs which have just been fertilized, spectacular cytoplasmic and/or membrane movements occur (Costello, 1948). These reorganizational events are relatively rapid. Frequently they are correlated both temporally and spatially with the establishment of one or more localizations of developmental potential (e.g. Reverberi, 1961 for ascidians). In certain cases the treatment of eggs with agents that prevent these reorganizational events also prevents the localization of developmental potential (Arnold and Williams-Arnold, 1974; Zalokar, 1974). One interpretation of these experiments is that these movements directly transport the determinants to the appropriate sites. Another interpretation argues that these movements cause local changes, such as the unmasking of a population of receptors at a given site in the egg which sequestors the determinants thereby giving them a local distribution. Because these cases are dramatic and accessible to experimental manipulation they have provided the main model for thinking about how determinants are localized. Much less is known about the establishment of localizations of developmental potential during oogenesis and cleavage stages of embryogenesis. There has been a tendency to generalize the model based on the reorganizational events which occur as a consequence of oocyte maturation and fertilization to these other two stages.

In several groups of animals the pattern of cell division during embryogenesis is quite precise; certain blastomere lineages typically inherit certain determinants. During the period when the notion of localized determinants was first being established as an explanatory mechanism in its present analytical sense (Gould, 1977), it was thought that cleavage played a major role in establishing particular localizations of developmental potential (Conklin, 1898; Wilson, 1896). However Lillie's (1906) demonstration that local differentiation can occur in uncleaved eggs of the annelid *Chaetopterus* and the failure to convincingly demonstrate that an alteration in cleavage pattern changed the pattern of localization in the eggs studied, led to the abandonment of this idea (Wilson, 1925; Morgan, 1927). In retrospect it is clear that these experiments did not test the hypothesis that cleavage plays a role in localizing developmental potential, because in the eggs that had been chosen for the experiments the localization events in question had been largely completed before cleavage was intiated.

The idea that cleavage plays a role in setting up localizations of developmental potential was replaced by the notion that the cleavages of early embryogenesis passively divide up the various determinants already localized in the egg into different blastomere lineages. In order to ensure that localized determinants are segregated to the appropriate blastomeres during embryogenesis, the argument was advanced that the promorphological scaffold also determines how the planes of the first cleavages will be oriented. This is probably a satisfactory explanation of what is going on when one is dealing with eggs in which most localization has occurred before cleavage is initiated. However it is not clear that these ideas provide an adequate picture of what is going on in those cases in which determinants are localized during the cleavage stages of development.

This review will examine a number of cases in which determinants become localized during cleavage stages of embryogenesis. Two topics that will receive particular attention are: 1) the progress that is made in localizing a determinant in the cells of a lineage as a function of cell division in that lineage, and 2) the conditions under which the plane of cleavage influences the sites at which localization will occur in the cells of a lineage.

II. THE ONTOGENY OF LOCALIZATION IN CLEAVAGE STAGE EMBRYOS

The period during development when a given determinant becomes localized can be established by identifying the site in the oocyte or early embryo where that determinant will be localized and then asking when the determinant is first found only at that site. Since determinants are defined by the kind of differentiation they specify, these measurements of the time course or determinant localization necessarily involve an assay which scores the ability of some suitable test embryo or embryo part to undergo a specific kind of differentiation.

A concrete example of how the process of determinant localization is measured in a case where localization occurs during cleavage, is the localization of the factors which specify the apical tuft and the gut for the pilidium larva of the nemertine *Cerebratulus*. The course of early embryogenesis and the larva that develops is indicated for this animal in Fig. 1. The unfertilized egg has a distinct region at one pole where the meiotic chromosomes are aligned in metaphase; when the egg is fertilized the first and second polar bodies are given off at this site. This region of the egg is referred to as the animal pole while the opposite pole of the egg

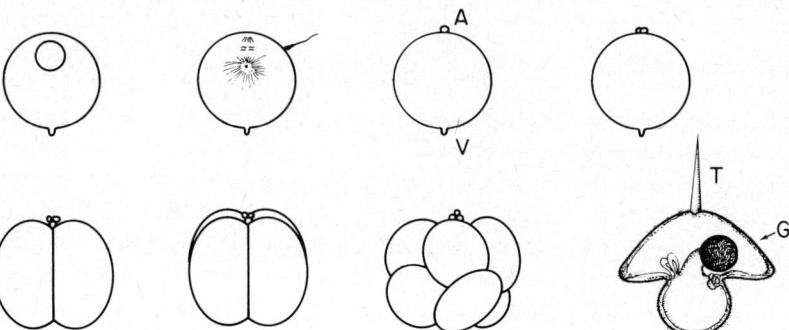

Fig. 1. Normal development in the nemertine *Cerebratulus*. The first panel shows the egg immediately after it has been shed; a prominent germinal vesicle is present; a small peduncle is frequently found at the vegetal pole of the egg. The second panel shows the egg being fertilized. The next set of panels show the stages when the polar bodies are given off and the first three cleavage stages of development. The animal (A) and vegetal (V) poles of the egg are marked at the first polar body stage. The last panel shows the pilidium larva. The apical tuft (T) and the gut (G) are marked. All views are from the side.

is referred to as the vegetal pole. Cleavage in this species is spiral. The first two cleavage planes pass at right angles to each other through the pole where the polar bodies are given off. The plane of the third cleavage is equatorial; the first quartet is derived from the animal hemisphere of the egg while the macromeres are derived from its vegetal hemisphere. The pilidum larva that forms is composed of several component parts including an apical tuft and a gut. Marking experiments with vital dyes for the purpose of establishing the fate of the blastomeres at the eight-cell stage have shown that the apical tuft forms from the blastomeres of the first quartet while the digestive tract forms from the macromeres (Hörstadius, 1937).

When the first quartet of blastomeres is separated from the macromeres at the eight-cell stage and each half is raised in isolation, the cells from the animal hemisphere give rise to an apical tuft but do not form a gut, while the cells from the vegetal hemisphere give rise to a gut but do not form an apical tuft. However, if the eight-cell stage embryo is divided into two vertical halves, each containing two macromeres and two first quartet cells, each half develops into a larva with both an apical tuft and a gut. When one takes an unfertilized egg after the germinal vesicle has broken down, cuts it equatorially into animal and vegetal halves, and then fertilizes each half, each fragment frequently develops into a normal larva with an apical tuft and a gut. Unfertilized eggs have also been cut into animal and vegetal halves at different time intervals after spawning in order to establish if the localization of these determinants is a function of the amount of time elapsed. These experiments indicate that the determinants which specify the apical tuft and the gut show no sign of localization in the unfertilized egg; however by the eight-cell stage they have become localized in the appropriate regions of the embryo (Hörstadius, 1971).

The exact period during which these determinants become localized in the animal and vegetal regions of the early embryo was defined by cutting the embryo into animal and vegetal halves at different stages between fertilization and the eight-cell stage (Freeman, 1978). The way in which these fragments differentiate was then assayed. As long as the cytoplasmic components that specify the apical tuft or the gut are not localized exclusively in one hemisphere, one would expect both of the isolated hemispheres to differentiate the respective structures, while the localization of one of these components in a given hemisphere would create a situation in which the other hemisphere would not be capable of differentiating that structure.

Table I presents the results of this study. The determinants that specify the apical tuft were localized primarily in the vegetal region of the unfertilized egg. During the period from fertilization to the first cleavage they became distributed throughout the egg. About the time of the first cleavage these factors began to disappear from the vegetal region of the embryo. They were localized in the animal hemisphere of the embryo at the eight-cell stage. The factors that specify the gut were initially distributed throughout the unfertilized egg. Between the first-polar body stage and the eight-cell stage these determinants were slowly localized at the vegetal pole of the embryo. As this localization process took place the size of the gut region that forms in animal hemisphere fragments became smaller while the size of the gut region that forms in vegetal hemisphere isolates increased. In control experiments where eggs were divided into fragments that contained both animal and vegetal hemisphere material, these lateral fragments differentiated both an apical tuft and a gut in the majority of cases, at all stages tested.

If the process of localization for these two determinants in the *Cerebratulus* embryo is compared with the localization events that occur during yellow crescent formation in ascidian eggs or grey crescent formation in amphibian eggs it is clear that the localization process is much slower in *Cerebratulus* and that it is not correlated with a visible cytoplasmic reorganization. The ctenophore *Mnemiopsis* is the only other form in which the dynamics of the localization process have been described in detail during cleavage stages of embryogenesis (Freeman, 1976a). Experiments that are similar to those described here for

TABLE I

The Differentiation Capabilities for Defined Regions of the Unfertilized Egg and the Embryo, at Different Developmental Stages.

stage	Animal hemisphere			Vegetal hemisphere			Lateral hemisphere		
	Number of cases	Percent[1] differentiation T	G	Number of cases	Percent differentiation T	G	Number of cases	Percent differentiation T	G
Unfertilized egg	57	28	96	38	71	85	49	67	94
Fertilized									
First polar body	33	64	79	24	63	83	15	93	80
Second polar body	33	79	73	11	91	91	21	81	91
Two-cell stage	44	87	29	31	45	97	20	100	90
Four-cell stage	36	83	29	34	12	94	19	100	95
Eight-cell stage	27	96	4	13	0	100	17	94	88

1. T, apical tuft; G, gut. Fig. 1 should be consulted for picture of these developmental stages. Complete data is in Freeman (1978).

Cerebratulus have been used to map the distribution of two determinants that specify different cell types in the cydippid larva, prior to their segregation to different blastomere lineages during the third cleavage in this form. In this case the localization of these two determinants is also a relatively slow process. In both *Cerebratulus* and *Mnemiopsis* there is evidence which suggests that the determinants studied continue to undergo localization at later cleavage stages of embryogenesis (Hörstadius, 1937; Freeman, 1976a).

III. THE COUPLING OF LOCALIZATION AND CLEAVAGE

The process of localization during cleavage stages of development may occur independently of cleavage. It is also conceivable that the cell cycle might regulate the process of localization in some way. The relationship between the establishment of localizations of developmental potential and the cell cycle during early embryogenesis can be investigated by reversibly inhibiting the cleavages that occur while a given determinant is being localized. This kind of experiment will be described here for the early embryo of the nemertine *Cerebratulus* and the ctenophore *Mnemiopsis*. Both studies have established the existence of mechanisms that regulate the localization of developmental potential during the cell cycle.

The localization of the determinants that specify apical tuft and gut differentiation following fertilization in the embryo of *Cerebratulus* has already been described. Cleavage can be reversibly inhibited during early embryogenesis in this form by using the drugs cytochalasin B and ethyl carbamate. Cytochalasin B appears to inhibit cleavage by preventing the formation of the contractile ring of microfilaments which mediates cleavage furrow formation; it does not affect aster formation (Schroeder, 1972). Ethyl carbamate inhibits cleavage by suppressing the formation of asters which are necessary for the induction of furrowing (Harvey, 1956; Rappaport, 1974). Fig. 2 outlines an experiment in which the first cleavage was reversibly inhibited. The experimental eggs were placed in a given cleavage inhibitor after the second-polar body had been given off and removed just before the second cleavage of their controls. Although the treated eggs began their first cleavage at the same time that their controls began their second cleavage, they still developed into normal larvae. Between second-polar body formation and the two-cell stage there is a marked localization of the determinants that specify the apical tuft and the gut (Table I). The distribution of the determinants that specify these cell types was mapped for eggs incubated in either of these inhibitors, by cutting them into animal and vegetal halves, at a time when the controls were at the two-cell stage. After the operation the inhibitor was removed and the fragments were allowed to develop. The eggs in which the first cleavage was artificially postponed with cytochalasin B showed a distribution of determinants that was typical of the two-cell stage, while the eggs inhibited with ethyl carbamate showed a distribution of determinants that was typical of the second polar body stage. This experiment suggests that aster formation must take place if progress is to be made in localizing the determinants that specify the apical tuft and the gut.

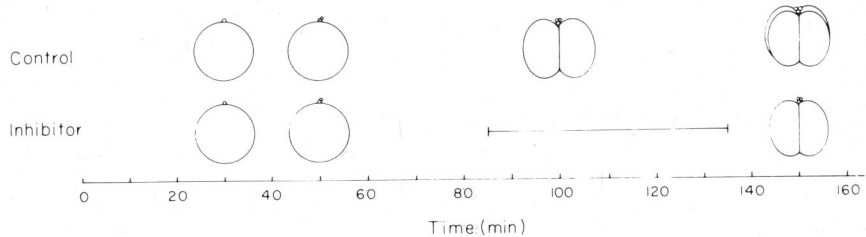

Fig. 2 Diagram comparing the cleavage patterns of normal *Cerebratulus* eggs and eggs in which the first cleavage was inhibited with cytochalasin B (.001M) or ethyl carbamate (.09M). The eggs were fertilized at time 0; development took place at 18-19°C. The line after inhibitor indicates the period of time the eggs were in these inhibitors. All views are from the side, (Freeman, 1978).

Another experiment that argues that the astral cycle is necessary for localization of determinants in *Cerebratulus* deals with the developmental period when the meiotic reduction divisions are occurring. The meiotic apparatus that mediates polar body formation contains a pair of asters. It is possible to remove the meiotic apparatus from the *Cerebratulus* egg prior to fertilization by cutting out this cytoplasmic region. When these eggs are fertilized they do not form polar bodies, however they initiate their first cleavage at the same time control eggs do and develop into normal larvae. Asters do not form in these eggs until just before the first cleavage. If eggs without a meiotic apparatus are operated on to produce animal and vegetal halves at a time period that is equivalent to the second-polar body stage, the isolates show a pattern of differentiation which reflects a distribution of apical tuft and gut determinants that is typical of the unfertilized egg. This result shows that in the absence of the two astral cycles which normally accompany polar body formation there has been no progress made in the localization of these determinants. If eggs without a meiotic apparatus are operated on to produce animal and vegetal halves following the first cleavage, the isolates show a pattern of differentiation which indicates a distribution of apical tuft and gut determinants that is typical for the two-cell stage. These results and other experiments on *Cerebratulus* suggest that the number of astral cycles do not determine the distribution of a given determinant; the distribution of the determinant is established by the stage specific state of the cell in which localization is occurring at the time when asters are present. This suggests that the astral cycle acts as a triggering mechanism.

There are a number of cases where there is suggestive evidence for aster mediated localization of developmental potential, either before cleavage begins or during cleavage stages of embryogenesis (Elinson and Manes, 1978; Guerrier, 1971; Kubota, 1967). It is not clear how asters might function to bring about the process of localization. Asters can promote cytoplasmic movements, either as a consequence of their presence or because of their presence at some prior time interval (Rebhun, 1975; Wolf, 1978). Asters can also act on the cell surface to cause a local change in its properties (Rappaport, 1974).

The initial cleavage stages for the *Mnemiopsis* embryo and its cydippid larva are depicted in Fig. 3. The cydippid larva contains several distinct cell types including comb plate cilia cells, which are found in rows on the outer surface of the larva over the radial canals, and photocytes, which contain a calcium activated photoprotein and are found in the radial canals. The localization process has been described for the determinants

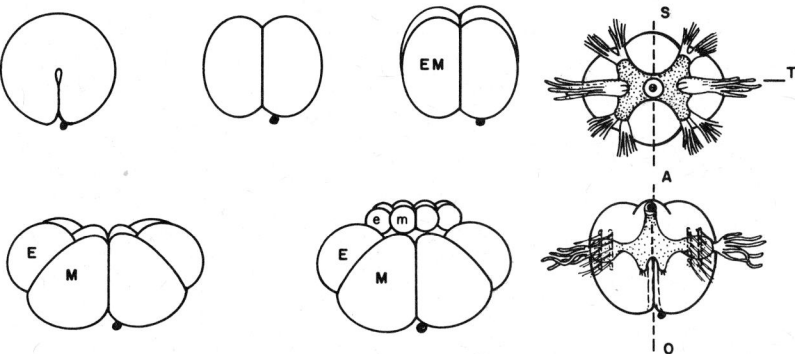

Fig. 3. Normal development in the ctenophore. The left half of the figure shows the first cleavage and the two-, four-, eight- and 16-cell stages of embryogenesis, viewed from the side. The right half of the figure shows the cydippid larva looking down on its aboral surface and from the side. The sagittal plane (S) and the tentacular plane (T) are marked in the aboral view, and the oral-aboral axis (A-O) is marked in the side view. The aboral view shows the eight rows of comb plates; the photocytes are in the radial canals under the comb plates. A mark which has been placed on the site where the first cleavage furrow originates is traced through the different cleavage stages to the larval stage of development where it resides in the oral region.

that specify these two cell types (Freeman, 1976a). During the third cleavage division of embryogenesis each blastomere forms an external E macromere and a central M macromere. This is the first division during development in which determinants become segregated into different blastomere lineages. If an E macromere is isolated after this division it continues to cleave and ultimately gives rise to comb plate cilia, however it does not form photocytes. When an M macromere is isolated after this cleavage it continues to cleave and subsequently forms photocytes but not comb plate cilia cells. At the two- and four-cell stages of development the region of each blastomere that will become the E and M macromeres at the eight-cell stage has been mapped. Localization measurements at the two-cell stage indicate that the determinants which specify comb plate cilia and photocytes are generally found in both the presumptive E and M macromere regions of each blastomere. At the four-cell stage the comb plate cilia determinant is found exclusively in the presumptive E macromere region of each blastomere while the photocyte determinants are generally found in both the presumptive E and M macromere regions.

Cytochalasin B and 2-4 dinitrophenol have been used to reversibly inhibit cleavage in *Mnemiopsis* embryos (In these experiments the kind of inhibitor used did not effect the experimental results.). Fig. 4B and C diagram the results of experiments in which selected cleavages were

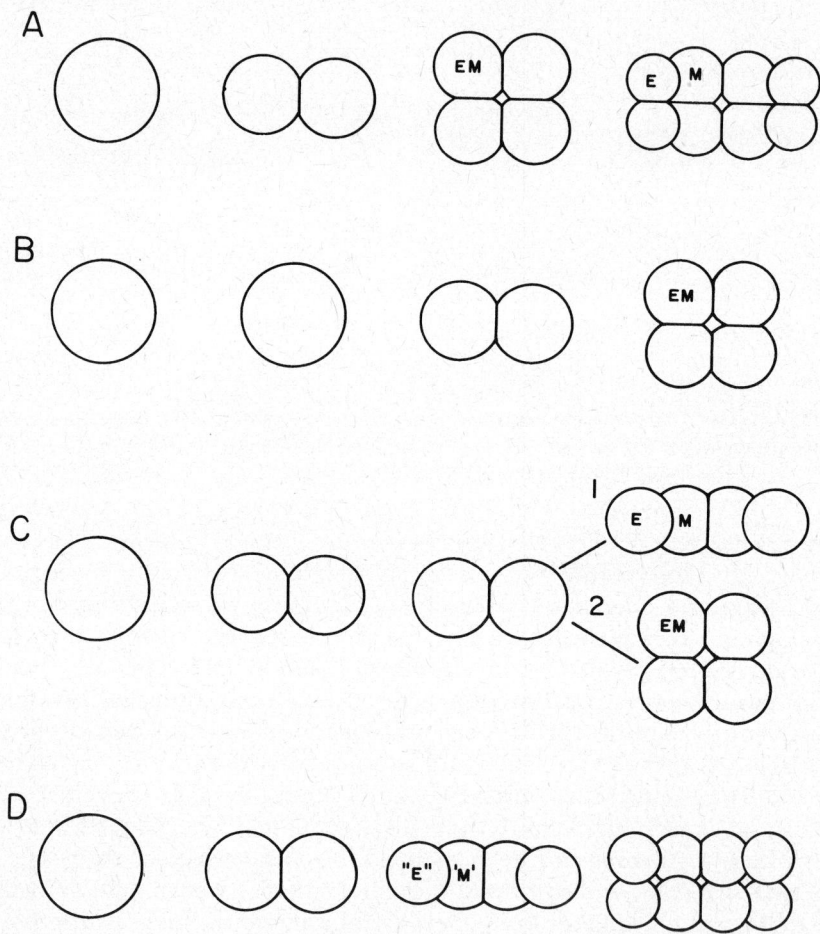

Fig. 4. Cleavage stages of normal and experimental ctenophore embryos viewed from the aboral pole. A) Uncleaved egg, two-, four-, and eight-cell stages of a normal embryo. B) Embryo in which the first cleavage has been reversibly inhibited. C) Embryo in which the second cleavage has been reversibly inhibited. (C1) shows the most common configuration that occurs after cleavage is initiated following the block; this configuration is similar to the configuration normally generated by the third cleavage. (C2) shows the blastomere configuration generated in the remaining cases; it is similar to the normal four-cell stage. D) Embryo in which the second cleavage occurred under compression, (Freeman, 1976, a & b).

postponed with these drugs. When a given cleavage was reversibly inhibited in *Cerebratulus* these embryos stay one cleavage stage behind their untreated controls. *Mnemiopsis* embryos behave in this way when the first cleavage is reversibly inhibited (Fig. 4B); however, if later cleavage stages are reversibly inhibited the cleavage that follows the lifting of the block usually has the character of the cleavage which is occurring in the untreated controls (Fig. 4C). The plane of the cleavage, the relative size of the blastomeres produced and their cytoplasmic composition corresponds to the control. These results show that the first cleavage sets up a timing system. Once this timing mechanism has been activated it determines the characteristics of a given cleavage without reference to the prior cleavage history of the blastomere. This situation is referred to as the cleavage clock phenomenon (Hörstadius, 1973).

Table II indicates how the reversible inhibition of a given cleavage affects the distribution of the determinants that specify comb plate cilia and photocytes when these determinants are mapped after cleavage has been resumed. In these experiments the first or the second cleavage was reversibly inhibited. At a given stage after cleavage had resumed the embryos were operated on to produce either blastomere fragments or blastomeres which correspond to the E and M macromeres at the eight-cell stage. The differentiation pattern of these isolates was used to map the distribution of the determinants. If the cleavage clock was not operating when cleavage was inhibited, the distribution of the determinants that specify the comb plate cilia and photocyte cells lagged one cleavage stage behind the control eggs. When this timing mechanism causes an embryo in which cleavage was inhibited to skip a cleavage stage, then the determinants that specify comb plate cilia and photocyte

TABLE II

The Differentiation Capabilities of Presumptive E and M Macromere Regions from Embryos in which the First or the Second Cleavage was Inhibited Compared with Their Unoperated Controls.

cleavage inhibited	stage blastomeres tested	E Macromere			M Macromere		
		Number of cases	Percent[1] differentiation		Number of cases	Percent differentiation	
			C	L		C	L
Control	two-cell stage	24	92	54	27	84	74
	four-cell stage	30	93	26	9	0	100
First Cleavage	two-cell stage	15	100	47	19	91	81
	four-cell stage	38	95	30	12	8	100
Second Cleavage	four-cell stage (clock C_1)	66	90	1	57	4	84
	four-cell stage (no clock C_2)	37	100	38	19	5	95

1. C, comb plate cilia, L. photocyte. Fig. 4 shows the cleavage inhibition experiments. Complete data is in Freeman (1976a).

differentiation show a pattern of localization which is typical of this stage.

This experiment shows that there is a correlation which relates the pattern of cleavage to the process of localizing the determinants that specify comb plate cilia and photocyte differentiation. There are two ways in which the pattern of cleavage could affect the localization of these determinants. 1) The plane in which a given cleavage occurs may determine the distribution of determinants. 2) Other mechanisms which are also controlled by the cleavage clock may determine the distribution of determinants. It is possible to alter the plane of the second cleavage in ctenophores by compressing two-cell embryos in a plane perpendicular to the first cleavage furrow (Zeigler, 1898). The blastomere configuration that forms at the next cleavage in these compressed embryos (Fig. 4D) is identical to the blastomere configuration generated in the majority of cases after reversible inhibition of the second cleavage (Fig. 4C). However these embryos differ from the latter because they are one cleavage stage younger. When the four blastomeres of these compressed embryos are raised in isolation both the "E" and "M" macromeres differentiate comb plate cilia; this determinant does not segregate until the next cleavage of these blastomeres (Freeman, 1976b). The experiment shows that some cleavage clock dependent function, other than the plane of cleavage determines the distribution of comb plate cilia determinants.

The cleavage clock phenomenon was discovered during the course of studies that tried to sort out how spindles are oriented and how aster size is controlled during the first cleavages of sea urchin eggs. Work has been done which demonstrates a cyclic change in the sulfhydryl content of KCl soluble and water soluble proteins in sea urchin embryos during early development that is correlated with the cell cycle (Sakai, 1960). There is evidence that this rhythmical fluctuation of the sulfhydryl contents of the egg proteins is the cleavage clock system (Dan and Ikeda, 1971). This hypothesis is based on studies in which different agents were used to reversibly inhibit a given cleavage in sea urchins and the effect of these inhibitors on the character of the next cleavage and the SH cycle was tested. This clock is thought of as a counting mechanism, at each cycle of the clock a unique set of cytoplasmic events that have been pre-programmed are activated. The effect of the program would depend on the cycle number and the cytoplasmic composition of a given blastomere. There is good evidence for a set of animalizing and vegetalizing determinants in the sea urchin embryo. These determinants appear to be localized in the appropriate regions of the egg prior to fertilization

(Hörstadius, 1973). This would preclude the cleavage clock from having a role in their localization (however see Dan, 1972 and Tanaka, 1976).

The cleavage clock phenomenon has only been demonstrated in echinoderm and ctenophore embryos. However in certain other kinds of embryos there are cytoplasmic movements associated with the localization of developmental potential that are coupled to cleavage but which can also occur in enucleated cytoplasmic fragments from fertilized eggs. In the mollusc *Dentalium* a polar lobe forms in the vegetal region of the egg during each of the first three cleavages (Fig. 5). As each cleavage occurs the contents of the lobe are segregated into only one of the blastomeres that form. This lobe-facilitated movement of cytoplasm is responsible for localizing a set of determinants primarily in one lateral quadrant of the embryo (Wilson, 1904; Geilenkirchen *et al.*, 1970). If a fertilized *Dentalium* egg is cut to produce an enucleated vegetal half prior to the first cleavage, this fragment does not cleave; however it forms polar lobes at the same time the nucleated animal fragment undergoes each of its first three cleavages (Wilson, 1904; Verdonk *et al.*, 1971).

These experiments show that when the localization of developmental potential occurs during cleavage, the progress made in localizing determinants is regulated by a mechanism which is cleavage stage specific. This coupling of localization and cleavage is of adaptive value in those forms in which cleavage occurs according to a set pattern because it helps to ensure that the various determinants are at the appropriate

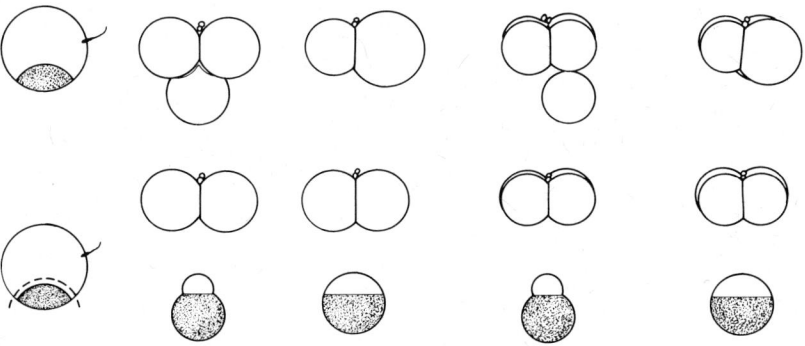

Fig. 5. Cleavage stages of normal and experimental embryos of a spiralian with a polar lobe viewed from the side. The top row shows the fertilization of the egg and polar lobe formation during the first and second cleavages of a normal embryo. The bottom row shows an egg which has been fertilized; after fertilization the vegetal region of the egg where the polar lobe (shaded) will form is cut out. The other panels show the behavior of this enucleated fragment and the animal half of the egg which contains the nucleus during the first two cleavages, (Verdonk *et al.*, 1971).

locations so that they will be segregated to the correct blastomere lineages. If the process of localization was out of phase relative to the process of cleavage this would not be the case.

IV. THE PLANE OF CLEAVAGE AND DETERMINANT LOCALIZATION

In some cases in which the localization of determinants occurs during cleavage, the plane of a given cleavage during the localization process has no effect on where determinants are localized. In other cases a given cleavage may provide directional information which establishes where localization will occur. A number of these situations will be examined in order to define the conditions that must be present in order for cleavage to play a directional role in setting up localizations of developmental potential.

The first case to be examined concerns the effects of changing the plane of the first cleavages on the sites where the determinants for the apical tuft and gut are localized in the *Cerebratulus* embryo. The planes of the first two cleavages in this embryo pass at right angles to each other along its animal-vegetal axis (Fig. 6A). The plane of the first or the second cleavage can be altered so that it is perpendicular to the animal-vegetal axis of the embryo by deforming the egg in the appropriate way prior to a given cleavage (Dederer, 1910) (Fig. 6B and C). Both of these alterations in the plane of cleavage create a condition in which the first two cleavage planes intersect along an axis in the equatorial plane of the egg rather than its animal-vegetal axis. In both kinds of embryos the third cleavage plane is perpendicular to this new axis. If this change in the plane of the first or second cleavage altered the sites at which the determinants for the apical tuft and gut are localized, one would expect the third cleavage to form two blastomere pairs in both the animal half and the vegetal half of the embryo in which only one pair of the daughter blastomeres differentiates an apical tuft while the other pair does not, or only one pair of the daughter blastomeres forms a gut while the other pair does not. This blastomere isolation experiment is outlined in Fig. 7. The results of the experiment are presented in Table III. These results show that both pairs of daughter blastomeres from either the animal or the vegetal half of the embryo differentiate the same structure in a significant number of cases.

In those cases in which the first cleavage is equatorial, quartets 2-4 appear to be formed along the animal-vegetal axis of the embryo. When the second cleavage is equatorial, quartets 2-4 appear to be formed

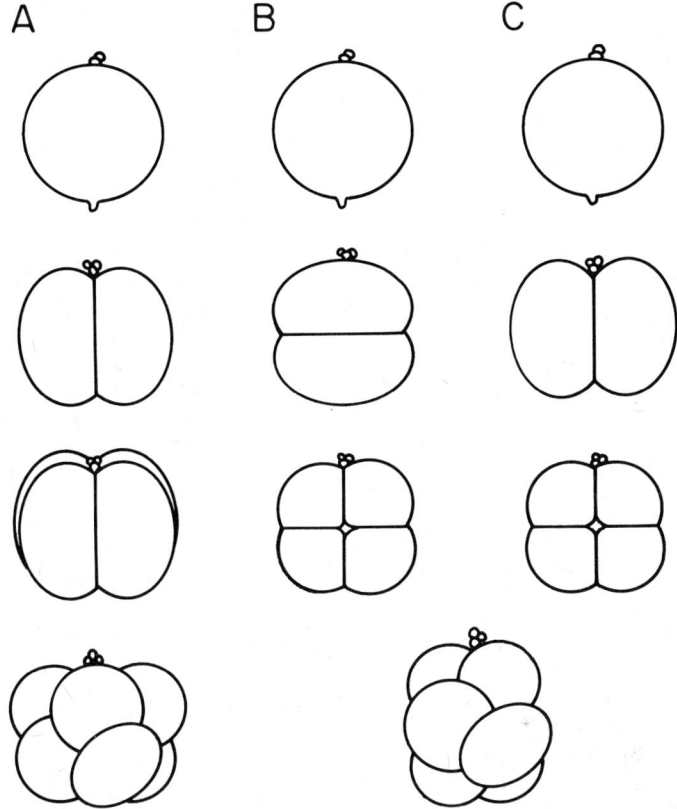

Fig. 6. Cleavage stages of normal and experimental *Cerebratulus* embryos viewed from the side. A) Uncleaved egg, two-, four-, and eight-cell stages of a normal embryo. B) Embryo in which the plane of the first cleavage is equatorial. C) Embryo in which the plane of the second cleavage is equatorial, (Freeman, 1978).

laterally. Both kinds of embryos develop into relatively normal pilidium larvae. When the vegetal poles of these embryos are marked at the four-cell stage, the marks always correspond to the gut forming region of the larvae. These observations show that the alteration of the plane of the first or second cleavage does not change the site where determinants are localized and does not change the symmetry properties of the embryo.

The next set of experiments in which the plane of cleavage is changed involves the ctenophore embryo. This embryo develops into a cydippid larva that is biradially symmetrical (Fig. 3). The larva has an oral-aboral axis which runs from the stomadeum to the apical organ. Around this axis there are two planes of symmetry: the sagittal plane and the

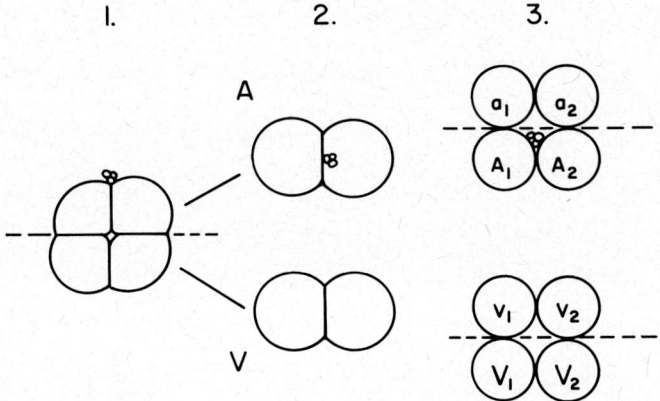

Fig. 7. Outline of the blastomere isolation experiment summarized in Table III. (1) Four cell stage embryo in which either the first or second plane of cleavage was equatorial, viewed from the side. The two animal blastomeres (A) are separated from the two vegetal blastomeres (V) at this stage. (2 & 3) The animal and vegetal blastomeres after their separation at the four- and eight-cell stages. At the end of the third cleavage the $a_1 a_2$ blastomere pair is separated from the $A_1 A_2$ blastomere pair along the third cleavage plane and the same blastomere isolation experiment is performed on the vegetal blastomeres.

TABLE III

The Differentiation Capabilities of Blastomere Pairs from Animal and Vegetal Halves of Embryos in which the Plane of the First or Second Cleavage was Equatorial.

Cleavage change	Animal half				Vegetal half			
	$\frac{++}{TG}$	$\frac{+-}{TG}$	$\frac{-+}{TG}$	$\frac{--}{TG}$[1]	$\frac{++}{TG}$	$\frac{+-}{TG}$	$\frac{-+}{TG}$	$\frac{--}{TG}$
First Cleavage								
$\frac{++}{TG}$	0				8			
$\frac{+-}{TG}$	1	9			4	1		
$\frac{-+}{TG}$	0	3	2		3	2	1	
$\frac{--}{TG}$	0	2	2	4	0	3	1	2
Second Cleavage	$\frac{++}{TG}$	$\frac{+-}{TG}$	$\frac{-+}{TG}$	$\frac{--}{TG}$	$\frac{++}{TG}$	$\frac{+-}{TG}$	$\frac{-+}{TG}$	$\frac{--}{TG}$
$\frac{++}{TG}$	1				0			
$\frac{+-}{TG}$	0	8			0	0		
$\frac{-+}{TG}$	0	2	0		0	1	8	
$\frac{--}{TG}$	0	4	1	8	0	0	6	4

1. T, apical tuft; G, gut; +, presence; −, absence.

tentacular plane. Early cleavage in this embryo is unipolar; the furrow is initiated at a circumscribed site on the cell surface and spreads out from there passing through the cell. Marking experiments have demonstrated that the site where the first cleavage is initiated corresponds to the oral pole of the embryo (Freeman, 1977). Marking experiments have also demonstrated that the plane of the first cleavage corresponds to the sagittal plane of the embryo while the second cleavage plane corresponds to the tentacular plane. When the blastomeres of the two-cell stage embryo are separated and raised in isolation, each half differentiates as a sagittal half embryo with a tentacle pouch and four rows of comb plates. When this operation is done at the four-cell stage, each blastomere differentiates a quadrant of the embryo containing two rows of comb plate cilia (Reverberi, 1966).

The ctenophore egg has a centrolecithal organization; it is composed of an inner endoplasmic zone and an outer layer of basophilic cytoplasm which surrounds the endoplasm. The zygote nucleus resides in the basophilic cytoplasm just underneath the cell membrane. In certain species of ctenophores one can predict where the first cleavage will be initiated with a high degree of certainty; for example, in *Pleurobrachia bachei* the first cleavage is almost invariably initiated where the polar bodies are given off. When one centrifuges *P. bachei* eggs just before the first cleavage, the eggs orient at random in the centrifuge tube and the basophilic cytoplasm takes up a centripetal position while the endoplasm takes up a centrifugal position in the egg. The zygote nucleus remains in the basophilic cytoplasm and is frequently transferred to a new site relative to the polar bodies. The movement of the zygote nucleus to a new location causes cleavage to be initiated at that site and the oral-aboral axis of the embryo forms at this new site. This experiment shows that cleavage sets up the oral-aboral axis of the embryo (Freeman, 1977).

In the two-cell ctenophore embryo there is a moderate localization of comb plate forming potential in the aboral half of each blastomere; 30% of the two-cell stage blastomeres in which this region is removed do not form comb plate cilia (Freeman, 1976a). If one measures the amount of localization for comb plate determinants in the aboral region of the egg which has just initiated cleavage where the furrow has not yet entered the aboral region, and in uncleaved eggs where one can use the site of polar body formation to predict where the aboral pole of the embryo will be, there is no evidence for localization of this determinant (Freeman, 1977). This means that localization of comb plate cilia determinants begins simultaneously with the cleavage which sets up the oral-aboral axis and that cleavage plays a role in establishing where localization will occur.

Another example showing that a cleavage which establishes a symmetry property also establishes where determinants will be localized concerns the ctenophore embryo in which the plane of the second cleavage has been altered by compression (Fig. 4D). The second cleavage normally occurs along the tentacular plane. When the plane of the second cleavage has been altered by compression the embryo develops into a monster which contains the cell types which a normal larva should have, however they are not arranged properly. While an oral-aboral axis can be detected, it is impossible to identify a sagittal or a tentacular plane. When the same blastomere configuration is created by reversibly inhibiting the second cleavage (Fig. 4C), the embryo has a set of symmetry properties that are relatively normal. This comparison suggests that the plane of cleavage which occurs at the developmental period when the embryo is normally undergoing its second division must have an appropriate orientation with reference to the first plane of cleavage if normal larval symmetry properties are to develop. The observation that embryos in which the second cleavage is reversibly inhibited have normal symmetry properties suggests that in the absence of a cleavage at a time equivalent to the four-cell stage a normal tentacular plane is established.

At the four-cell stage in a normal ctenophore embryo virtually all of the comb plate determinants are localized in the region of each blastomere that will become the E macromere at the next cleavage (Freeman, 1976a). However when the second cleavage occurs along a plane that mimics the plane of the third cleavage, instead of creating a situation in which only the "E" macromere inherits all of the comb plate cilia determinants, both the "E" and the "M" blastomeres inherit the determinant indicating that a change in the plane of this cleavage has changed the sites where comb plate potential is localized.

The last case to be considered concerns polar lobe mediated localization of determinants in spiralians. In many annelids and molluscs the polar lobe serves as a localizing device for moving determinants from the vegetal region of the egg to a lateral position in one quadrant of the developing embryo. The presence of these determinants in the vegetal region of the uncleaved egg reflects a localization process that presumably occurred prior to fertilization along the animal-vegetal axis of the oocyte (for example see Verdonk *et al.*, 1971). The process of polar lobe mediated localization accompanies the first two cleavages (Fig. 5); these events set up the dorsal-ventral axis of the embryo. If the polar lobe forming region or the polar lobe proper is removed from the embryo a characteristic set of larval structures fail to develop and the resulting larva lacks a dorsal-ventral axis of symmetry (see Reverberi, 1971 for reviews involving different spiralians with polar lobes). One can argue that the lateral quadrant of the embryo where the determinants in the vegetal region of the egg will become localized is predetermined. This argument assumes that the promorphological scaffold which specifies the lateral quadrant that will inherit the polar lobe contents also controls the meridian in which the plane of the first cleavage is positioned. It is also possible to argue that the plane of the first cleavage plays a role in determining the lateral site where the contents of the polar lobe will be shunted.

In the annelid *Sabellaria*, Guerrier (1970) has demonstrated that the position of the first cleavage spindle can be changed just before cytokinesis by rotating it so that it comes to lie in any one of a number of new positions around the animal-vegetal axis of the egg. This change causes the meridian where the first cleavage furrow will form to change. After the meridional rotation of the first plane of cleavage the polar lobe is still segregated to only one blastomere. In many cases these meridional shifts cause the lobe contents to be segregated to a lateral quadrant of the embryo in which they would not normally be segregated. Since subsequent cleavage and polar lobe formation are normal, and the larva that develops is normal, one is able to argue that the lateral quadrant which become the site where the contents of the lobe become localized orginates as a consequence of cleavage.

Guerrier (1970) has also done experiments in which the first or the second plane of cleavage is shifted so that it is perpendicular to the animal-vegetal axis of the embryo. Both of these shifts create a condition in which the first two cleavage planes intersect along an axis in the equatorial plane of the egg rather than its animal-vegetal axis. When the first cleavage plane is shifted so that it is perpendicular to the animal-vegetal axis of the embryo the larva that develops is abnormal and lacks a dorsal-ventral axis. When the second cleavage plane is shifted so that it is perpendicular to the animal-vegetal axis of the embryo the larva that develops is normal. This situation can be explained by arguing that, at the time of the first cleavage, events take place which program the dorsal-ventral axis of the embryo; and a cleavage in an inappropriate orientation with reference to the animal-vegetal axis of the embryo causes abnormal morphogenesis because the appropriate determinants receive a set of localization cues that are not appropriately placed.

An analysis of these experiments in which the plane of cleavage is changed during the localization process shows that when the promorphological scaffold, along which a given determinant is localized, is set up prior to cleavage, the plane of cleavage during determinant localization does not effect where localization occurs. However, when a cleavage occurs at a time when an axis of symmetry is being set up, the plane of cleavage plays a directional role in establishing where determinants which sort out along that axis are localized.

There are several kinds of embryos in which axes of symmetry are set up during cleavage stages of development. In many of these cases the plane of a given cleavage determines where an axis of symmetry will be laid down (see Guerrier, 1971 for an important review of this literature for spiralians). The literature suggests that there are defined time periods of relatively short duration when a given axis of symmetry can form. One would like to know much more about the special metabolic properties of the embryo during these time periods. One would also like to know how the position of a plane of cleavage functions to define where an axis of symmetry will form when cleavage takes place at one of these special times.

V. CONCLUDING REMARKS

This review shows that during the cleavage stages of embryogenesis, determinants can be localized at specific sites in the blastomeres of the embryo, and that these localization events are regulated by the cell cycle. Each cell cycle allows a certain amount of progress to be made in localizing a given determinant in a definite region of a blastomere; the amount of progress made depends on the specific cell cycle and the cytoplasmic composition of the cells involved. Certain cell cycles also establish where a given determinant will be localized; at these cell cycles an axis of symmetry is set up.

I do not think that the coupling reported here between localization events which occur during cleavage and the cell cycle represents a special case. These findings form part of the larger literature on the relationship between the cell cycle and cell differentiation (Reinert and Holtzer, 1975).

ACKNOWLEDGEMENTS

I want to thank Drs. David Miyamoto, Klaus Kalthoff and Antone Jacobson for reading this manuscript. This work was supported by research grant GM 20024 from the U.S. Public Health Service.

REFERENCES

Ancel, P. and Vintemberger, P. (1948). *Bull. Biol. France et Belg.* Suppl. **31**, 1-182.
Arnold, J. and Williams-Arnold, L. (1974) *J. Embryol. Exp. Morphol.* **31**, 1-25.
Baxter, A. (1974). "Edmund Beecher Wilson and the Problem of Development: From the Germ Layer Theory to the Chromosome Theory of Inheritance". Doctoral Dissertation, Yale University.
Bonner, J.T. (1952). "Morphogenesis, An Essay On Development". Princeton University Press, Princeton.
Conklin, E.G. (1898). *In* "Biological Lectures from the Marine Biological Laboratory, Woods Hole, Mass.", pp. 17-43. Gin and Company, Boston.
Costello, D. P. (1948). *Ann. N.Y. Acad. Sci.* **49**, 663-683.
Dan, K. (1972). *Exp. Cell Res.* **72**, 69-73.
Dan, K. and Ikeda, M. (1971) *Develop. Growth and Differen.* **13**, 285-301.
Davidson, E. (1976). "Gene Activity in Early Development". Academic Press, New York.
Dederer, P. (1910). *Arch. Entwicktangsmech.* **29**, 225-242.
Elinson, R. and Manes, M. (1978). *Develop. Biol.* **63**, 67-75.
Freeman, G. (1976a). *Develop. Biol.* **49**, 143-177.
Freeman, G. (1976b). *Develop. Biol.* **51**, 332-337.
Freeman, G. (1977). *J. Embryol. Exp. Morphol.* **42**, 237-260.
Freeman, G. (1978). *J. Exp. Zool.* **206**, 81-107.
Geilenkirchen, W.L.M., Verdonk, N. and Timmermans, L. (1970). *J. Embryol. Exp. Morphol.* **23**, 237-243.
Gould, S. J. (1977) "Ontogeny and Phylogeny". Harvard University Press, Cambridge.
Guerrier, P. (1970). *J. Embryol. Exp. Morphol.* **23**, 639-665.
Guerrier, P. (1971). *Ann. Biol.* **10**, 152-192.
Harvey, E.B. (1956). "The American *Arbacia* and other Sea Urchins". Princeton University Press, Princeton.
Hillman, N., Sherman, M. I. and Graham, C. F. (1972). *J. Embryol. Exp. Morphol.* **28**, 263-278.
Hörstadius, S. (1937). *Biol. Bull.* **73**, 317-342.
Hörstadius, S. (1971). *In* "Experimental Embryology of Marine and Freshwater Invertebrates" (G. Reverberi, ed.), pp. 164-174. North-Holland Pub. Co., Amsterdam.
Hörstadius, S. (1973). "Experimental Embryology of Echinoderms". Oxford University Press, London.
Jacobson, A. (1966). *Science* **152**, 25-34.
Jaffe, L. F. (1969) *Develop. Biol.* Suppl. **3**, 83-111.
Kubota, T. (1967). *J. Embryol. Exp. Morphol.* **17**, 331-340.
Kühn, A. (1971). "Lectures on Developmental Physiology". Springer-Verlag, New York.
Lillie, F. R. (1906). *J. Exp. Zool.* **3**, 153-268.
Morgan, T.H. (1927). "Experimental Embryology". Columbia University Press, New York.
Quatrano, R.S. (1972). *Exp. Cell Res.* **70**, 1-12.
Raff, R. (1977). *Bio. Sci.* **27**, 394-401.
Rappaport, R. (1974). *In* "Concepts of Development" (J. Lash and J. Whittaker, eds.), pp. 76-98. Sinauer Assoc., Stamford.
Rebhun, L. (1975). *In* "Molecules and Cell Movement" (S. Inoue' and R. Stephens, eds.), pp. 233-238. Raven Press, New York.
Reinert, J. and Holtzer, H. (eds.), (1975). "Cell Cycle and Cell Differentiation". Springer-Verlag, New York.
Reverberi, G. (1961). *Adv. in Morphogen.* **1**, 55-101.

Reverberi, G. (1966). *Ann. Biol.* **5,** 375-390.
Reverberi, G. (ed.), (1971). "Experimental Embryology of Marine and Freshwater Invertebrates". North-Holland Pub. Co., Amsterdam.
Sakai, H. (1960). *J. Biophys. Biochem. Cytol.* **8,** 609-615.
Schroeder, T. (1972). *J. Cell Biol.* **53,** 419-434.
Tanaka, Y. (1976). *Develop. Growth and Differn.* **18,** 113-222.
Verdonk, N., Geilenkirchen, W.L.M. and Timmermans, L. (1971). *J. Embryol. Exp. Morphol.* **25,** 57-63.
Wilson, E.B. (1896). "The Cell in Development and Inheritance". 1st ed., Macmillan, New York.
Wilson, E.B. (1904). *J. Exp. Zool.* **1,** 1-74.
Wilson, E.B. (1925). "The Cell in Development and Heredity". 3rd ed., Macmillan, New York.
Wolf, R. (1978). *Develop. Biol.* **62,** 464-472.
Zalokar, M. (1974). *Wilhelm Roux' Arch.* **175,** 243-248.
Zeigler, H. (1898). *Arch. Entwicklungsmech.* **7,** 34-64.

The Control of the Polar Deposition of a Sulfated Polysaccharide in *Fucus* Zygotes

Ralph S. Quatrano, Susan H. Brawley[*]
and William E. Hogsett

Department of Botany and Plant Pathology
Oregon State University
Corvallis, OR 97331
and
Department of Botany[*]
University of California
Berkeley, CA 94720

I. Introduction 78
II. Polar Axis Fixation 80
    A. Requirement for a Cytochalasin B-Sensitive Process 80
    B. Fucoidin Localization is Prevented by CB 82
    C. Ion Accumulation and Cortical Clearing Precedes Axis Fixation 84
    D. Model for Axis Fixation 85
III. Localization of Rhizoid Specific Products 87
    A. Evidence for Transport of Fucoidin 87
    B. Dependence Upon Sulfation for Transport of Fucoidin .. 91
IV. Mechanism of Localization 94
    A. Contractile 94
    B. Electrophoretic 94
    References 96

I. INTRODUCTION

The mechanisms underlying the temporal and quantitative regulation of gene activity must be integrated with spatial controls for a complete understanding of differentiation. One of the major questions involving spatial control is how specific components are localized within definite areas of a cell. Cytoplasmic localization is initiated in response to the extracellular microenvironment, usually in the form of a gradient. The result is the segregation of certain subcellular components to ends of an axis established with respect to the gradient. The result is a polar cell. These local cytoplasmic regions can serve as a site for intracellular differentiation or as an unique microenvironment for genetically identical nuclei when they are partitioned into these areas, e.g. blastula formation from a polar egg cell. Thus, the local divergence of gene expression resulting from these unique nucleo-cytoplasmic interactions in the developing blastula is a consequence of the unequal distribution of cytoplasmic components during oogenesis (c.f. Davidson, 1976; Quatrano, 1978).

The mechanism whereby cytoplasmic components are directed to a predetermined location is difficult to directly approach in most systems. However, zygotes of the brown alga *Fucus* are ideally suited to answer questions pertaining to cytoplasmic localizations. Within 14 hours after fertilization, the zygote forms a localized protuberance or rhizoid which represents the first sign of asymmetric development. Within the cytoplasm of the rhizoid one can detect the accumulation of subcellular components which are the earliest signs of polarity in the previously apolar, homogeneous cytoplasm of the spherical zygote. When the rhizoid and its unique contents are separated from the remaining zygote by the first cell plate, the resulting two cells of the embryo are different from one another in structure, biochemical composition and developmental fate (Fig. 1). The rhizoid and its progeny contribute to the holdfast portion of the mature plant while the thallus cell and its derivatives form the frond. The polarity acquired by the zygote is the basis for cellular differentiation in the two-celled embryo and the developmental axis of the entire plant (Jaffe, 1968, 1970; Quatrano, 1974, 1978).

A wide variety of external gradients can orient the polar axis of the zygote (Jaffe, 1968). For example, rhizoid-specific components accumulate on the shaded side of a unilateral light gradient, the low end of a pH gradient, the positive pole of a voltage gradient, and the more concentrated side of a Ca^{2+} or K^+ gradient. The polar axis remains labile

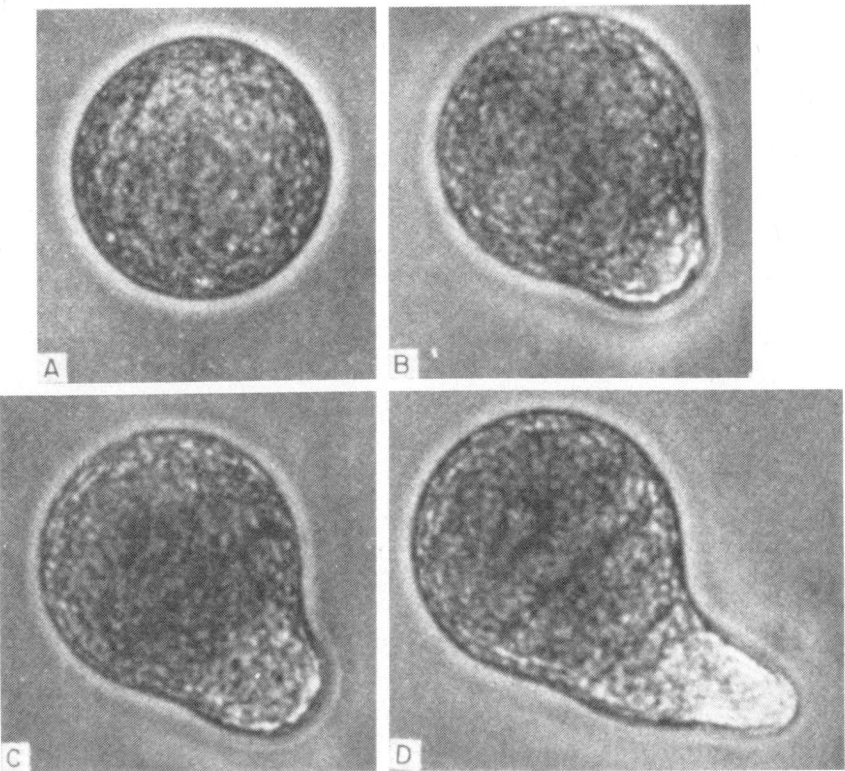

Fig. 1. Photomicrographs of zygote development in *Fucus*. The apolar zygote (A) forms a localized protuberance (rhizoid) at 14 hr after fertilization (B), which is subsequently partitioned from the rest of the cell by the first division at about 20 hr (C) and the second division at about 24 hr (D). (Photomicrographs courtesy of Dr. G. Benjamin Bouck.)

until a few hours before rhizoid formation, and each vector is perceived by the zygote at different but overlapping periods after fertilization (Jaffe, 1968; Quatrano, 1973). Hence, an axis oriented by unilateral light at one time can be reoriented at a later time by light (or by another vector) from a different direction. However, none of the above gradients are required for a polar zygote to form. Once fertilized, zygotes of a number of fucoid species become polar in a gradient-free environment. Sperm entry, however, appears to be required (Knapp, 1931; Quatrano, 1978). The point of sperm entry is the subsequent site of polar development unless a subsequent vector is imposed across the cell. This data and the demonstration by Jaffe (1958) that polarized light produces bipolar embryos, argues strongly for a polar axis which is initially labile and "arises in some more epigenetic manner than through the directed rotation of some preformed asymmetric structure." Therefore, the main advantages of the *Fucus* embryos are: (a) the structural and biochemical polarity exhibited in the zygote occurs in a previously homogeneous egg cytoplasm, (b) the cytoplasmic site of the localizations can be experimentally determined and controlled by a variety of externally applied gradients. Thus, in a synchronously developing unicellular system one can describe and analyze the cytological and biochemical changes occurring at a cytoplasmic site which was predetermined by an extracellular gradient.

The rhizoidal area is characterized by the intracellular accumulation of mitochondria, Golgi and associated vesicles, as well as the localization of sulfated polysaccharides in the cell wall (Quatrano, 1972, 1974, 1978). One possible role of the dense accumulation of these mucilagenous polysaccharides at this site is for the adhesion of the developing embryo to the substratum by the "root-like" rhizoid (Crayton et al., 1974). During the past several years we have used the sulfated polysaccharide *fucoidin* as a localized, cell-specific product to study the events required for its incorporation into the cell wall at a site determined by an extracellular gradient. I will present data and some speculation bearing on three questions: What are the cytological events which lead to the establishment of a fixed polar axis oriented with respect to an extracellular light gradient? What characteristics of the sulfated polysaccharide are essential for its accumulation at one end of the fixed axis? What is the mechanism for its localization?

II. POLAR AXIS FIXATION

A. *Requirement for a Cytochalasin B-Sensitive Process*

Orientation of the polar axis is accomplished by unilateral light pulses (60 min) between 4 and 11 hours after fertilization. The axis is no longer labile after 11 hours and is fixed, i.e. light or any other gradient from a different direction cannot reorient the axis (Quatrano, 1973). Colchicine does not block axis fixation or photopolarization, but prevents the first oriented cell division (perpendicular to the rhizoid axis). Some cytoskeletal basis other than microtubules and the spindle apparatus is responsible for the original polar axis. It follows that the plane of the first cell division is predetermined by the previously set rhizoid axis. Unfortunately, polar axis fixation and the localized accumulation of macromolecules experimentally directed by light or other gradients cannot be separated in time from each other, even in synchronously developing cultures of zygotes. In an attempt to separate these events we treated zygotes exposed to unilateral light with reversible inhibitors of rhizoid formation. By temporarily inhibiting the localization processes, we asked whether fixation was likewise delayed or if these two processes could be uncoupled. To assay whether the polar axis was fixed, we rotated the source of orienting light after untreated controls had established a fixed axis. In such cultures, rhizoids forming from the shaded side of the first orienting light indicated that a fixed axis was established in the presence of the rhizoid inhibitor. Cycloheximide and sucrose-treated zygotes responded in this manner i.e. these inhibitors uncoupled fixation from localization (Fig. 2). Cultures treated with cytochalasin B (CB), however, responded to the second orienting light indicating that this inhibitor prevented axis fixation (Quatrano, 1973). CB has several other interesting effects on polar axis induction by light. If CB is present only during one unilateral light pulse of 60 min, the resulting rhizoids are normal but not oriented with respect to light (Nelson and Jaffe, 1973; Quatrano, 1973). When CB is introduced for only 4 hours *after* a single

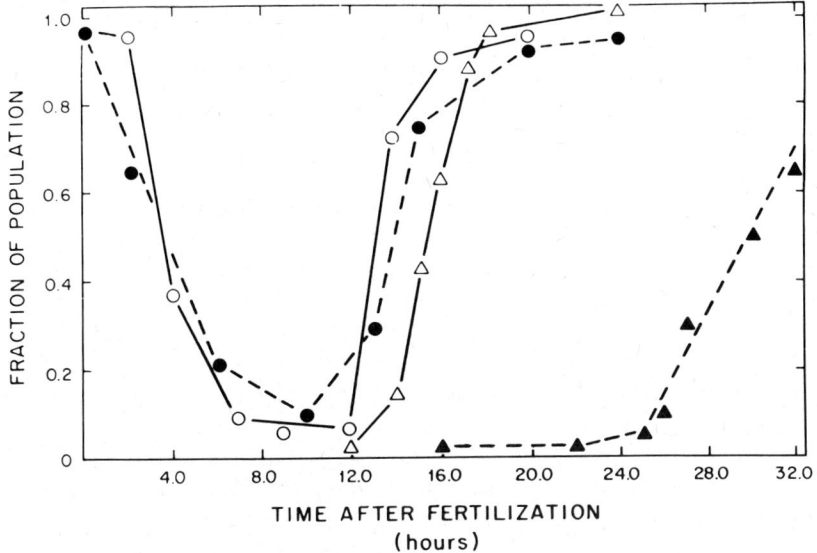

Fig. 2. Time course of sensitivity to polarity-inducing light and the initiation of rhizoid outgrowth in *Fucus distichus*, with (----) and without (——) cycloheximide treatment (1.0 µg/ml from 7 to 15 hr after fertilization). Each point represents the average of more than 200 zygotes that were exposed to a 2-hr pulse of unilateral light beginning 60 min before and ending 60 min after the designated time. Circles (O) represent fraction of population lacking photosensitivity, and triangles (Δ) represent fraction of the population possessing rhizoids. Notice how treatment with cycloheximide (----) effectively separated the fixation of a stable polar axis (●) from the formation of the rhizoid (▲). Similar results were obtained with sucrose. (From Quatrano, 1973).

light pulse, but without a second orienting light from a different direction, rhizoids are oriented with respect to the light. Thus, although CB prevents axis establishment and axis fixation by light, it does not disrupt the orientation of a previously induced axis (Quatrano, 1973).

B. *Fucoidin Localization is Prevented by CB*

One would expect that, in the absence of a fixed site of rhizoid formation fucoidin would not become localized. If cell walls were isolated after 16 hours and treated with Toluidine Blue O (TBO) at a pH below 2 (specific stain for sulfated polysaccharides), only a localized region of the wall was stained, that area forming the rhizoid. Walls isolated prior to 8 hours did not stain with TBO, indicating that very little if any sulfated fucoidin was present in the wall at that time. When walls from 16 hour embryos incubated with CB were stained, fucoidin was not localized, but the stain was evenly distributed throughout the wall surface (Novotny and Forman, 1974; Quatrano and Stevens, 1976). Hence, CB does not interfere with the process of fucoidin sulfation (to be discussed later), secretion or its incorporation in the cell wall, but apparently uncouples these processes from the *directed deposition* of fucoidin to the fixed site. These results are consistent with CB selectively interfering with the processes of axis fixation.

In an attempt to determine the nature of the CB block, zygotes treated with CB continually from fertilization were examined cytologically and stained with TBO to localize fucoidin. Rhizoid formation was prevented but some zygotes divided, which is consistent with previously published reports (Nelson and Jaffe, 1973; Quatrano, 1973). All treated zygotes exhibited an abnormally large accumulation of Golgi and associated vesicles in the perinuclear area (Brawley and Quatrano, 1979). These areas stained metachromatically with TBO, were capable of incorporating $^{35}SO_4$ into fucoidin, and persisted as long as CB was present, even after cell division (Fig. 3 A-D). The cell wall stained weakly, but uniformly with TBO but did not exhibit a localized accumulation of fucoidin (Brawley and Quatrano, 1979). Apparently those vesicles in the peripheral cytoplasm containing fucoidin were not inhibited from depositing their contents into the cell wall in the presence of CB. Not only did CB interfere with the directed movement of these vesicles to a predetermined location, but CB also prevented Golgi and associated vesicles in the perinuclear region from migrating to a predetermined site in the cytoplasm. An earlier EM study of untreated zygotes indicated that if a median section was viewed, the perinuclear region was highly

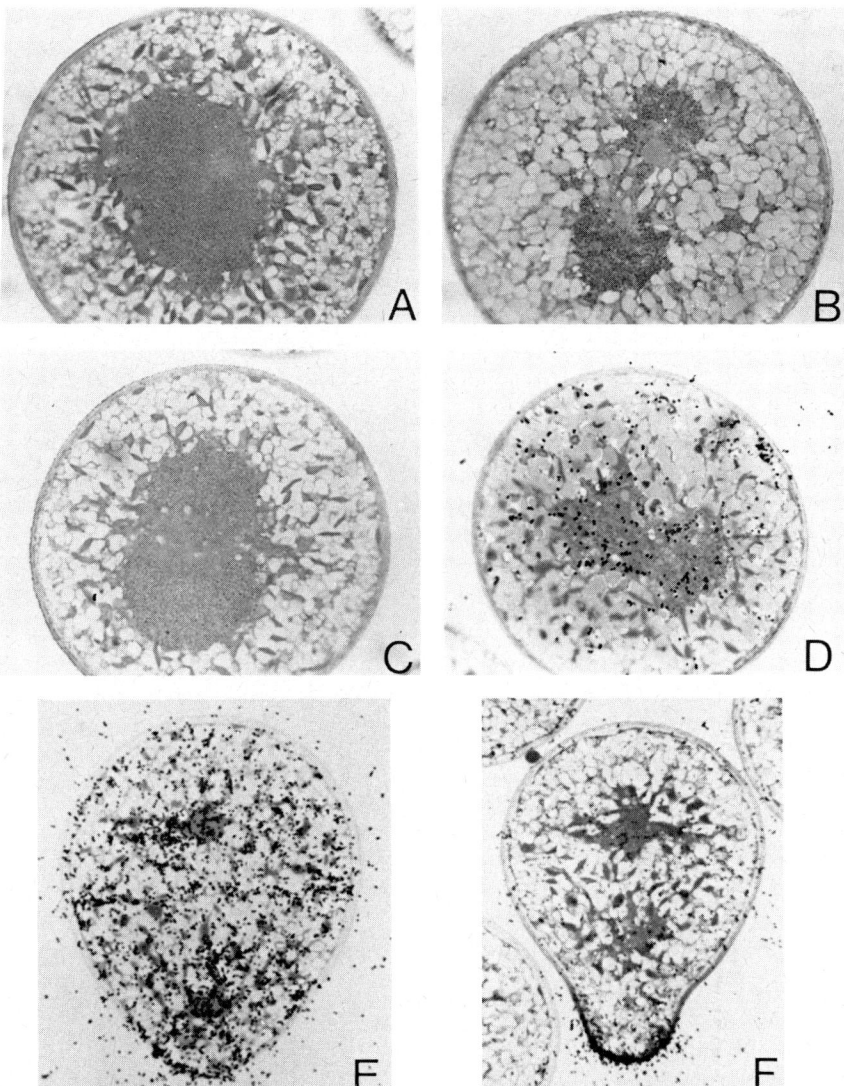

Fig. 3. A-D (X70) Sections of zygotes of *F. distichus* treated with CB (50 μg/ml) continually from fertilization and stained in TBO. In the different stages of cell division, notice the large accumulation of the metachromatic material in the perinuclear region, most probably large numbers of Golgi. The cell wall is also stained uniformly with TBO. Photograph D is an autoradiograph of a $^{35}SO_4$ pulse demonstrating that fucoidin sulfation is not inhibited and occurs within the Golgi.

Photos E and F (X60) are autoradiographs of embryos grown in the absence (E) and presence (F) of sulfate in the medium. $^{35}SO_4$ was given for 5 min at the same developmental stage (2-celled embryo) and then chased with 0.1 mM cold sulfate for 2 hours before fixing. Notice the random location of Golgi (sulfating sites) in E and the precise localization of the sulfated fucoidin in only the tip of the rhizoid wall.

polarized with fingerlike projections of cytoplasm containing mitochondria and Golgi radiating only toward the rhizoid pole (Quatrano, 1972). This localized accumulation of cytoplasmic components also stained metachromatically with TBO. In summary, the prevention of axis fixation by CB results in the lack of fucoidin localization due to an apparent disruption of the directed transport of Golgi and associated vesicles.

C. *Ion Accumulation and Cortical Clearing Precedes Axis Fixation*

What other localized events occur at the presumptive site of rhizoid formation before and during axis fixation, and what is the event blocked by CB? Nuccitelli (1978) has recently shown that the influx of ions and the secretion of wall material occur at the presumptive rhizoid site *before* fixation of a light-induced polar axis. *Pelvetia fastigiata* zygotes exhibit both an accumulation of cytoplasmic vesicles and a clear area between the plasma membrane and cell wall at the rhizoid pole several hours prior to rhizoid emergence. This "cortical clearing" is most likely material for cell wall formation accumulating intracellularly and being secreted. Consistent with the latter point, Peng and Jaffe (1976) have described by freeze-fracture techniques the deposition of new membrane patches covering the rhizoid region at the time of "cortical clearing". By moving the source of unilateral light 180° just after the first signs of clearing appeared, Nuccitelli (1978) observed a second cortical clearing area 180° from the first. *Both* areas persisted for a few hours, with the rhizoid finally emerging from the region of the latest formed clearing. This points to a slightly later event, not cortical clearing, that finalizes or fixes a particular site for subsequent rhizoid extension.

In addition to cortical clearing, local fluxes of inward directing ions clearly precede rhizoid formation in *Pelvetia,* and can be localized in the plasma membrane regions of the presumptive rhizoid site before axis fixation. An ultra-sensitive vibrating probe (c.f. Jaffe and Nuccitelli, 1977) was utilized to detect currents entering and leaving various surface regions. Using the probe, Nuccitelli (1978) demonstrated that shortly after fertilization the spatial current pattern around the *Pelvetia* zygote is shifting between several inward current regions. However, as the time of fixation approaches, these inward currents were concentrated at the presumptive site of rhizoid formation and represents intracellular areas of ion accumulation. Using the same material, he then observed cortical clearing occurring at this site and always at the region of largest inward current. If the orienting light was reversed, the spatial current pattern changed with an inward current on the new dark side within 40 minutes, followed in about an hour by a second region of cortical clearing at the new site of inward current. If the light reversal occurred after fixation, neither the reversal of the inward current nor the second clearing area was observed. The main points of this important study seem clear: (a) the appearance of a second cortical clearing was preceded by a local inward current, (b) both events occur prior to fixation and *predict* the final site of rhizoid formation, and (c) only when the electrical polarity was reversed did the cortical clearing and localized growth change.

When summarized, then, these results lead to the following time course of events within the zygote:

1. The detection of inward directing currents in the plasma membrane following fertilization.

2. Inward currents directed toward the presumptive site of rhizoid formation by an external gradient e.g. light. This represents the first sign of localization.

3. Accumulation and secretion of vesicles containing cell wall material (cortical clearing).

4. A process sensitive to CB and insensitive to CH and sucrose which fixes this site of intracellular ion accumulation and vesicle secretion for subsequent localizations of particles and organelles needed for wall extension.

5. The incorporation of fucoidin and other polysaccharide material into the rhizoid wall.

D. *Model for Axis Fixation*

The primary role of inward directing current patches of the plasma membrane in determining the site of rhizoid formation suggests that their stabilization within the membrane, or to the underlying cytoplasm, is the critical factor in axis fixation. Given the above information, what can be proposed as a model or working hypothesis for the fixation process?

Although no direct information is available in *Fucus* or other plant systems on the stabilization of membrane components, an extensive literature is available implicating a cytoskeletal basis for restricting movement of membrane components in animal cells by microtubules and/or microfilaments (Albertini and Anderson, 1977; Edelman, 1976; Nicholson, 1976; Pollard and Wiehing, 1974). Plant cells do contain microfilaments and actin filaments (Condeelis, 1974; Hepler and Palevitz, 1974; Forer and Jackson, 1976; Clark and Spudich, 1977). *Fucus*

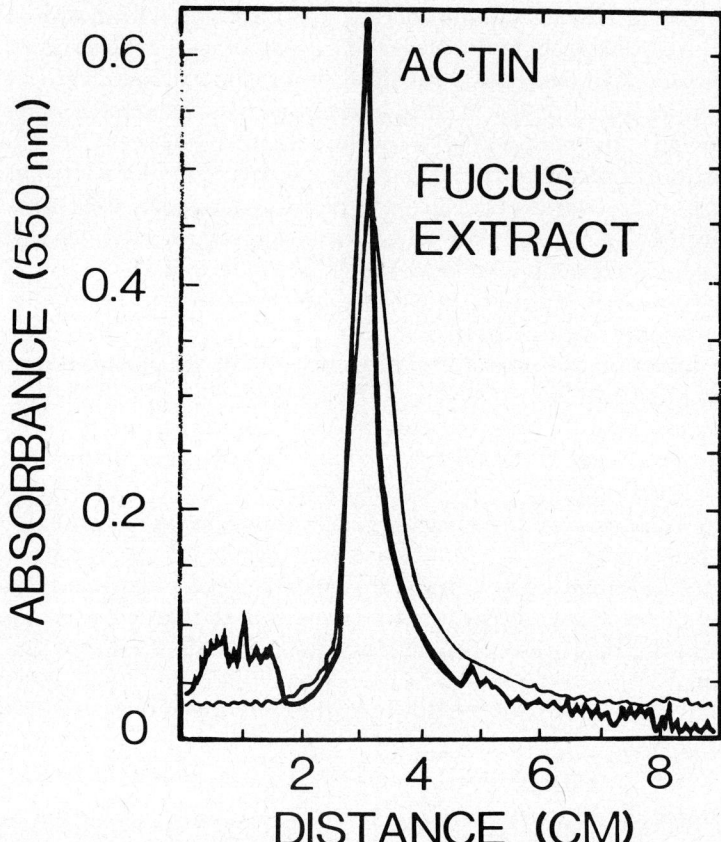

Fig. 4 Densitometer tracings of two acrylamide gels stained with Coomassie Blue containing authentic rabbit muscle F-actin and an extract from an acetone powder prepared from 2-celled *Fucus* embryos.

might also contain microfilaments (Brawley *et al.*, 1977) and a protein has been isolated in our lab which co-electrophoresis with F-actin (Fig. 4). The fact that CB reversibly prevents polar axis fixation in *Fucus* could be used as indirect evidence implicating a plasma membrane/microfilament association as the basis for fixation. Attachment of microfilaments to the sol-gel interface in *Nitella* is prevented by CB (Chen, 1973). Great care must be used in linking CB effects to microfilaments without direct evidence, but in view of the information now available (primarily from animal cells) it could be helpful to propose a model using this association as a basis to stimulate and focus future thinking and testing.

A working model is as follows. Membrane components in *Fucus* involved in transporting ions (e.g. Ca^{2+}) inward are translocated to the dark side of the light gradient. This movement is mediated by a CB sensitive process. As a result of this local increase in intracellular Ca^{2+} caused by the localized membrane patches in the rhizoid region, many cellular processes including enzyme activation, vesicle secretion (Cochrane and Douglas, 1974) and microtubule assembly could be regionally activated or suppressed thereby amplifying the existing polarization. The accumulation of these membrane patches at the presumptive rhizoid site are stabilized in the membrane or to the underlying cytoplasm by a cytoskeletal component, possibly microfilaments. This corresponds to polar axis fixation. The microfilaments and/or other cytoskeletal components (e.g., microtubules), may serve to provide a "track" for the accumulation of vesicles needed for cell wall extension, and serve as a point to organize and orient the spindle apparatus (Franke et al., 1972). Such processes in this region may be further modified by the local increase of specific ions caused by the accumulation of specific membrane components directing inward current fluxes. The same could be said for events occurring at the opposite pole.

Certain parts of this model are testable: Can microfilaments, actin bundles, microtubules, and/or colchicine binding proteins be localized at the fixed site of rhizoid formation? Are they associated with the plasma membrane, and is their structure or polymerization influenced by ions such as Ca^{2+}? Are membrane patches directing ions inward prevented from becoming localized in the presence of CB (like the deposition of fucoidin and the polar axis)? Are the vesicles which are localized in the rhizoid region associated with any cytoskeletal component?

III. LOCALIZATION OF RHIZOID SPECIFIC PRODUCTS

A. *Evidence for Transport of Fucoidin*

Previous reports indicated that a new and different sulfated polysaccharide is deposited in only that region of the zygote cell wall which forms the rhizoid (Quatrano and Stevens, 1976). The localized polysaccharide is an α-1,2 linked fucan characterized by an ester-sulfate bond to the C-4 of the fucose residues. Although fucose is the predominant sugar, xylose, mannose, galactose and glucuronic acid are also found but in an unsulfated form. At least a portion of the galactose is found as the terminal sugar in the chain(s), and various chemical fractionation schemes have described a range of polymers with differing amounts of these sugar and uronic acid residues (Mian and Percival,

1973; Percival and McDowell, 1967). However, upon electrophoresis in two different buffers, the sulfated polysaccharides from each of the three major chemical fractions displayed essentially the same two subfractions (Hogsett and Quatrano, 1975; Quatrano and Stevens, 1976). This group of polysaccharides containing fucose sulfate as the predominant monomer unit will be referred to as *fucoidin*, while the same sugar chains that lack sulfate will be termed *fucan*. The two subfractions observed upon electrophoresis are fucoidin 1 and 2 (F_1 and F_2). F_1 has a lower electrophoretic mobility, is not as heavily sulfated and contains less fucose and more uronic acid moieties than F_2. At least three "isoforms" of F_2 are observed in certain extracts (Fig. 5).

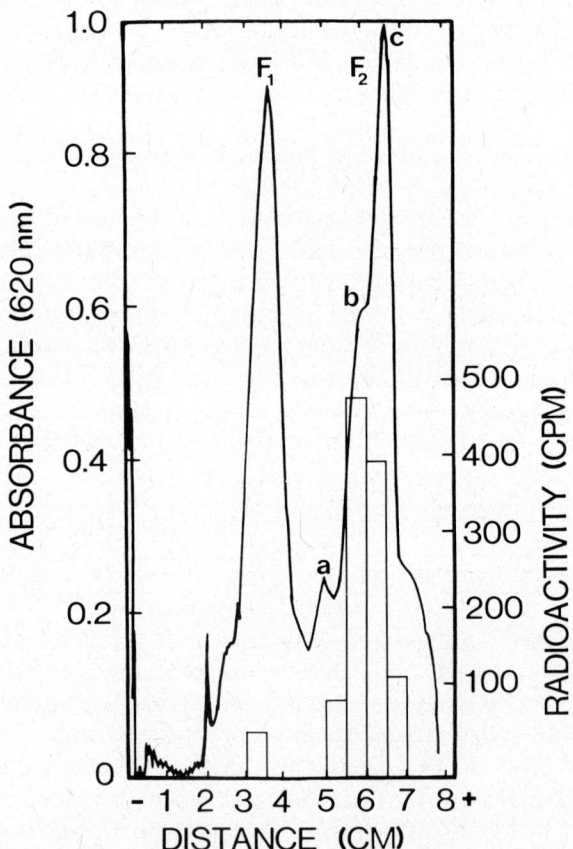

Fig. 5. Densitometric tracing of a cellulose acetate strip showing the two major fucoidin bands (F_1 and F_2) when stained with TBO. F_2 can be shown to have three "isoforms" (a,b,c). When pulsed for 60 min with $^{35}SO_4$ during rhizoid formation, the majority of the sulfate incorporation (bars) appears in the F_2 fraction.

Fucoidin comprises about 20% of the polysaccharides found in the cell wall at 4 hours after fertilization and remains at that level throughout early development. Alginic acid (60%) and cellulose (20%) are the other cell wall polysaccharides present during early embryogenesis (Quatrano and Stevens, 1976). When extracts of walls from 4 hour zygotes are subjected to cellulose acetate electrophoresis at pH 7, a single fucan (F_1) can be detected. By 12 hours the developing wall acquires the F_2 fucan which appears heavily sulfated due to its rate of migration in an electric field and upon staining with TBO. The color and timing coincide precisely with the cytochemical detection of sulfated polysaccharide at the rhizoid site. From this data we believe that the fucoidin component in the rhizoid wall is F_2 and differs from fucoidin at other wall sites (Quatrano and Stevens, 1976). Alginate and cellulose distribution appears to be uniform as evidenced by various cytochemical tests (unpublished observations).

How does fucoidin (F_2) become localized at the rhizoid site? Since TBO is staining the negatively charged sulfate groups, several possibilities exist: a fully-sulfated fucoidin could be unmasked, a fucan could be sulfated, or, the entire molecule may be synthesized *de novo*. Whether one of these possibilities occurs throughout the cell at random (with the resulting product transported to the site of rhizoid formation), or whether a specific localized region is the site of de novo synthesis, sulfation, or unmasking of fucoidin is the main question to be asked.

A considerable amount of data (c.f. Quatrano and Crayton, 1973; Quatrano, 1974, 1978) supports the conclusion that there does not appear to be de novo synthesis of the entire molecule nor the unmasking of the sulfate groups on fucoidin, but rather the sulfation of a previously existing fucan. For example, when zygotes were pulsed with $^{35}SO_4$ for 60 min at different times after fertilization, the enzymatic sulfation of fucoidin was initiated at 10 hours after fertilization (Quatrano and Crayton, 1973), the same time this polymer was detected in the rhizoid region by cytochemical and autoradiographic studies (McCully, 1970; Quatrano *et al.*, 1979). Most of the ^{35}S is found in F_2 after these relatively long pulses (Fig. 5). The pattern of sulfate accumulation into fucoidin cannot be accounted for on the basis of changes in the pool size or permeability of the zygotes to sulfate. Sulfation is apparently dependent upon new protein synthesis as judged by the ability of cycloheximide to prevent incorporation of ^{35}S into fucoidin (Quatrano and Crayton, 1973).

Previous autoradiographic and cytochemical evidence cited above indicated that sulfation occurs during rhizoid formation, and the sulfated fucoidin is detected in the rhizoid region. However, because of the long pulses (several hours) of $^{35}SO_4$ used in the autoradiographic experiments

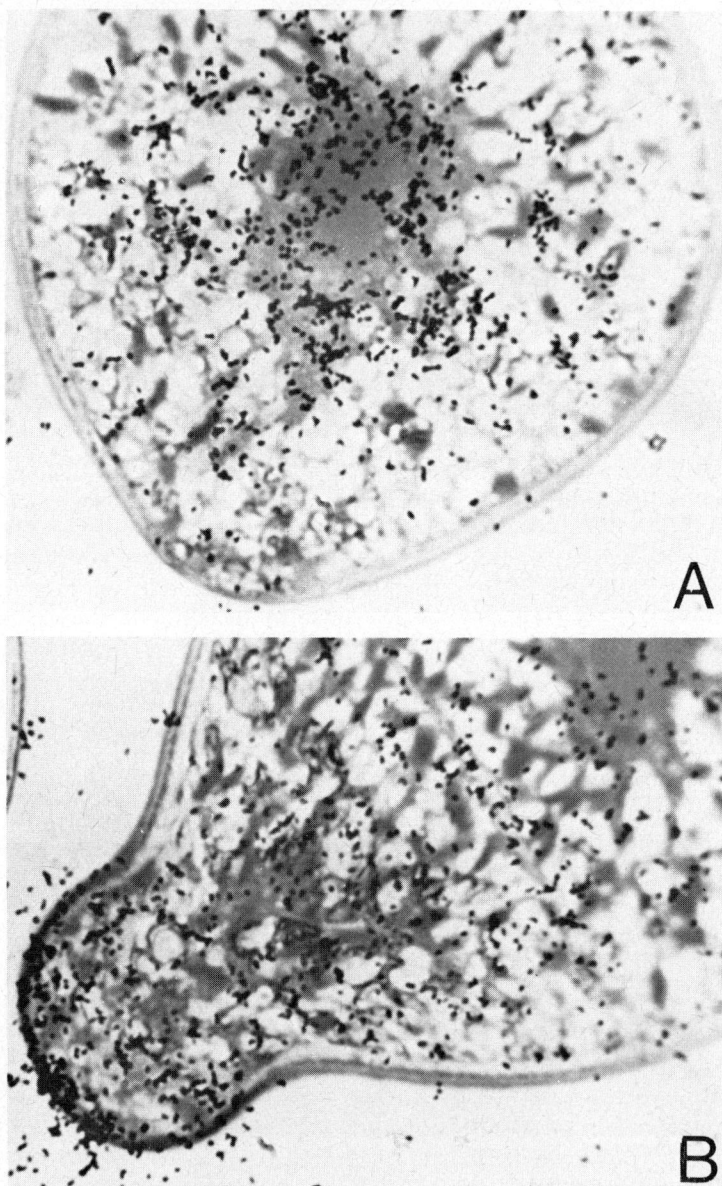

Fig. 6. Autoradiographs of an 18 hour zygote (A) pulsed for 5 min with $^{35}SO_4$ with no chase period. Notice the lack of localized grains in the wall or in the peripheral cytoplasm and the accumulation of sulfated fucoidin in the perinuclear region. In B, notice the localization of label in the rhizoid cytoplasm and cell wall after a 2 hour chase (X 150).

(McCully, 1970), it could not be determined if sulfation was initially random and then localized, or if the sulfating system was unevenly distributed initially, i.e. polar. In other words, is the cell-specific rhizoid product (fucoidin) transported to a predetermined intracellular site or is it sulfated only at the site? Using zygotes of *Fucus distichus* exposed to 5 min pulses of $^{35}SO_4$ followed by various chase periods, we recently found (Quatrano et al., 1979, Brawley and Quatrano, 1979) that at the time of rhizoid initiation, cytoplasmic sulfation of the fucoidin was initially random. During subsequent chase periods, the ^{35}S-polysaccharide was transported and secreted selectively into the rhizoid wall (Fig. 6). In more developmentally advanced embryos, sites of sulfation appeared sequestered into the rhizoid area since short pulses resulted in incorporation of the ^{35}S preferentially into the rhizoid region. These results are consistent with the hypothesis that the sulfating sites of fucoidin (shown by Evans and collaborators (1974) to be the Golgi) are randomly distributed in the cytoplasm at the time of rhizoid formation and gradually transported to the rhizoidal area during rhizoid development.

B. *Dependence Upon Sulfation for Transport of Fucoidin*

The formation of the rhizoid itself (i.e. a polar cell) is not dependent upon this sulfation however. Crayton *et al.*, (1974) found that zygotes grown in sea water lacking sulfate but containing methionine (necessary for protein synthesis), form rhizoids and two-celled embryos which do not stain metachromatically with TBO. Apparently there are no endogenous pools of SO_4^{2-} and hence, fucan sulfation can be controlled by the amount of exogenously added SO_4^{2-}. Although fucoidin is not needed for rhizoid formation, we were in a position to ask if the enzymatic sulfation is required for localization of the polymer, i.e. its assembly into the rhizoid cell wall. Two different approaches were used; one involving a fluorescent probe, the other, autoradiography. If the sulfating sites (Golgi) and fucans are localized in the absence of sulfation we should be able to detect this accumulation by short pulses of ^{35}S, as well as by a specific stain for the unsulfated fucan.

1. *FITC-Ricin as a Probe for Fucan Localization.* TBO depends upon the charged sulfate groups for its specificity and hence cannot distinguish between a rhizoid wall lacking fucoidin or a wall containing fucan. We approached this problem by determining if certain proteins or glycoproteins that bind specifically to certain sugar moieties in polysaccharides (i.e. lectins) could be used as a tag for fucoidin. We

demonstrated *in vitro* that the lectin, ricin (RCAι), complexed with both sulfated and desulfated fucoidin but not with other brown algal polysaccharides. The binding of RCAι to fucoidin was inhibited by galactose, indicating that the complex is formed through terminal galactose units on the fucoidin (Hogsett and Quatrano, 1978).

With the *in vitro* specificity of RCAι for fucoidin demonstrated, we conjugated this lectin with FITC to determine if it could be used as a cytological marker for fucoidin *in vivo*. When two-celled *Fucus* embryos are treated with the FITC-RCAι, a very intense fluorescence at the rhizoid tip was observed confirming the specificity of the probe for the fucoidin polymer *in vivo* (Fig. 7). Under the same conditions, the rhizoid wall stains metachromatically with TBO at pH 1.5, indicating the presence of a sulfated polymer. A similar localization of the fluorescent conjugate was

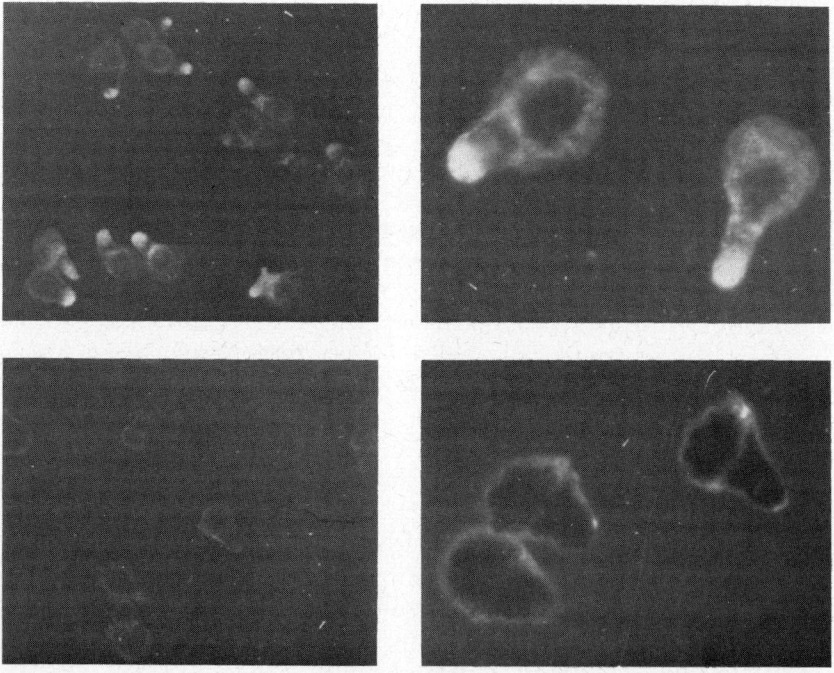

Fig. 7. Two-celled F. *distichus* embryos after various treatments with FITC-RCAI. The top two photos (X60 on left and X240 on right) represent embryos grown in the presence of sulfate. In both these Figures a concentration of the FITC-RCAI is observed in the region of expected fucoidin deposition (i.e. the rhizoid cell wall). The lower left photo (X60) shows the lack of FITC-RCAI binding when embryos grown in the presence of sulfate are incubated in O.1M galactose. The lower right photo (X240) represents embryos grown in the absence of sulfate and no localization of FITC-RCAI in the rhizoid area is observed. (from Hogsett and Quatrano, 1978).

observed in the two-celled embryos desulfated with methanolic-HCl. However, no metachromatic staining was observed in these desulfated embryos. This indicated that if an unsulfated fucan was incorporated into the wall, the FITC-RCAı complex would bind (as predicted by the *in vitro* results), whereas the TBO would not detect the unsulfated fucan. However, when embryos grown in the absence of sulfate are treated with FITC-RCAı or with TBO, no localization of the markers is evident at the rhizoid tip (Fig. 7). If unsulfated fucoidin had been incorporated into the rhizoid wall, localized fluorescence would have been evident. Extracts from zygotes grown in the absence of sulfate had the same amount of fucan and RCAı binding material as extracts from sulfate-grown zygotes. However, if an unsulfated fucan is not localized, yet detectable *in vitro* by RCAı precipitation, why can we not detect it randomly distributed in the cytoplasm? Two explanations are possible. Unlike the highly localized concentration of fucoidin in the rhizoid cell wall or in the underlying cytoplasm, the random localization of the unsulfated fucan throughout the cytoplasm may not bind sufficient FITC-RCAı within a given area to be detected by U.V. microscopy. Secondly, the fixative used may selectively extract or redistribute the unsulfated fucan while preserving the localization of fucoidin. Both of these explanations are not mutually exclusive and are presently being tested.

2. *Demonstration of Sulfating Sites by Autoradiography.* If fucoidin is not localized in the rhizoid area of embryo grown without sulfate, the Golgi and its vesicles should be randomly distributed. We tested this by pulsing such embryos with ^{35}S for 5 min to determine the localization of the sulfating sites. Whereas two-celled embryos grown in the presence of sulfate exhibited localized sulfating sites when pulsed for 5 min, randomly distributed Golgi are evident in two-celled embryos grown in the absence of sulfate (Fig. 3 E,F). These results along with the FITC-RCAı data are both consistent with the conclusion that the enzymatic sulfation of fucoidin is required for its ultimate accumulation in the rhizoid. We cannot determine at this time if sulfation is also required for secretion into the rhizoid wall.

The lack of fucoidin in the rhizoid wall when grown in the absence of sulfate also points to a role of the sulfated fucans in adhesion. It has been shown that embryos grown without sulfate but containing methionine form rhizoids but do not adhere to the substratum (Crayton *et al.*, 1974). When sulfate is added to the medium, the sulfated polysaccharide is detected in the rhizoid wall (by TBO staining the FITC-RCAı binding) and the embryos adhere to the substratum.

IV. MECHANISM OF LOCALIZATION

It is clear in *Fucus* that certain particles and macromolecules are synthesized throughout the cell and then redistributed to some predetermined site specified by the polar axis. Two possible mechanisms could control this type of segregation: a cytoplasmic electrical potential gradient or field which could segregate particles/molecules on the basis of their net charge, and, a mechanical, contractile mechanism involving actin and microfilaments.

A. *Contractile*

Numerous reviews have focused on the relationship between movement and a contractile system involving actin/myosin and associative proteins organized into a structure such as microfilaments (Pollard and Wiehing, 1974; Clark and Spudich, 1977; Edelman, 1976; Nicholson, 1976). In a previous section the possible role of microfilaments in polar axis fixation in *Fucus* was discussed in terms of directing and stabilizing components within the fluid matrix of the membrane. In view of the above studies, microfilaments may also play a role in the directional transport of subcellular particles via a contractile mechanism. In *Fucus* zygotes (which contain actin) the directed deposition of fucoidin into the rhizoid cell wall is prevented by CB. Mollenhauer and Morré (1976) have shown in maize root tips that a CB-sensitive subcellular component "is involved with the vectorial movement of secretory vesicles from sites of formation at dictyosomes to sites of fusion at the cell surface."

All of these studies are at best suggestive that a mechanism of intracellular transport may involve the interaction of vesicle binding to filamentous structures that possess contractile properties. However, with present techniques and methodology, the following approaches can be undertaken: (a) Is a contractile mechanism operative during the expression of polarity in *Fucus*? (b) If so, is it localized at the fixed site of rhizoid or cap formation? (c) Are vesicles which contain the localized macromolecules and have been shown to be transported to the fixed site bound to filamentous structures?

B. *Electrophorectic*

It was shown earlier that the *Fucus* zygote (c.f. Jaffe and Nuccitelli, 1977) drive an electrical current through their cytoplasms. The demonstrated gradient of electrical potential in these systems may serve to orient a cytoskeletal component such as microfilaments or microtubules, or act as a force to localize charged components.

Jaffe's (1966) elegant demonstration over a decade ago of an electrical current passing through the *Fucus* zygote rekindled interest in this as a mechanism for subcellular localization. He proposed in *Fucus* that "in traversing the cytoplasm the current will generate a field that may significantly localize negatively-charged molecules or particles toward the growth point (or, if there are any, positively charged ones towards its antipode)." The rhizoid pole of the current (positive) represents the site of entering cations such as Ca^{2+}. The magnitude of an electrical field generated is in large part dependent upon the ions carrying the current. If the entering cations have a low cytoplasmic mobility and are immobilized locally by an anionic gel in the cytoplasm, local binding would initiate a fixed charge gradient and thus a field. Jaffe has calculated, based on the current measurements and the fact that Ca^{2+} carries at least a portion of the current, that a field on the order of 100 mV/cm could be generated across the *Fucus* zygote. This is sufficient, even in view of the leveling action of diffusion, to localize large macromolecules and small particles. Is there any evidence that negatively charged elements are localized in the rhizoid area? As was previously discussed, we have shown that fucoidin must be sulfated in order to be localized in the rhizoid cell wall. This enzymatic addition of sulfate to fucoidin results in a net negative charge on the polymer. We also demonstrated that the amount of enzymatic sulfation *in vivo* is proportional to its electrophorectic mobility *in vitro* (Quatrano and Crayton, 1973). Although the electrical potential gradient is sufficient to account for the localization of free fucoidin, most of the sulfated molecules in the cytoplasm are found in Golgi-derived vesicles. Secretory vesicles which have been isolated from other systems are found to be negatively charged (Mathews *et al.*, 1972). It is not known whether the enzymatic sulfation that occurs within the Golgi apparatus of *Fucus* results in an increase in the net negative charge on the vesicles that accumulate at the rhizoid pole. Since these vesicles that are localized in the rhizoid region can be isolated (unpublished observations), the surface and electrophoretic properties can now be investigated before and after sulfation. We hope these approaches will lead us to a more direct determination of the role of electrical fields in localizing macromolecules at predetermined sites within the zygote.

REFERENCES

Albertini, D.F. and Anderson, E. (1977). *J. Cell Biol.* **73**, 111-127.
Brawley, S.H. and Quatrano, R.S. (1979). in preparation.
Brawley, S.H., Quatrano, R.S. and Wetherbee, R. (1977). *J. Cell Sci.* **24**, 275-294.
Chen, C.W. (1973). *Protoplasma* **77**, 427-435.
Clarke, M. and Spudich, J.A. (1977). *Ann. Rev. Biochem.* **46**, 797-822.
Cochrane, D.E. and Douglas, W.W. (1974). *Proc. Nat. Acad. Sci. U.S.* **71**, 408-412.
Condeelis, J.S. (1974). *Exp. Cell Res.* **88**, 435-439.
Crayton, M.A., Wilson, E. and Quatrano, R.S. (1974). *Develop. Biol.* **39**, 164-167.
Davidson, E.H. (1976). "Gene Activity in Early Development". (2nd ed.), pp. 452. Academic Press, New York.
Edelman, G.M. (1976). *Science* **192**, 218-226.
Evans, L.V., Simpson, M. and Callow, M.E. (1974). *Planta* **117**, 93-95.
Forer, A. and Jackson, W.T. (1976). *Cytobiology* **12**, 199-214.
Franke, W.W., Herth, W., van der Woude, W.J. and Morré, D.J. (1972). *Planta* **105**, 317-341.
Hepler, P.K. and Palevitz, B.A. (1974). *Ann. Rev. Plant Physiol.* **25**, 309-362.
Hogsett, W.E. and Quatrano, R.S. (1975). *Plant Physiol.* **55**, 25-29.
Hogsett, W.E. and Quatrano, R.S. (1978). *J. Cell Biol.* **78**, 866-873.
Jaffe, L.F. (1958). *Exp. Cell Res.* **15**, 282-299.
Jaffe, L.F. (1966). *Proc. Nat. Acad. Sci. U.S.* **56**, 1102-1109.
Jaffe, L.F. (1968). *Adv. Morphogen.* **7**, 295-328.
Jaffe, L.F. (1970). *Develop. Biol. Suppl.* **3**, 83-111.
Jaffe, L.F. and Nuccitelli, R. (1977). *Ann. Rev. Biophys. Bioeng.* **6**, 445-476.
Knapp, E. (1931). *Planta* **14**, 731-751.
Mathews, E.K., Evans, R.J. and Dean P.J. (1972). *Biochem. J.* **130**, 825-832.
McCully, M.E. (1970). *Ann. N.Y. Acad. Sci.* **175**, 702-711.
Mian, A.J. and Percival, E. (1973). *Carbohydr. Res.* **26**, 146-161.
Mollenhauer, H.H. and Morré, D.J. (1976). *Protoplasma* **87**, 39-48.
Nelson, D.R. and Jaffe, L.F. (1973). *Develop. Biol.* **30**, 206-208.
Nicholson, G.L. (1976). *Biochim. Biophys. Acta* **457**, 57-108.
Novotny, A.M. and Forman, M. (1974). *Develop. Biol.* **40**, 162-173.
Nuccitelli, R. (1978). *Develop. Biol.* **62**, 13-33.
Peng, H.B. and Jaffe, L.F. (1976). *Planta* **133**, 57-71.
Percival, E. and McDowell, R.H. (1967). In "Chemistry and Enzymology of Marine Algal Polysaccharides", pp. 219. Academic Press, New York.
Pollard, T.D. and Wiehing, R.R. (1974). *CRC Rev. Biochem.* **2**, 1-65.
Quatrano, R.S. (1972). *Exp. Cell Res.* **70**, 1-12.
Quatrano, R.S. (1973). *Develop. Biol.* **30**, 209-213.
Quatrano, R.S. (1974). In "Experimental Marine Biology" (R. Mariscal, ed.), pp. 303-346. Academic Press, New York, pp.373.
Quatrano, R.S. (1978). *Ann. Rev. Plant. Physiol.* **29**, 487-510.
Quatrano, R.S. and Crayton, M.A. (1973). *Develop. Biol.* **30**, 29-41.
Quatrano, R.S., Hogsett, W.E. and Roberts, M. (1979). *Proc. Int. Seaweed Symp., 9th, 1977,* Santa Barbara, in press.
Quatrano, R.S. and Stevens, P.T. (1976). *Plant Physiol.* **68**, 224-231.

Analysis of a Morphogenetic Determinant in an Insect Embryo (Smittia Spec., Chironomidae, Diptera)

Klaus Kalthoff
Department of Zoology
University of Texas at Austin
Austin, Texas 78712

I. Introduction 97
II. Double Cephalons and Double Abdomens in Insects 101
III. Analysis of an Anterior Determinant 103
   A. The Smittia Embryo as a Test System 103
   B. Topographical Localization 107
   C. Cellular Localization 108
   D. Tentative Molecular Characterization 110
   E. Masked Messenger RNA Hypothesis 115
IV. Discussion 117
   A. Messenger RNA-Containing RNP Particles in Insect Eggs 117
   B. Formal Concepts of Insect Embryogenesis 118
   C. Metamerization and Antero-Posterior Decision 120
   D. Relation to Homoeotic Mutations and Compartments ... 122
   References 124

I. INTRODUCTION

Insect embryogenesis may be subdivided very broadly into initial periods of nuclear multiplication and cell proliferation, and subsequent periods of increasingly complex regional differentiation (see Fig. 1). The zygote nucleus undergoes a series of mitotic divisions within the yolk-rich endoplasm of the egg. This period is usually referred to as *"intravitelline cleavage"*, although the egg cell is not cleaved. Rather, the embryo develops in a plasmodial state, containing eventually hundreds of *energids*, i.e. nuclei with jackets of cytoplasm. Some energids remain as vitellophages in the endoplasm, while most energids enter the yolk-free periplasm at the surface. There the nuclei become enclosed by infoldings of the *oolemma*, i.e. the plasmalemma of the egg cell. The resulting *blastoderm cells*, however, may not always be entirely separated; at least in *Drosophila*, cytoplasmic connections persist between blastoderm cells and the yolk endoplasm (Rickoll, 1976). Following formation of blastoderm cells, and usually further mitotic divisions, the period of visible regional differentiation begins. Part of the blastoderm cells build the originally unsegmented *germ anlage* while the remainder form amnion and serosa, i.e. the embryonic covers. After gastrulation and segmentation, the embryo reaches the *germ band* stage which already reflects the basic organization of the larva (Fig. 1). Typically, we can distinguish the procephalon (A), the gnathocephalon (B), the thorax (C), and 8 to 12 abdominal segments (D and E). These regions may be regarded as elements of a *basic longitudinal body pattern*.

The formation of a spatial body pattern depends upon recognizable differences, in structure or arrangement, between cells or groups of cells.

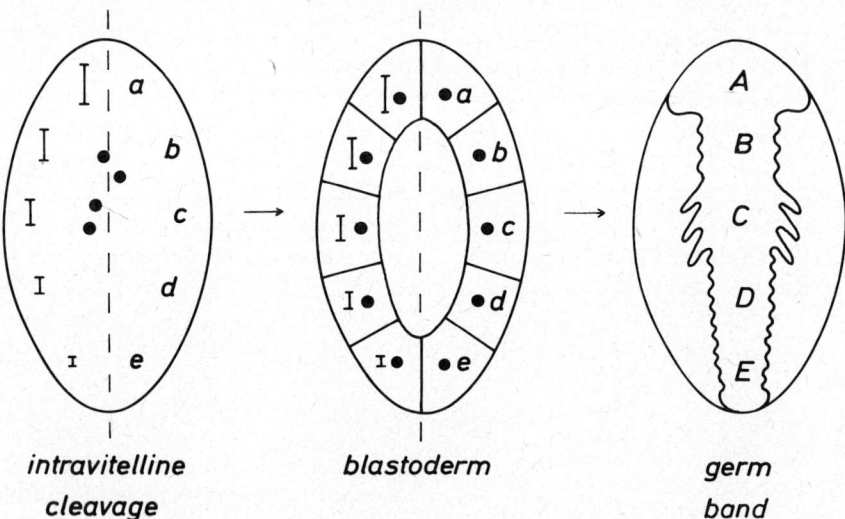

Fig. 1. Development of a generalized insect; stages are described in the text. Black discs represent nuclei. Pattern elements A (procephalon), B (gnathocephalon), C (thorax), D and E (abdomen) can be identified at the germ band stage. Cytoplasmic determinants thought to exist at earlier stages are represented by small letters a — e (mosaic model), or bars of different lengths (gradient model).

Such differences in turn are ascribed to *differential gene activity*. Each cell nucleus, however, seems to inherit the same complement of genetic information. This has been shown by transplantation of nuclei from differentiated *Xenopus* cells (Gurdon *et al.*, 1975) and *Drosophila* gastrula cells (Illmensee, 1973) back into enucleated or unfertilized eggs. These experiments have left us with the problem of explaining how, in a spheroid layer of blastoderm cells with totipotent nuclei, subgroups of cells can be programmed for different gene activities. Moreover, our explanations have to account for the *spatial order* in which differential gene activation must occur.

A fundamental concept in developmental biology ascribes the programming of differential gene activities to interactions of nuclei with *cytoplasmic determinants*. This idea is especially suggestive in cases such as in some ascidians where several egg regions can be distinguished by their unique pigmentation, and where the inheritance of these particular cytoplasmic sectors by certain blastomeres seems to determine the kinds of tissue that they will ultimately form (see Davidson, 1976, for review). Histochemical studies with ascidian embryos have indicated that the synthesis of messenger RNA for specific enzymes is triggered differentially in certain cell lineages at specific stages (Whittaker, 1973; Whittaker, *et al.*, 1977; Whittaker, 1979). While some cytoplasmic determinants are apparently prelocalized during oogenesis, others do not become localized until early cleavage stages (Freeman, 1976, 1979). In these cases, a scaffold of localizing organelles rather than the determinants themselves seems to be prelocalized during oogenesis.

In insect eggs, the comparatively late subdivision of periplasm into blastoderm cells could facilitate the incorporation of various localized determinants which might then trigger differential gene activity in a spatial order. A conspicuous increase of nuclear volume in *Drosophila* blastoderm cells obviously involves a transfer of cytoplasmic components into the nuclei (Jacobson and Fullilove, 1973). However, there is little direct evidence for localized components in insect eggs except for the *oosome*. The posterior pole region of the eggs of many Coleoptera, Diptera, and Hymenoptera contains basophilic, granular organelles known as *polar granules*. The cells that originate at this site and incorporate polar granules are referred to as *pole cells*. They usually bud off prior to the formation of blastoderm cells, and they display a different morphology. Pole cells in *Drosophila* give rise to germ cells, midgut cells, and possibly vitellophages (Counce, 1973; Illmensee *et al.*, 1976). The presence of germ cell determinants in the oosome region of *Drosophila* eggs has been demonstrated by transplanting this material to ectopic locations and showing that the pole cells induced at the transplantation site can give rise to germ cells (Illmensee and Mahowald, 1974; see also Mahowald et al., 1979).

The ultrastructure of insect eggs has been scrutinized in search for localized organelles that might be related to the determination of embryonic regions other than the pole cells, but little if any clues have been found (Okada and Waddington, 1959; Mahowald, 1972; Zissler and Sander, 1973, 1977). Conspicuous accumulations of mitochondria have been observed near the posterior pole of *Melanosoma populi* eggs (Jura et al., 1957), and near the anterior pole of *Smittia* eggs (Zissler and Sander, 1973). The idea that mitochondria could determine the antero-posterior polarity, however, was not supported by the results of relevant experiments with *Smittia* embryos (Kalthoff et al., 1975, 1977). The anterior pole region of newly deposited *Smittia* eggs also contains a *cytaster* which seems unrelated to the meiotic apparatus (Zissler and Sander, 1973). However, the function of this structure remains to be examined. Further, biochemical and immunological techniques have been used to detect region specific proteins in insect embryos (Koch and Heinig, 1968; Nünemann and Moser, 1970; Graziosi and Roberts, 1975; Roberts and Graziosi, 1977), but their role in the formation of the body pattern has not been investigated.

In the absence of ultrastructural or biochemical markers that could be correlated with cytoplasmic determinants, their localization and other characteristics have to be inferred entirely from the results of experiments causing abnormal development. The methodology of such experiments is therefore crucial. Methods causing the incomplete formation or lack of certain parts of the body do *not* allow unequivocal conclusions about cytoplasmic determinants because there are usually several possible explanations why a given part of an embryo may be defective or missing. Cells that normally would have built this part may have failed to pick up their instructions from specific morphogenetic determinants but may also have failed in some way to express their determination. Methods allowing less ambiguous interpretations include those which cause *switches from one well defined morphological pathway into another*, i.e. the formation of body parts in places where they do not normally occur. At present, we have no idea of *how many* cytoplasmic determinants are present in any egg, and what their *molecular nature* and *modes of action* might be. Is there a *"mosaic"* of many different and independent determinants, each one responsible for a particular element of the body pattern, or are there only a few *"gradients"* of determinants, the local levels of which cause differential gene activation in a spatial order (Fig. 1)? To which degree are cytoplasmic determinants similar in different animal species?

In the pursuit of these questions, we may try to carry the analysis of a few cases to the molecular level; appropriate *experimental systems* should have the following properties. The morphogenetic program of embryonic cells should be switched into a well defined pathway not normally open to these cells. The switch should occur, under suitable conditions, with a yield of virtually 100%. At the same time, the test system should respond to specific rather than unspecific experimental treatments. A test system meeting these criteria fairly well has been developed, based on the work of earlier investigators. The system has allowed the analysis of a cytoplasmic determinant which has a major influence on the antero-posterior polarity and segment pattern in the embryo of a chironomid midge. Evidence for a topographical and cellular localization of the determinant, and its tentative molecular characterization, will be reviewed and discussed in relation to other work in the field.

II. DOUBLE CEPHALONS AND DOUBLE ABDOMENS IN INSECTS

Distinct types of abnormal embryos providing clues to the role of cytoplasmic determinants in the specification of the body pattern have been observed in several dipterans. The spontaneous occurrence of a characteristic *pattern aberration* in the pitcher plant mosquito, *Wyeomyia smithii*, was reported by Price (1958). In 15 out of 1,570 fertile eggs, he observed embryos representing longitudinal mirror image duplications of the abdomen without heads and thoracic segments. A longitudinal mirror duplication of head and thoracic segments without abdominal structures was observed in a *Drosophila* embryo (Lohs-Schardin and Sander, 1976). Such *double abdomens* and *double cephalons* were also produced experimentally, by centrifugation or partial UV irradiation, in eggs of the harlequin fly, *Chironomus dorsalis* (Yajima, 1960, 1964). Under certain experimental conditions, the resulting monster embryos were perfectly symmetrical in their external and internal morphology (see Fig. 3, k, l) except that pole cells occurred only in the posterior halves of both double abdomens and double cephalons (Yajima, 1970).

Phenotypes resembling the double abdomens in *Wyeomyia* and *Chironomus* were found also in the *bicaudal* mutant of *Drosophila melanogaster* (Bull, 1966). Because the *embryonic phenotype* is controlled by the *maternal genotype*, the pattern aberrations must be ascribed to a defective oogenetic

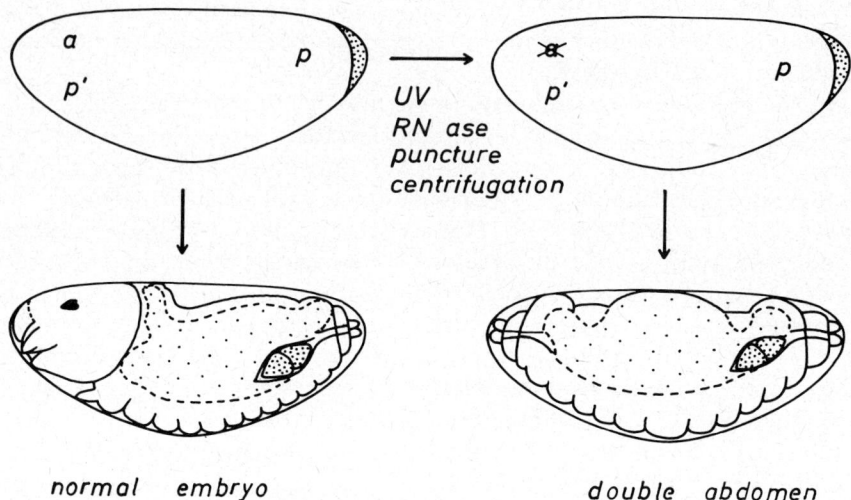

Fig. 2. Diagrammatic representation of double abdomen induction in *Smittia* by various types of experimental interference, all of which appear to inactivate or displace anterior determinants designated as *a*. These are thought to cooperate with other factors *p'*, in the anterior half of the embryo, so as to allow the formation of head and thorax. Upon inactivation or displacement of *a*, *p'* is assumed to cause abdomen formation in the anterior half, as the formation of the posterior abdomen is ascribed to similar or identical factors *p*. Note that germ cells (shaded) are present only in the posterior abdomen.

condition. The mutation is recessive, and maps at a single locus on the second chromosome close to *vg* (67.0 ± 0.1). Eggs from homozygous mutant females apparently inherit insufficient amounts of the wild-type gene product, which functions as a cytoplasmic determinant (Nüsslein-Volhard, 1977, 1979).

Besides the dipterans, double abdomens also have been produced in embryos of the pea beetle, *Callosobruchus maculatus* (Van der Meer, 1978). In this case, the monsters developed in posterior egg fragments from which a smaller anterior region had been separated by a complete but temporary ligation. Since the first instar larval cuticle of *Callosobruchus* provides sufficient landmarks to recognize almost each segment individually, the *polarity reversal* associated with double abdomen formation could be studied in detail. Both the segment order (sequence polarity) and the arrangement of markers on individual segments (element polarity) were occasionally found to be inconsistent with the overall antero-posterior polarity reversal in the anterior abdomen. In about half of the cases in which polarity reversal occurred, it was restricted to one lateral half of the embryo while the contralateral half showed a more or less normal segment pattern. Other asymmetrical phenotypes were found in the *bicaudal* mutant syndrome; the anterior abdomen was often shorter than its counterpart. Extremely asymmetrical phenotypes consisted of a complete and normally oriented abdomen with a pair of rudimentary spiracles at the anterior end (Nüsslein-Volhard, 1977). This type of pattern aberration was also observed upon UV irradiation of anterior regions of wild-type *Drosophila* eggs (Bownes and Kalthoff, 1974; Bownes and Sander, 1976).

Combined ligation and translocation experiments with eggs of the leaf hopper, *Euscelis plebjus*, also have led to the formation of double abdomens (Sander, 1960, 1961, 1975a, 1976). In this case, the pattern abnormality was apparently caused by the translocation of posterior pole material, while the other cases of double abdomen formation seem to result from the removal or inactivation of *anterior determinants*.

III. ANALYSIS OF AN ANTERIOR DETERMINANT

A. *The Smittia Embryo as a Test System*

In a chironomid midge of the genus *Smittia*, the aberrant pattern double abdomen can be produced by several unrelated types of experimental interference, including UV irradiation of anterior embryonic regions (Kalthoff and Sander, 1968), centrifugation (Kalthoff *et al.*, 1977), puncture of the embryo at the anterior pole (Schmidt *et al.*, 1975), and application of RNase to the anterior pole region (Kandler-Singer and Kalthoff, 1976). It seems improbable that all these different procedures could *de novo* generate specific determinants for the formation of an abdominal end. It is much more likely that the different methods have in common the *displacement or inactivation* of some crucial anterior component(s), designated as *a* in Fig. 2. This view is strongly supported by the fact that the UV induction of double abdomens is photoreversible, photoreversal being commonly ascribed to a light-dependent, enzymatic *repair* of UV damage to nucleic acids (see below, section III D). *The switch from the normal to the abnormal developmental pathway and vice versa is therefore ascribed to the activity or inactivity of crucial components localized in the anterior pole region of the egg cell.* Since the activity of the components is apparently required for head and thorax formation, they are referred to as *anterior determinants*.

The cells giving rise to the anterior half of a double abdomen are the ones that normally would have formed head and thoracic segments; this can be observed directly (see Fig. 3) and in time lapse films (Kalthoff, 1975a). Moreover, the formation of the anterior abdomen does not even

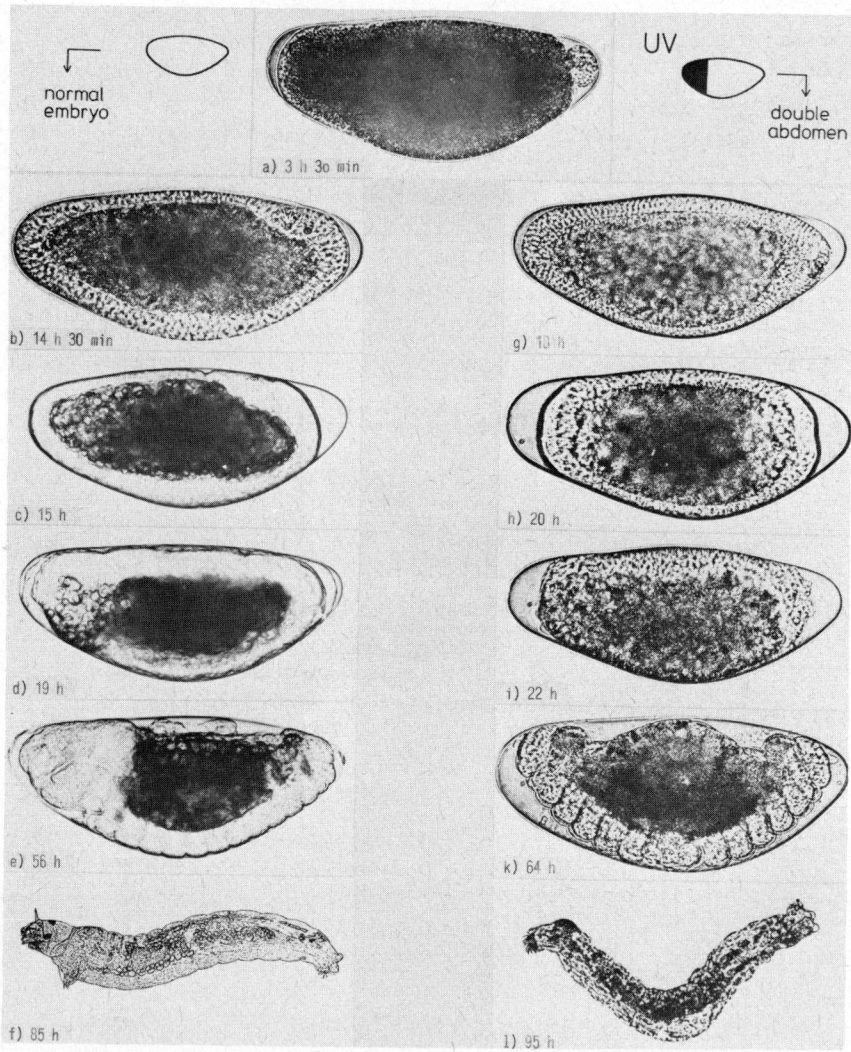

Fig. 3. Development of a normal larva (b-f) and a double abdomen (g-1) in *Smittia*. The aberrant pattern was produced by UV irradiation of the anterior quarter of the embryo during intravitelline cleavage (a). The developmental pathways differ markedly during germ anlage formation (b-d versus g-i); the double abdomen germ anlage develops by fusion of two thick layers of blastoderm cells formed near the pole regions (h,i). The two abdomens develop in strict symmetry and synchrony unless one partner is handicapped by the lack of space in the egg shell. The age of the embryos is given in hours after deposition at 19°C (Kalthoff and Sander, 1968).

require interactions with the posterior half of the embryo, as shown by a combination of UV irradiation with ligation experiments (Sander, unpublished; Ritter, 1976). Upon transverse ligation, anterior embryonic fragments produced head structures while the posterior fragments gave rise to a set of abdominal segments (Fig. 4b). However, if the anterior pole region was UV irradiated after ligation, the anterior fragment instead of a head produced another set of abdominal segments with reversed antero-posterior polarity (Fig. 4d). In order to check whether the development of abdomens in anterior fragments was due to a transfer of components from posterior fragments through accidentally persisting cytoplasmic bridges, posterior fragments were ruptured after ligation. Irradiated anterior fragments were still capable of forming abdomens with reversed polarity. Therefore, the conditions required for abdomen formation (designated as p and p' in Fig. 2) exist not only in the posterior but also in the anterior half of the embryo. Conditions p and p' which may be identical, apparently allow the formation of an abdomen in the absence of, and the formation of head and thorax in the presence of additional anterior determinants. This hypothetical situation would not be unique. The *bicaudal* and other *homoeotic mutations* (see Ouwennel, 1976) demonstrate that the activity or inactivity of a single gene may cause dramatic switches in developmental pathways.

The yield of double abdomens, upon UV irradiation of *Smittia* embryos under appropriate conditions (Kalthoff, 1971a) is virtually one hundred percent. Since this experiment can be carried out in a simple bulk operation, hundreds of embryos programmed for double abdomen development can be obtained rather easily.

The specificity of the experimental procedures causing double abdomen formation is also encouraging. Irradiation with UV inactivates nucleic acids, proteins, and a number of other biologically important molecules. However, the *photoreversal*, i.e. the light-dependent mitigation, of a UV effect can frequently be ascribed to the light-dependent *repair* of UV damaged nucleic acids (Harm, 1976; Gordon et al., 1976). The UV induction of double abdomens is photoreversible, i.e. light (300 – 500 nm) after UV (240 – 300 nm) causes an increase in the fraction of normal embryos at the expense of double abdomens (see below, section III D). Double abdomens also can be produced by application of RNase but not by various other enzymes (Kandler-Singer and Kalthoff, 1976). Both UV and RNase must be applied near the anterior pole in order to cause double abdomen formation. The apparent specificity of double abdomen induction as indicated by these experiments is not invalidated by the possibility of producing double abdomens by either centrifugation or

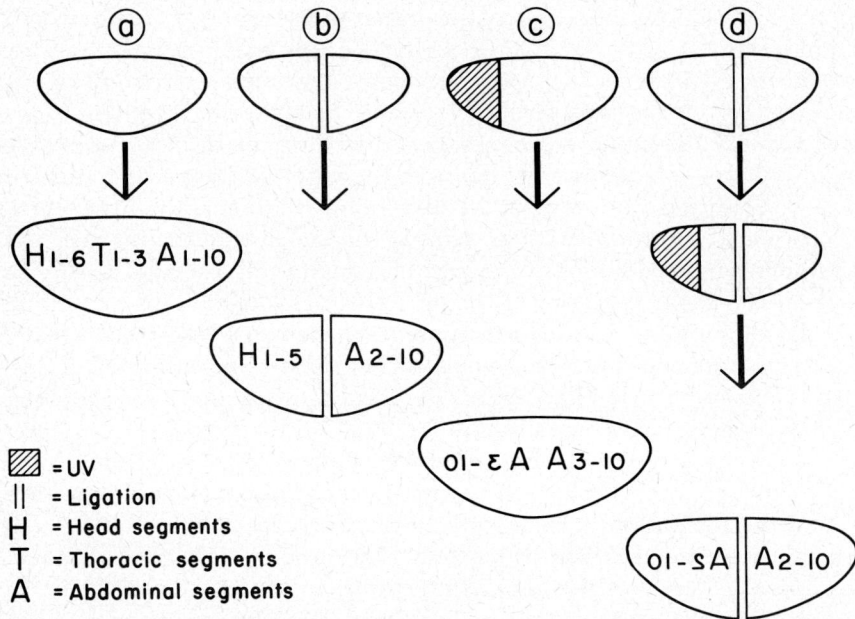

Fig. 4. Combined ligation and UV irradiation of *Smittia* embryos. (a) The normal segment pattern comprises 6 head segments, 3 thoracic and 10 abdominal segments. (b) Upon ligation, the anterior embryonic fragment develops an incomplete head while the posterior fragment forms abdominal segments. Some intermediate segments are missing from both partial embryos (gap phenomenon). (c) UV irradiation of the anterior quarter causes the formation of double abdomens, most of which represent a mirror image duplicaton of the posterior 8 abdominal segments, (d) Ligation and subsequent UV irradiation of the anterior fragment shows that an anterior abdomen with reversed polarity may develop even if interactions with the posterior fragment are inhibited. The isolated abdomens comprise, on an average, one more segment than the halves of a joined double abdomen (Data from Sander, 1975a, and Ritter, 1976).

puncture of the embryo. Although centrifugation causes the displacement of many constituents in the egg cell, double abdomen formation may still result from the changed localization of a few specific components. Likewise, double abdomen induction by puncture (Schmidt *et al.*, 1975) may result from leakage of specific components or from the release of endogenous RNase. The following observation is also important with respect to specificity. After induction of double abdomens by UV irradiation at an early intravitelline cleavage stage, the photoreverting treatment with light can be postponed for several hours without detriment to its efficiency (Fig. 9, top). Thus, double abdomen formation appears to result from the inactivation of an embryonic component that must function at a specific stage, i.e., shortly before blastoderm formation, in order to support normal development (see below, section III E).

B. *Topographical Localization*

In an attempt to determine the topographical localization of the anterior determinants, various areas of *Smittia* embryos were exposed to UV from a germicidal lamp, or irradiated with a UV microbeam (Kalthoff, 1971a). The effective targets for the UV induction of double abdomens are apparently distributed symmetrically to the long egg axis, since irradiation of dorsal, ventral, or lateral areas of the anterior pole regions were equally effective. The response to UV decreased as the target area was shifted from the anterior pole toward more posterior locations within the anterior half of the embryo. In a semiquantitative evaluation of these data, the figure E was calculated as

$$E = \frac{1}{D \times F}$$

where F represents the size of the exposed embryonic area and D the UV dose required for a standard percentage of double abdomens. Regarding E as an indicator for the "efficiency" of double abdomen induction by UV irradiation of the respective area, the results can be illustrated as shown in Fig. 5, where the density of hatching indicates higher or lower efficiency. The diagram can be thought to represent a corresponding distribution in the concentration of the effective targets of UV, i.e., the anterior determinants. However, shielding of the targets by yolk or other components would also be reflected by the results. Also, "subcritical" quantities of anterior determinants which normally are insufficient to cause head and thorax formation, may exist in the posterior half of the embryo undetectable by the method used here. Moreover, Fig. 5 represents the efficiency of double abdomen induction during and after nuclear migration which may differ from the situation during earlier stages.

A change in the localization of the anterior determinants from a more central position in the anterior pole region to a more peripheral localization seems to occur during the migration of nuclei from the endoplasm into the periplasm of the embryo. Microbeam irradiation of a small target area close behind the anterior pole caused high double abdomen yields during intravitelline cleavage but became ineffective thereafter (Kalthoff *et al.*, 1977). Conversely, when embryos were irradiated with the entire anterior pole region facing the UV beam, low UV doses had no effect prior to nuclear migration but produced almost

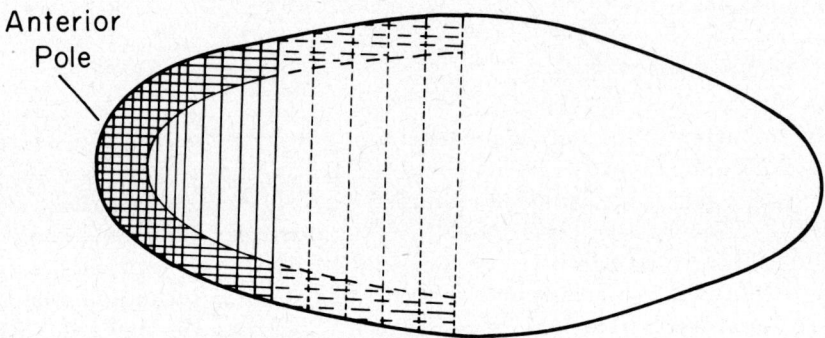

Fig. 5. Diagrammatic representation of the efficiency with which double abdomens can be induced by UV irradiation of various embryonic areas (Kalthoff, 1971a). The topographical distribution of the efficiency does not necessarily reflect a corresponding distribution of anterior determinants in *Smittia* eggs (see text).

100% double abdomens thereafter (Fig. 9, middle). Changes in the localization of cytoplasmic determinants are also observed during early development of embryos undergoing holoblastic rather than superficial cleavage (see Davidson, 1976, and Freeman, 1979).

The synthesis of anterior determinants in *Smittia* seems to occur during oogenesis, since double abdomens can be induced readily by UV irradiation of newly deposited eggs (Kalthoff, 1971a). Since the anterior determinants apparently consist of ribonucleoprotein particles (see below, section III D), and because there is little if any RNA synthesis in insect eggs during intravitelline cleavage (Zalokar, 1976), it seems very unlikely that the anterior determinants are synthesized during a few minutes elapsing between egg deposition and irradiation. This view is in line with the observation that the double abdomen phenotype produced by the *bicaudal* mutation in *Drosophila* is also a maternal effect.

C. *Cellular Localization*

The cellular localization of the anterior determinants in *Smittia* embryos is clearly extranuclear, since double abdomens can be produced with high efficiency by UV irradiation of target regions containing no nuclei (Kalthoff, 1971a). To further determine the extranuclear fraction containing the anterior determinants, embryos were centrifuged and various stratified components were irradiated with a UV microbeam. Experimental conditions were selected so that the stratification alone caused little or no pattern aberrations. After centrifugation at 30,000 g for 20 min with the centrifugal force perpendicular to the long egg axis, 90% of the embryos survived, and all survivors developed the normal body segment pattern. These centrifugation conditions caused an accumulation of proteid spheres on the centrifugal, and lipid droplets on the centripetal side of the egg. These yolk components, however, remained still embedded in a matrix of cytoplasm. Between the proteid and lipid layers, a zone of cytoplasm, approximately 18 μm thick, was cleared of yolk components and therefore appeared translucent in the light microscope. Direct observation of the centrifuged embryos suggested, and electron microscopy confirmed that mitochondria were stratified in the middle of the clear cytoplasmic zone (Fig. 6). Between the mitochondrial and the lipid layer, abundant ER cisternae and vesicles, and the nuclei were collected. However, between the mitochondrial and the proteid layer, few if any organelles larger than ribosomes could be

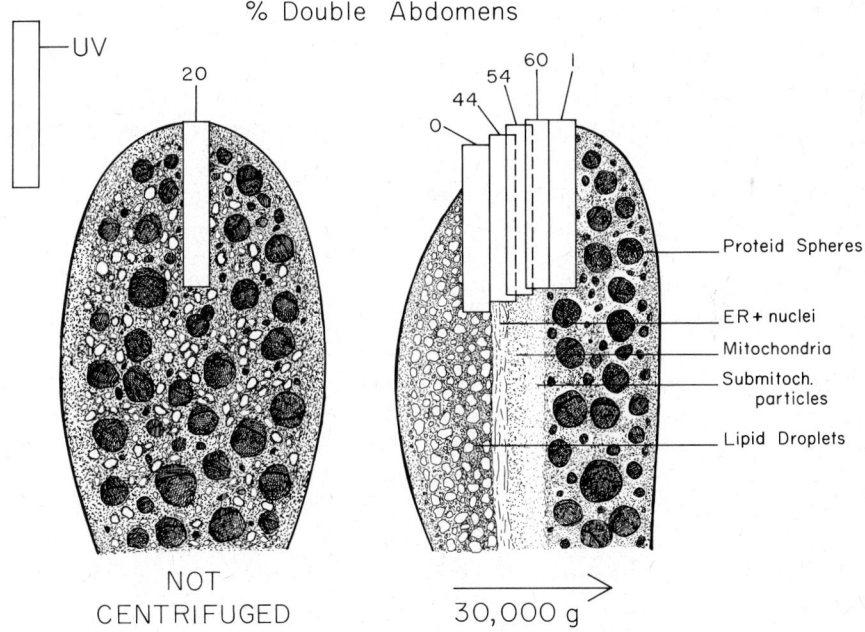

Fig. 6. Analysis of the cellular localization of anterior determinants in *Smittia* by microbeam irradiation of centrifuged embryos. Irradiation under conditions causing 20% double abdomens in uncentrifuged embryos has virtually no effect if proteid spheres or lipid droplets are accumulated in the target area. By contrast, removal of yolk components from the irradiated region significantly enhances double abdomen induction. Stratification within the yolk-free zone of endoplasmic reticulum plus nuclei, and mitochondria, is not paralleled by corresponding differences in the double abdomen yields obtained upon irradiation of the respective layer. A maximum double abdomen yield is found after UV irradiation of a layer containing little, if any, organelles larger than ribosomes (Kalthoff *et al.*, 1977).

detected, after fixation with either glutaraldehyde-osmium or $KMnO_4$.

Microbeam irradiation of stratified egg components with a standard UV dose caused double abdomen formation with different yields. Irradiation of the lipid or proteid layers produced practically no double abdomens, while irradiation within the clear cytoplasmic layer led to significantly higher double abdomen yields than irradiation of an equivalent but yolk containing zone in uncentrifuged embryos (Fig. 6). The anterior determinants are therefore not associated with either of the yolk components. Within the cytoplasmic zone, there was no correlation between the accumulation of mitochondria and ER and the double abdomen yield resulting from irradiation of the respective layers. These data corroborated an earlier study in which mitochondria were ruled out as anterior determinants in Smittia embryos (Kalthoff et al., 1975).

Our interpretation of the results shown in Fig. 6 and of other experiments (Kalthoff et al., 1977) is that the anterior determinants were *not directly stratified* under our conditions, but only *displaced to a certain degree as other components became stratified*. This was also indicated by the fact that the small differences in the yields resulting from irradiation of the three layers within the clear cytoplasm became even smaller upon milder or shorter centrifugation. The results also show that the anterior determinants were *not attached to the oolemma*, since removal by centrifugation of yolk components from the cytoplasm underneath the oolemma increased the yield of double abdomens after UV irradiation while the accumulation of yolk components decreased it (Fig. 6). This cannot be ascribed to changes in the exposure of supposedly oolemma-bound targets turned away from the microbeam, since both yolk-containing and yolk-free cytoplasm absorb UV very strongly (Kalthoff, 1973, and unpublished data). The same kind of rationale would exclude the notion that the anterior determinants were exclusively localized in the yolk-free periplasm ("cortex") which is about 5 μm thick in Smittia eggs (Kalthoff et al., 1977).

The conclusion seems therefore hard to escape that the anterior determinants are associated with *components of submitochondrial size in the cytoplasmic matrix*. This interpretation leaves us with the problem of explaining why the anterior determinants apparently remain localized for many hours and do not diffuse all over the egg. A possible explanation is that they might be bound to cytoskeletal structures, as has been suggested to be the case for polysomes in HeLa cells (Lenk et al., 1977).

D. Tentative Molecular Characterization

To obtain a clue to the molecular nature of the anterior determinants in Smittia eggs, an *action spectrum* for the UV induction of double abdomens

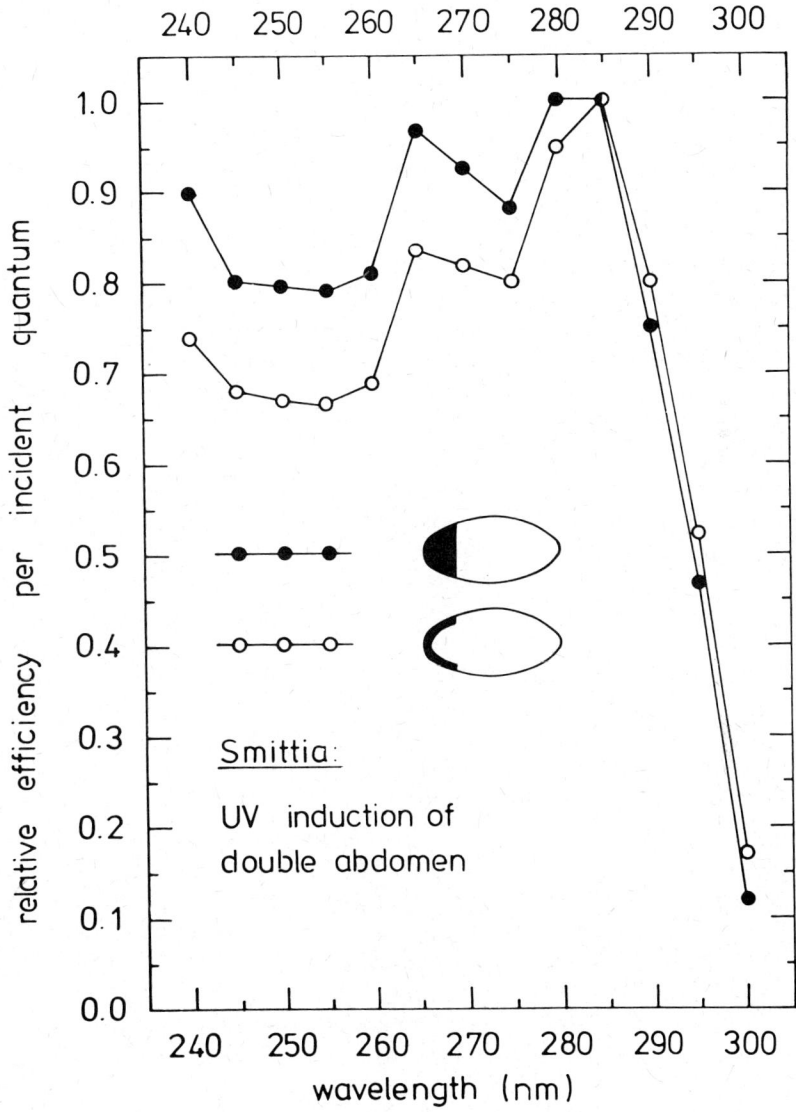

Fig. 7. Action spectrum for UV induction of double abdomens in *Smittia* embryos. The relative efficiencies per incident quantum have been determined by dose-response curves at each wavelength. Wavelength-dependent shielding is taken into account assuming target localization throughout the irradiated anterior quarter of the embryo (●) or only within a superficial layer of periplasm (○). Peaks at 265 and 285 nm indicate a nucleic acid-protein complex as effective targets (Kalthoff, 1973).

was established (Kalthoff, 1973). Using UV from a monochromator, dose response curves were obtained at wavelengths between 240 and 300 nm at 5 nm increments. At each wavelength, the dose required for 50% double abdomens was determined and the *relative quantum efficiency* computed. The computation included corrections for the wavelength dependence of quantum energies as well as shielding of the effective targets by the chorion and components within the egg. Transmittance spectra of the chorion and homogenized embryos were determined experimentally, and the effective targets were assumed to be distributed homogeneously either in the irradiated anterior quarter, or in a 5 μm thick superficial layer (periplasm) of this area. Under either of these assumptions, the maximum efficiency per incident quantum was found between 280 and 285 nm, indicating a *protein moiety* in the effective targets. A minor peak was also found at 265 nm, suggesting the involvement of a *nucleic acid* (Fig. 7).

Further independent evidence for the involvement of a nucleic acid is the *photoreversibility* of the UV induction of double abdomens (Fig. 8; see also Kalthoff, 1971b, 1973). Photoreversal is defined as the mitigation of UV effects by subsequent irradiation with light of longer wavelength. In our case, light was effective only after but not before UV, and it had to be received by the same egg area. These results, together with the wavelength dependence, temperature dependence, and dose rate saturation of photoreversal suggest that the underlying mechanism belongs to the "direct" or "enzymatic" type of photoreactivation (Kalthoff, 1973; Kalthoff et al., 1978), which is ascribed to light-dependent, enzymatic splitting of UV induced pyrimidine dimers in DNA (Harm, 1976) or RNA (Gordon et al., 1976). In fact, we have recently observed that UV irradiation of *Smittia* embryos caused formation of pyrimidine dimers in the (largely ribosomal) RNA, and that the dimers disappeared when the embryos where exposed to light after UV (Fig. 8). The production of dimers by UV apparently occurred in a photosensitized reaction, and after UV inactivation of the embryos at different wavelengths, the photoreactivable sector was correlated with the amount of pyrimidine dimers produced (Jäckle and Kalthoff, 1978). Other data suggest that the light-dependent repair of UV damage to RNA enables *Smittia* embryos to survive under bright sunlight without need for pigmentation or shading (Kalthoff, 1975b).

Taken together, the evidence from the action spectrum for double abdomen induction and the localization experiments indicate that some nucleic acid-protein complexes of submitochondrial size should function as effective targets of UV. Components of insect embryos meeting these

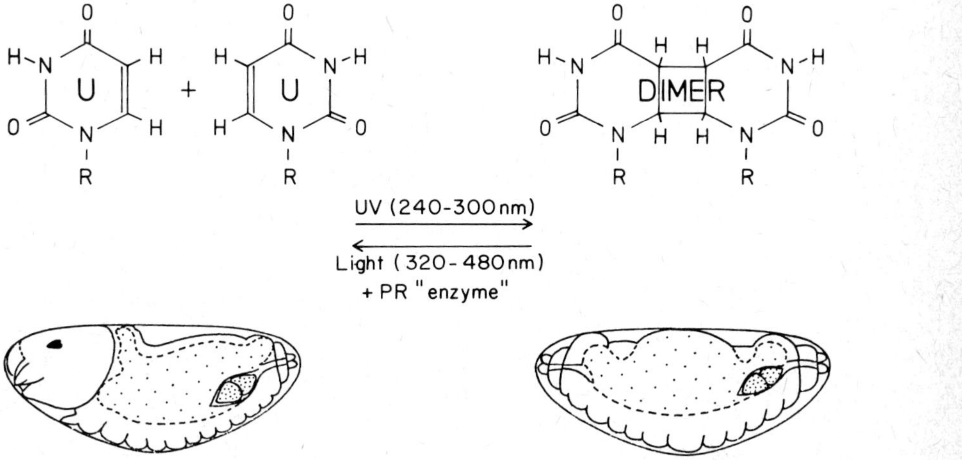

Fig. 8. Photoreversal of UV effects in *Smittia* embryos. Irradiation with UV (240 — 300 nm) during intravitelline cleavage causes the formation of pyrimidine dimers in the RNA of the embryos, and leads to the formation of double abdomens. Subsequent exposure of the UV irradiated region to light (320 — 480 nm) causes the disappearance of pyrimidine dimers from RNA, and the formation of normal embryos at the expense of double abdomens. The light-driven reaction(s) appear(s) to be associated with a temperature-dependent and light-independent step which may represent a complex formation of pyrimidine dimers with a "photoreactivating enzyme" (Kalthoff, 1973; Jäckle and Kalthoff, 1978; Kalthoff *et al.,* 1978).

characteristics include ribosomes, ribosomal subunits, and subribosomal ribonucleoprotein particles, all of which contain RNA but not DNA. This result is in line with the correlation between the photoreversibility of the UV induction of double abdomens and the light-dependent disappearance of pyrimidine dimers from RNA of UV irradiated embryos (Fig. 8).

Independent evidence for the involvement of RNA in the anterior determinants of *Smittia* has come from the *application of enzymes* to various embryonic regions. Application of RNase (ribonuclease I, ribonucleate 3'-pyrimidino-oligonucleotidohydrolase, EC 3.1.4.22) to the anterior pole of *Smittia* embryos caused the same switch in the developmental program as UV irradiation (Kandler-Singer and Kalthoff, 1976). By puncturing embryos during submersion in a hypotonic solution of pancreatic RNase A, double abdomens could be produced with a maximum yield of 29% (Fig. 9). However, application of RNase in concentrations causing double abdomen formation (0.5 to 0.8 µg/ml) inevitably killed a major fraction of the treated embryos. Control embryos punctured in water, boiled RNase, or oxidized RNase did not produce double abdomens, and only a few of them died (Table 1). Another control experiment made use of the complementation of inactive RNase S fragments to form active RNase S (Richards and Vithayathil, 1959). The S peptide which has no enzymatic activity produced no double abdomens, while the complementary S protein which has residual enzymatic activity produced a few double abdomens. The recombination of the fragments which results in an almost complete restoration of the enzymatic activity also caused double abdomen formation with a yield comparable to that of the native enzyme (Table I). These control experiments are regarded as proof that *the switch in the pattern formation of Smittia embryos resulted from the RNase activity* and not from other stimuli associated with the experiment. Neither application of other enzymes to the anterior pole nor application of RNase to other embryonic regions produced double abdomens in significant yields (Kandler-Singer and Kalthoff, 1976).

The data obtained so far do not prove, but suggest that both UV and RNase inactivate one type of anterior cytoplasmic determinant. Both agents have to be applied at the same site, i.e. the anterior pole region, and during the same period of development, i.e. from deposition until blastoderm formation, in order to cause double abdomen formation (Fig. 9). As discussed above, the effective UV targets also appear to contain RNA. The idea that both UV and RNase inactivate one type of anterior

TABLE I

Production of the aberrant pattern "double abdomen" in Smittia eggs (5.5 to 7.5 hours after deposition) by puncturing at the anterior pole during submersion in RNase.

agent	Conc. (μg/ml)	batches	eggs (total)	NL	ud (numbers)	DA	DA (% tot.)	DA (% surv.)
RNase A	0.8	19	382	95	203	84	22	47
RNase A denatured	0.5-1.0	23	476	436	40	—	0	0
RNase A oxidized	10.0	4	93	93	—	—	0	0
H_2O		8	158	151	7	—	0	0
RNase S	1.0	9	189	76	63	50	26	40
S-Peptid	0.16	9	205	195	10	—	0	0
S-Protein	0.84	4	81	77	3	1	1	1
	1.68	7	170	151	15	4	2	3
S-Peptid plus S-Protein	0.16 + 0.84	8	184	69	77	38	21	36

NL = normal larvae, ud = undifferentiated eggs, DA = double abdomens, % tot. = percentage of total, % surv. = percentage of survivors.

cytoplasmic determinant is therefore adopted as the simplest interpretation so far of our data. Under this provision, the available evidence indicates that *ribonucleoprotein particles function as anterior determinants in Smittia embryos.*

E. *Masked Messenger RNA Hypothesis*

Possible candidates for anterior determinants in *Smittia* embryos include ribosomes, ribosomal subunits, and subribosomal ribonucleoprotein (RNP) particles. Is there any further evidence supporting one of these candidates rather than the others? Spirin (1966) has proposed that messenger RNA can exist in a "masked" form in eggs and early embryos, in which "masking" proteins serve to inhibit both the premature translation of the messenger RNA and its precocious digestion by endogeneous RNase. The possibility of gene regulation by localized messenger like RNA has been suggested by Davidson and Britten (1971) in a theoretical model of pattern formation during early embryogenesis. Experimental evidence for *localized* maternal messenger RNA has come from studies of protein synthesis in embryos of *Ilyanassa* (Donohoo and Kafatos, 1973; Newrock and Raff, 1975), and a histochemical study on the synthesis of a specific enzyme in an ascidian embryo (Whittaker, 1977, 1979).

It is tempting to speculate that masked messenger RNAs might function as anterior determinants in *Smittia* embryos. Polyadenylated RNA of heterogeneous size (8-45 S) has been found in subribosomal (15-60 S) RNP particles from newly deposited eggs (Jäckle, 1977). The fraction possibly serving as anterior determinants may not be translated until nuclear migration. This assumption would explain why, after UV inactivation of anterior determinants, their photoreactivation can be delayed until stage M_2 without becoming less effective (Fig. 9, top). From nuclear migration stages on, the supposed messenger RNAs serving as anterior determinants might be translated and, at the same time, become RNase sensitive (Fig. 9, bottom). The conformation changes rendering them both translatable and RNase sensitive, may be related to the apparent changes in the localization of anterior determinants during nuclear migration (Fig. 9, middle, Section III B). After blastoderm formation, the hypothetical translation of the anterior determinants may be completed to a degree sufficient for the formation of the normal body pattern. Therefore, the effects of UV irradiation, photoreverting treatment, and RNase on the double abdomen yield would cease at this

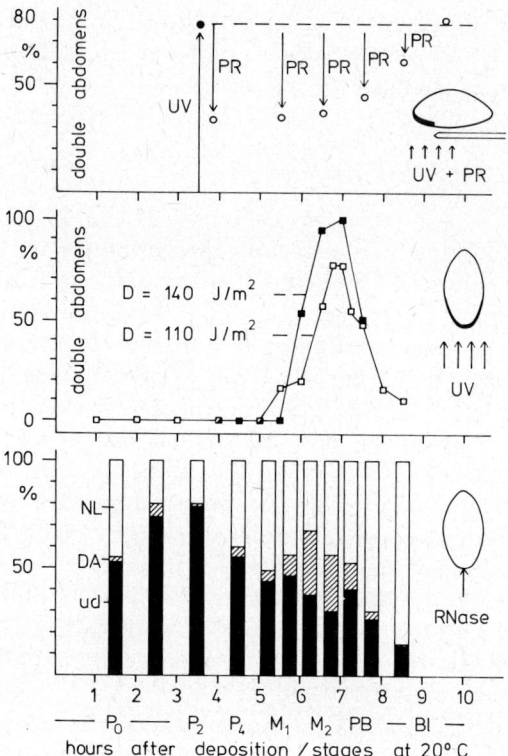

Fig. 9. Sensitive phases for induction of double abdomens in *Smittia* by RNase (bottom), by UV irradiation with the anterior pole facing the beam (middle), and for photoreversal after UV irradiation (top). Age of embryos is indicated in hours after deposition, and by the stages reached upon incubation at 20°C for the indicated period after deposition. P_0: no pole cells; P_2 and P_4: two or four pole cells, resp.; M_1 and M_2: migration of energids into periplasm; PB; preblastoderm; Bl: blastoderm.

Bottom: A maximum yield of double abdomens (DA) was found upon application of RNase around stage M_2. NL: normal larvae; ud: undifferentiated eggs (Data from Kandler-Singer and Kalthoff, 1976).

Middle: When embryos were irradiated with the anterior pole facing the UV beam, small doses had no effect prior to nuclear migration, when the double abdomen yield rose sharply to virtually 100% at stage M_2. Thereafter, the efficiency of UV irradiation decreased again until stage B1. D: UV dose (Kandler-Singer, unpublished data).

Top: After irradiation of embryos at stage P_2 with a UV dose causing 78% double abdomens, exposure to light immediately after UV reduces the percentage of double abdomens to 33. When UV irradiated eggs are incubated in the dark for 2 or 3 hours before exposure to light, photoreversal is still equally effective. The efficiency of the photoreverting treatment decreases only after nuclear migration, until it becomes ineffective beyond blastoderm formation (Kalthoff *et al.*, 1975).

stage. Further experiments are needed to either corroborate or disprove these speculations.

IV. DISCUSSION

A. *Messenger RNA-Containing RNP Particles in Insect Eggs*

Subribosomal RNP particles with polyadenylated RNA or RNA showing messenger activity have been found in various eggs and embryos (see Davidson, 1976). In insects with meroistic oogenesis, polyadenylated RNA is provided to the oocyte by the nurse cells. In the telotrophic ovary of a cotton bug, *Dysdercus intermedius*, subribosomal RNP particles (10-45S) were found to be synthesized by the nurse cells in the trophic chamber, and deposited in oocytes. The RNA extracted from these particles was heterodisperse and polyadenylated, and showed messenger activity in a cell free system. The native RNP particles contained substantial amounts of various proteins, some of which inhibited while others stimulated protein synthesis (Winter, 1974). Corresponding work with silkmoths (Paglia *et al.*, 1976) and *Smittia* (Jäckle, 1977), both of which have polytrophic ovaries, has led to similar or complementary results. In addition, postvitellogenetic oocytes of *Dysdercus* synthesize a short-lived messenger RNA themselves (Winter *et al.*, 1977). Material entering the oocyte via the follicular epithelium has been described as a low molecular weight RNA or nucleotide fraction associated with yolk proteins (Duspiva *et al.*, 1973; Schmidt and Jäckle, 1978). In situ hybridization data also indicate the presence of large amounts of polyadenylated RNA in follicle cells of ovaries in the milkweed bug, *Oncopeltus fasciatus* (Capco and Jeffery, personal communication).

In *Drosophila* oocytes and early embryos, heterodisperse (7-40 S) RNA with poly (A) segments (50 – 200 nucleotides long) was found in 30 - 70S RNP particles. The proportion of polyadenylated RNAs outside the polysome region of sucrose gradients decreased from 58% in mature oocytes and 51% in 1 hour old embryos to 30% in 7 hours old embryos (Lovett and Goldstein, 1977). The two fractions of polyadenylated RNA in polysomes and outside contained essentially the same sequences as indicated by complementary DNA hybridization (Goldstein, 1978). A 7-methylguanosine 5′ triphosphate "cap" was found at the 5′ end of polyadenylated RNA from *Drosophila* embryos but not oocytes. However, the significance of this structure for the initiation of protein synthesis in this case remains to be shown (Kastern and Berry, 1976; Berry, personal communication).

Although messenger RNAs clearly exist in insect eggs, evidence for their localization is so far indirect and limited to the posterior pole plasm in *Drosophila*. A stable association of polar granules with ribosomes was observed in fertilized eggs but not in oocytes; this may suggest that the granules contained messenger RNA released for protein synthesis at fertilization (Mahowald, 1977). Except for this case, there seem to be no further observations with insects that could support our hypothesis of localized masked messenger RNA in the anterior cytoplasm of *Smittia* embryos.

B. *Formal Concepts of Insect Embryogenesis*

A thorough review of "classical" work on pattern formation in insect embryogenesis including a reinterpretation of Seidel's (1961) differentiation center has been provided by Sander (1976).

Considering mosaic versus gradient models, one must specify the developmental stage under consideration and the level of resolution at which the body pattern is analyzed. This discussion will be limited to insect embryos before blastoderm formation, and the longitudinal body pattern will be thought of as comprising at least five somatic elements (A to E as illustrated in Fig. 1). Within these limitations our data do not support the idea that the longitudinal body pattern of the *Smittia* embryo is specified by a mosaic of different and independent determinants. If this were the case, UV irradiation of different areas of the embryo should have caused aberrant patterns of the types XBCDE, AXCDE, and ABXDE (with X standing for missing or abnormal pattern elements). Instead, virtually all surviving embryos showed either the normal pattern (ABCDE) or a symmetrical double abdomen pattern (ƎᗡDE, where E stands for the 5 posterior abdominal segments, while D represents the 5 anterior abdominal segments in the normal embryo, or 1-5 anterior abdominal segments in double abdomens).

The limited determination range of the oosome has sometimes been taken to support a mosaic model of insect embryogenesis. However, germ cells and somatic cells differ in several respects (see Sander, 1975a). The pole cells giving rise to germ cells tend to segregate early from the somatic part of the body that still develops as a plasmodium. The early insulation of the pole cells may help to preserve the germ line cells in a non-specialized state while somatic cells undergo progressive specialization. In *Drosophila*, different maternal effect mutations affect the formation of pole cells and somatic blastoderm cells separately (see Gehring, 1976). Moreover, pole cells are not duplicated in double abdomens (Bull, 1966; Yajima, 1970; Gollub, 1970; Nüsslein-Volhard, 1977), although a genital imaginal disc is formed in both halves (Gehring, personal communication). The determinations of germ cells and somatic cells are thus clearly dissociated, and there is no evidence so far that formal or molecular aspects of germ cell determination will be similar to the determination of somatic cells.

A gradient model of insect embryogenesis has been proposed by Sander (1960, 1961), originally on the basis of his ligation and transplantation experiments with *Euscelis* embryos. Components localized near the anterior and posterior poles are thought to build up, during early embryonic development, gradients of cytoplasmic determinants. The main lines of evidence supporting the model were (1) the formation of double abdomens after translocation of posterior pole material to equatorial periplasm, (2) the increased segment forming capacities of anterior partial embryos after providing them with posterior pole material, and (3) the observation that a given set of segments could differ in length considerably. Moreover, a requirement for interactions between anterior and posterior egg parts for the determination of intermediate segments is suggested by the "gap phenomenon". In ligation experiments, the segment patterns of developing partial embryos are often not complementary because some segments are missing from both fragments (see Fig. 4b). In several holometabolous insects, ligation of early embryos causes large gaps while embryos ligated at later stages show progressively smaller gaps (see Herth and Sander, 1973). Moreover, Schubiger et al., (1977) have found that a major gap in the segment pattern observed after temporary ligation of *Drosophila* embryos disappeared upon rupture of the transverse cellular barrier formed as a result of the ligation. However, differences in the ligation technique and/or the animal strain used have led to different gap sizes and hence different conclusions (compare Schubiger and Wood, 1977, and Vogel, 1977).

Double abdomens in *Smittia* embryos apparently do not result from interference with a *static* gradient of anterior determinants supposed to be transformed directly into a sequence of segments. If this were the case, parameters affecting the shape of the presumed gradient, such as UV dose or irradiated area, should have an impact on the segment pattern displayed by double abdomen embryos. Actually, the double abdomens are always symmetrical, and the number of duplicated segments is not well related to any of the experimental parameters examined so far (see the following section). Therefore a *dynamic* gradient that has a monotonous profile in normal development, but can assume a symmetrical profile upon experimental interference, would seem more appropriate for an explanation of our results.

A mathematical model of pattern formation based on lateral inhibition has been devised (Gierer and Meinhardt, 1972) and used to explain (Meinhardt, 1977) Sander's ligation and translocation experiments with *Euscelis* eggs, and the UV induction of double abdomens in *Smittia* eggs. Although the model in fact accounts for many of the experimental results, other data obtained with *Euscelis* (Vogel, 1978) and *Smittia* (Kalthoff, 1978) are at variance with Meinhardt's model. Part of the difficulties can be alleviated by a modified application of the model (Meinhardt, 1978). An alternative interpretation involving *short range inductive events* between cephalic, thoracic, and abdominal determinants has been proposed by Vogel (1977, 1978) on the basis of segment patterns observed in anterior and posterior fragments of *Drosophila* embryos, and *Euscelis* embryonic fragments isolated from both polar regions. The formation of double abdomens in *Callosobruchus* (see Section II) has been explained along the lines of Meinhardt's model (Van der Meer, 1978). This has required the introduction of a number of additional assumptions. To account for the unilateral double abdomens, where in the anterior part the segments are normal on one lateral side but of abdominal character on the other, Van der Meer has postulated the existence of barriers which allow the diffusion of morphogenetic substances in antero-posterior direction but sometimes prevent diffusion between lateral halves of the embryo. Such difficulties could be avoided if sequences of anterior and posterior segments were formed under the control of one common determinant superimposed on a second determinant which accounts only for the gross distinction between anterior and posterior.

C. *Metamerization and Antero-Posterior Decision*

I propose that the formation of the antero-posterior body pattern in dipteran and possibly other insect embryos involves two processes termed "metamerization" and "antero-posterior decision" (Kalthoff, 1978). While metamerization is visualized as a process generating segmental groups of cells with determinations specified from either end to the mid-body region, the antero-posterior decision is regarded as a binary switch between the specification of either the anterior or the posterior half of the body pattern. The metamerization process appears compatible with the idea that cells receive positional information (Wolpert, 1971) about their distance from the egg poles. This information might be carried by two gradients that are symmetrical to each other (p and p' in Fig. 2). The antero-posterior decision is regarded as an independent threshold mechanism superimposed on the metamerization. If the activity of the anterior determinants (a in Fig. 2) exceeds a critical level, a and p' are thought to cooperate so that the anterior half of the segment pattern is specified. If a critical activity level of a is not reached at the appropriate stage of development, p' alone specifies the posterior half of the segment pattern. Since the terminal abdominal segment is then formed in the place of the terminal head segment, the proximo-distal polarity (i.e., two arrows pointing in opposite directions from the middle to the terminal segments) is identical in normal embryos and double abdomens.

Experimental interference causing double abdomen formation in *Smittia* embryos such as UV irradiation or application of RNase, is thought to affect primarily the antero-posterior decision and only to a minor extent the metamerization. Some parameters like the UV wavelength, or light and temperature conditions after UV irradiation, had a strong impact on the percentage of double abdomens formed (i.e. on the antero-posterior decision in terms of the model) but no significant influence on the metamerization as expressed by the number of segments per double abdomen. Only the application of extremely high UV doses or irradiation of unnecessarily large egg areas caused a slight decrease in the number of segments in double abdomens (Bathe, 1977). Some influence of the UV irradiation on the metamerization is indicated also by the fact that the number of segments in double abdomens is usually 14 or 16 as compared to 19 in normal embryos. However, the separated double abdomens obtained after combined ligation and UV irradiation together comprised an average of 18 segments (Fig. 4d, data of Ritter, 1976).

Some advantages and difficulties of the interpretation proposed here have been discussed elsewhere (Kalthoff, 1976, 1978). The production of double cephalons by UV irradiation of posterior egg regions as reported by Yajima (1964) is hard to explain on the basis of either Meinhardt's (1977) or my propositions. However, the maximum frequency of apparent double cephalons in Yajima's experiments was less than 7 per cent, and this pattern aberration was scored only after UV irradiation of the posterior two-thirds of *Chironomus* embryos. The infliction of UV damage to such extended areas, however, may cause abnormalities that are very hard to analyze. Our own rather extensive attempts to produce double cephalons by irradiation of posterior embryonic regions of several chironomids have failed completely. Another discrepancy in the results obtained after centrifugation has been discussed previously (Kalthoff,

1976; Kalthoff *et al.*, 1977). Transverse stratification of embryos prior to blastoderm formation results in the development of *both* double cephalons and double abdomens. This may be ascribed to a faulty redistribution of anterior determinants which could occur in two different ways. Above-threshold quantities of anterior determinants may end up in both anterior and posterior embryonic halves causing double cephalon formation, or major fractions may be "trapped" rendering sub-threshold anterior determinant activities in both halves with double abdomen formation ensuing.

D. *Relation to Homoeotic Mutations and Compartments*

While the process of metamerization is included among the phenomena usually explained with a gradient model, the idea of an antero-posterior decision may require additional justification. Major switches in developmental pathways are not restricted to the pattern aberrations discussed so far. *Transdeterminations* and *homoeotic mutations* which are well known in *Drosophila* but occur also in other insects (see Gehring and Nöthiger, 1973) represent formally analogous phenomena with possibly similar underlying mechanisms. They result in the transformation of one body part into another not normally present in that place, e.g. the replacement of an antenna by a leg. The mere existence of homoeotic loci demonstrates that the presence or absence of the corresponding gene products determines the developmental pathway of the cells of an imaginal disc or part thereof. Most of the known transformations are intersegmental, but some cause the intrasegmental replacement of part of an imaginal disc by a mirror image duplication of the complementary part. The *engrailed* mutation transforms posterior wing structures into what appear to be anterior wing structures (Garcia-Bellido and Santamaria, 1972). In *wingless* flies the wings are replaced by mirror image duplications of the dorsal thorax (Morata and Lawrence, 1977a). The formal analogy to the formation of double abdomens is obvious.

The realms of homoeotic mutations often coincide with *compartments*, i.e. regions with species- and stage-specific boundaries which are not transgressed by marked cell clones (Garcia-Bellido, 1975; Lawrence and Morato, 1979). Each compartment is filled by a polyclone, i.e. the descendents of a small number of founder cells (Crick and Lawrence, 1975). Compartments may become subdivided progressively during development, each subdivision involving the activation of another homoeotic gene in one subcompartment but not in the other (Lawrence and Morata, 1976). The specific combination of activated homoeotic genes not only accounts for the developmental pathway taken by the cells in a given compartment but also confers on them an ability to associate with cells that have the same set of active homoeotic genes, and to dissociate from other cells (Nöthiger, 1964; Garcia-Bellido and Lewis, 1976). The activation of homoeotic genes thus appears to be the basis of what is operationally defined as determination (Morata and Lawrence, 1977b). In order to distinguish the controlling role of homoeotic genes from the more executive functions of other genes, Garcia-Bellido (1975) has coined the terms "selector genes" for the former and "realisator genes" for the latter.

The mechanism of selector gene activation is therefore a key problem in developmental biology. Since the subdivision of compartments does not normally follow clonal boundaries (Crick and Lawrence, 1975), the demarcation apparently follows some regional cues. Kaufman (1977) has proposed that the lines between compartments are drawn by "nodal" threshold levels of chemical patterns dictated by the geometry of the developing system. Alternatively, or in addition to such a mechanism, cytoplasmic determinants are apparently involved, at least during early stages of embryogenesis. Sander (1975b) has suggested that an initial system of longitudinal gradients may cause the activation of segment specific selector genes. In terms of this model, he has interpreted two mutations of the *bithorax* series as resulting from a faulty "read off" by the mutant blastoderm cells of some of the levels of the original gradient system. An alternative interpretation involving a regulatory allosteric protein coded for by the *bithorax* gene complex has been proposed by Kiger (1973) and used to explain determinative decisions observed in a *Drosophila* strain carrying paradoxical *bithorax* mutations (Kiger, 1976). In *Drosophila*, blastoderm cells are determined, according to operational criteria, to form either anterior or posterior structures (Chan and Gehring, 1971; Illmensee, 1976). The evidence discussed here indicates that the specification of the anterior half of the body pattern in dipterans requires the activity of cytoplasmic determinants. The anterior determinants in *Smittia* eggs have been tentatively characterized as ribonucleoprotein particles. It is hoped that further experiments with this system will contribute to a better understanding of the mechanisms underlying the activation of selector genes in a spatial order.

ACKNOWLEDGEMENTS

I wish to thank Drs. G. Freeman, A.G. Jacobson, H. Pianka, K. Sander, and O. Vogel for their comments on the manuscript. Work in the author's previous laboratory at Freiburg (FRG) has been supported by the Deutsche Forschungsgemeinschaft, SFB 46.

REFERENCES

Bathe, M.M. (1977). Staatsexamensarbeit, Fak. f. Biologie, Univ. Freiburg.
Bownes, M. and Kalthoff, K. (1974). *J. Embryol. Exp. Morphol.* **31**, 329-345.
Bownes, M. and Sander, K. (1976). *J. Embryol. Exp. Morphol.* **36**, 394-408.
Bull, A. (1966). *J. Exp. Zool.* **161**, 221-242.
Chan, L-N. and Gehring, W. (1971). *Proc. Nat. Acad. Sci. U.S.* **68**, 2217-2221.
Counce, S.J. (1973). *In* "Developmental Systems" (S.J. Counce and C.H. Waddington, eds.), Vol. II, pp. 1-156. Academic Press, New York.
Crick, F. and Lawrence, P.A. (1975). *Science* **189**, 340-347.
Davidson, E.H. (1976). "Gene Activity in Early Development". Acad. Press, New York.
Davidson, E.H. and Britten, R.J. (1971). *J. Theor. Biol.* **32**, 123-130.
Donohoo, P. and Kafatos, F.C. (1973). *Develop. Biol.* **32**, 224-229.
Duspiva, F., Scheller, F., Weiss, D. and Winter, H. (1973). *Wilhelm Roux' Arch.* **172**, 83-130.
Freeman, G. (1976). *Develop. Biol.* **49**, 143-177.
Freeman, G. (1979). This volume.
Garcia-Bellido, A. (1975). *In* "Cell Patterning" (Ciba Foundation Symposium 29), pp. 241-263. Associated Scientific Publishers, Amsterdam.
Garcia-Bellido, A. and Santamaria, P. (1972). *Genetics* **72**, 87-104.
Garcia-Bellido, A. and Lewis, E.B. (1976). *Develop. Biol.* **48**, 400-410.
Gehring, W.J. (1976). *Ann. Rev. Genet.* **10**, 209-252.
Gehring, W.J. and Nöthiger, R. (1973). *In* "Developmental Systems" (S.J. Counce and C.H. Waddington, eds.), Vol. II, pp. 211-290. Acad. Press, New York.
Gierer, A. and Meinhardt, H. (1972). *Kybernetik* **12**, 30-39.
Goldstein, E.S. (1978). *Develop. Biol.* **63**, 59-66.
Gollub, G. (1970). Staatsexamensarbeit, Fak. f. Biologie, Univ. Freiburg.
Gordon, M.P., Huang, C.W., and Hurter, J. (1976). *In* "Photochemistry and Photobiology of Nucleic Acids" (S.Y. Wang, ed.), Vol. II, pp. 265-308. Acad. Press, New York.
Graziosi, G. and Roberts, D.B. (1975). *Nature* **258**, 157-159.
Gurdon, J.B., Laskey, R.A. and Reeves, O.R. (1975). *J. Embryol. Exp. Morphol.* **34**, 93-112.
Harm, H. (1976). *In* "Photochemistry and Photobiology of Nucleic Acids" (S.Y. Wang, ed.), Vol. II, pp. 219-263. Acad. Press, New York.
Herth, W. and Sander, K. (1973). *Wilhelm Roux' Arch.* **172**, 1-27.
Illmensee, K. (1973). *Wilhelm Roux' Arch.* **121**, 331-343.
Illmensee, K. (1976). *In* "Insect Development" (P.A. Lawrence, ed.), pp. 76-96. Blackwell, Oxford.
Illmensee, K. and Mahowald, A.P. (1974). *Proc. Nat. Acad. Sci. U.S.* **71**, 1016-1020.
Illmensee, K., Mahowald, A.P. and Loomis, M.R. (1976). *Develop. Biol.* **49**, 40-65.
Jacobson, A.G. and Fullilove, S.L. (1973). *Develop. Biol.* **33**, f-1.
Jäckle, H. (1977). Dissertation, Fak. f. Biologie, Univ. Freiburg.
Jäckle, H. and Kalthoff, K. (1978). *Photochem. Photobiol.* **27**, 309-315.

Jura, C., Krzysztofowicz, A. and Weglarska, B. (1957). *Zool. Polon.* **8,** 201-215.
Kalthoff, K. (1971a). *Wilhelm Roux Arch.* **168,** 63-84.
Kalthoff, K. (1971b). *Develop. Biol.* **25,** 119-132.
Kalthoff, K. (1973). *Photochem. Photobiol.* **18,** 355-364.
Kalthoff, K. (1975a). *Encyclop. Cinematograph.* (Göttingen), Film E2158.
Kalthoff, K. (1975b). *Oecologia* **18,** 101-110.
Kalthoff, K. (1976). *In* "Insect Development" (P.A. Lawrence, ed.), pp. 53-75. Blackwell, Oxford.
Kalthoff, K. (1978). *J. Cell Sci.* **29,** 1-15.
Kalthoff, K. and Sander, K. (1968). *Wilhelm Roux Arch.* **161,** 129-146.
Kalthoff, K., Kandler-Singer, I., Schmidt, O., Zissler, D. and Versen, G. (1975). *Wilhelm Roux Arch.* **178,** 99-121.
Kalthoff, K., Hanel, P. and Zissler, D. (1977). *Develop. Biol.* **55,** 285-305.
Kalthoff, K., Urban, K. and Jäckle, H. (1978). *Photochem. Photobiol.* **27,** 317-322.
Kandler-Singer, I. and Kalthoff, K. (1976). *Proc. Nat. Acad. Sci. U.S.* **73,** 3739-3743.
Kastern, W.H. and Berry, S.J. (1976). *Biochim. Biophys. Res. Comm.* **71,** 37-44.
Kauffman, S. (1977). *Amer. Zool.* **17,** 631-648.
Kiger, J.A., Jr. (1973). *J. Theor. Biol.* **40,** 455-467.
Kiger, J.A., Jr. (1976). *Develop. Biol.* **50,** 187-200.
Koch, P. and Heinig, S. (1968). *Wilhelm Roux Arch.* **161,** 241-248.
Lawrence, P.A. and Morata, G. (1976). *In* "Insect Development" (P.A. Lawrence, ed.), pp. 132-149. Blackwell, Oxford.
Lawrence, P.A. and Morata, G. (1979). This volume.
Lenk, R., Ransom, L., Kaufman, Y. and Penman, S. (1977). *Cell* **10,** 67-78.
Lohs-Schardin, M. and Sander, K. (1976). *Wilhelm Roux Arch.* **179,** 152-162.
Lovett, J.A. and Goldstein, E.S. (1977). *Develop. Biol.* **61,** 70-78.
Mahowald, A.P. (1972). *In* "Developmental Systems: Insects" (S.J. Counce and C.H. Waddington, eds.), Vol. I, pp. 1-49. Acad. Press, New York.
Mahowald, A.P. (1977). *Amer. Zool.* **17,** 551-563.
Mahowald, A.P., Allis, C.D., Karrer, K.M., Underwood, E.M., and Waring, G.L. (1979). This volume.
Meinhardt, H. (1977). *J. Cell Sci.* **23,** 117-139.
Meinhardt, H. (1978). *Rev. Physiol. Biochem. Pharmacol.* **80,** 47-104.
Morata, G. and Lawrence, P.A. (1977a). *Develop. Biol.* **56,** 227-240.
Morata, G. and Lawrence, P.A. (1977b). *Nature* **265,** 211-216.
Newrock, K.M. and Raff, R.A. (1975). *Develop. Biol.* **42,** 242-261.
Nöthiger, R. (1964). *Wilhelm Roux Arch.* **155,** 269-301.
Nünemann, H. and Moser, J.G. (1970). *Zool. Anz. Suppl.* **33,** 113-120.
Nüsslein-Volhard, C. (1977). *Wilhelm Roux Arch.* **183,** 249-268.
Nüsslein-Volhard, C. (1979). This volume.
Okada, E. and Waddington, C.H. (1959). *J. Embryol. Exp. Morphol.* **7,** 583-597.
Ouweneel, W. (1976). *Adv. Genetics* **18,** 179-248.
Paglia, L.M., Berry, S.J., and Kastern, W.H. (1976). *Develop. Biol.* **51,** 173-181.
Price, R.D. (1958). *Ann. Entomol. Soc. Amer. 51,* 600-604.
Richards, F.M. and Vithayathil, P.J. (1959). *J. Biol. Chem.* **234,** 1459-1465.
Rickoll, W.L. (1976). *Develop. Biol.* **49,** 304-310.
Ritter, W. (1976). Staatsexamensarbeit, Fak. f. Biologie, Univ. Freiburg.
Roberts, D.B. and Graziosi, G. (1977). *J. Embryol. Exp. Morphol.* **41,** 101-110.
Sander, K. (1960). *Wilhelm Roux Arch.* **151,** 660-707.

Sander, K. (1961). *In* "Symposium on Germ Cells and Development", pp. 338-353. Institut Intern. d'Embryologie Pallanza.
Sander, K (1975a). *In* "Cell Patterning" (Ciba Foundation Symposium 29), pp. 241-263. Associated Scientific Publishers, Amsterdam.
Sander, K. (1975b). *Verh. Dtsch. Zool. Ges. 1974*, 58-70.
Sander, K. (1976). *Adv. Insect Physiol.* **12**, 125-238.
Schmidt, O., Zissler, D., Sander, K. and Kalthoff, K. (1975). *Develop. Biol.* **46**, 216-221.
Schmidt, O. and Jäckle, H. (1978). *Wilhelm Roux Arch.* **184**, 143-154.
Schubiger, G., Moseley, R.C., and Wood, W.J. (1977). *Proc. Nat. Acad. Sci. U.S.* **74**, 2050-2053.
Schubiger, G. and Wood, W.J. (1977). *Amer. Zool.* **71**, 565-576.
Seidel, F. (1961). *Zool. Anz. Suppl.* **24**, 121-142.
Spirin, A.S. (1966). *Curr. Topics Develop. Biol.* **1**, 1-38.
Van der Meer, J.M. (1978). Ph.D. thesis, Faculteit der Wiskunde en Natuurwetenschappen, Katholieke Universiteit te Nijmegen.
Vogel, O. (1977). *Wilhelm Roux Arch.* **182**, 9-32.
Vogel, O. (1978). *Develop. Biol.* **67**, 357-370.
Whittaker, J.R. (1973). *Proc. Nat. Acad. Sci. U.S.* **70**, 2096-2100.
Whittaker, J.R. (1977). *J. Exp. Zool.* **202**, 139-153.
Whittaker, J.R. (1979). This volume.
Whittaker, J.R., Ortolani, G., and Farinella-Ferruzza, N. (1977). *Develop. Biol.* **55**, 196-200.
Winter, H. (1974). *Wilhelm Roux Arch.* **175**, 103-127.
Winter, H., Wiemann-Weiss, D. and Duspiva, F. (1977). *Wilhelm Roux Arch.* **182**, 39-58.
Wolpert, L. (1971). *Curr. Topics Develop. Biol.* **6**, 183-224.
Yajima, H. (1960). *J. Embryol. Exp. Morphol.* **8**, 198-215.
Yajima, H. (1964). *J. Embryol. Exp. Morphol.* **12**, 89-100.
Yajima, H. (1970). *J. Embryol. Exp. Morphol.* **24**, 287-303.
Zalokar, M. (1976). *Develop. Biol.* **49**, 425-437.
Zissler, D. and Sander, K. (1973). *Wilhelm Roux Arch.* **172**, 175-186.
Zissler, D. and Sander, K. (1977). *Wilhelm Roux Arch.* **183**, 233-248.

Germ Plasm and Pole Cells of *Drosophila*

A.P. Mahowald, C.D. Allis, K.M. Karrer,
E.M. Underwood, and G.L. Waring

*Program in Molecular, Cellular and Developmental Biology
and the Department of Biology
Indiana University
Bloomington, Indiana 47401*

I. Introduction 127
II. *Drosophila* Germ Plasm 128
    A. Properties of Germ Plasm 128
    B. Properties of Polar Granules 132
    C. Characteristics of Grandchildless (*gs*) Mutations 137
III. Pole Cells 138
    A. General Properties 138
    B. Pole Cell Specific Nuclear Body 140
    C. Pole Cells in Culture 141
IV. Perspective 144
    References 145

I. INTRODUCTION

One of the central issues in cellular determination concerns the sequence of events within a cell by which it becomes restricted to a specific fate. Extensive analyses, both classic (Wilson, 1928) and recent (this symposium), provide ample examples of the presence of determinants in the egg cytoplasm. In some instances the determinants or morphogens are present as discrete localizations (e.g., germ plasms; reviewed in Beams and Kessel, 1974; Eddy, 1975; Smith and Williams, 1975), as regions (e.g., polar lobe of molluscs and annelids, cf. Dohmen and Verdonk, this symposium; gray crescent of amphibians, Curtis, 1962; Chung and Malacinski, 1975) or gradients (e.g., Childs, 1940;

Sander 1975; Schubiger and Wood, 1977). However in each instance the cellular or nuclear events which occur subsequent to the interaction with the morphogen may be similar. For example, we can reasonably suppose that there is a response by the cell (or cytoplasmic region) which induces a specific pattern of chromosomal activiation and repression. This, in turn, leads to a series of gene products which both determine the cell and gradually lead to a stable nuclear state.

The successful analysis of the processes involved in determination will almost certainly require the possibility of identifying the molecules involved in establishing the determined state. This requirement, in turn, demands that cells be analyzed as they are becoming restricted in developmental potential. The germ plasm, as a specific determinant, and the pole cells, as examples of newly determined cells, appear to be an especially favorable system for the systematic investigation not only of the nature of a morphogenetic determinant, but also for how the cell responds to this determinant. In this paper we will review our current knowledge of both aspects of this system. Although we certainly have many gaps in our knowledge, the prospects for a complete picture of the critical developmental phenomenon of determination appears feasible.

The general biology of germ cell formation has been frequently reviewed (Beams and Kessel, 1974; Eddy, 1975; Smith and Williams, 1975; Mahowald, 1977) and thus we will not attempt to be comprehensive. Instead, we will concentrate on the recent studies that have been carried out on the polar plasm of *Drosophila*. First, we will analyze the work characterizing properties of the germ plasm as a whole, and then we will consider the unique organelles, the polar granules of the germ plasm. Then we will describe the properties of the pole cells and their unique organelles; finally we will attempt to give a prospectus of future possibilities for the analysis of the function of germ plasm.

II. *DROSOPHILA* GERM PLASM

A. *Properties of Germ Plasm (Summarized in Table I).*

The first complete cells which form in a number of insect groups bud off from the posterior pole of the early syncytial insect embryo. Metschnikoff (1866) was the first to show that these cells in *Miastor* (a cecidomiid) were the precursors to the primordial germ cells. Hegner (1908) showed that the removal or destruction of this polar plasm prevented the formation of these pole cells, as they came to be called, and caused subsequent sterility. Since then, workers have shown the obligate requirement for this region for the formation of germ cells. The favorite approach has been to irradiate the polar plasm prior to pole cell formation with sufficient dose of ultra-violet (UV) light so that no pole cells form (Geigy, 1931). Analogous experiments with UV-irradiation have clearly identified a germ plasm in amphibians in the vegetal hemisphere (Bounoure, 1934; 1937). Smith (1966), in a series of important pioneering experiments showed that the UV-induced lesion of the germ plasm in *Rana pipiens* could be corrected by a transplantation of unirradiated vegetal cytoplasm. Furthermore, he obtained a spectrum of the UV-sensitivity of this region which suggested that nucleic acids were the important components.

TABLE I

Properties of Posterior Pole Plasm

1. Included in pole cells which are precursors to germ cells[1]
2. UV sensitive[2]
3. Functional in ectopic locations: both anterior tip and mid-ventral[3]
4. Functional without fertilization and prior to completion of oogenesis[4]
5. Functional in heterospecific combinations

[1]Huettner, 1923
[2]Geigy, 1931; Graziosi and Marzari, 1976; Okada *et al.*, 1974
[3]Illmensee and Mahowald, 1974; 1976
[4]Illmensee *et al.*, 1976
[5]Mahowald *et al.*, 1976

Recently, analyses of UV-irradiation and transplantation of polar plasm in *Drosophila* have extended our knowledge of the biological properties of polar plasm. Analysis of the effect of UV on the polar plasm of *Drosophila* eggs has been complicated by the considerable lethality following UV-irradiation. For example, in recent experiments utilizing UV, the dose sufficient to produce 60% sterility caused 30% lethality (Illmensee *et al.*, 1976) and a dose sufficient for 98% sterility caused 90% embryonic lethality (Okada *et al.*, 1974). A systematic study of the dose relationship between these two effects has been made by Graziosi and Marzari (1976) and their results suggest different targets responsible for the two effects. Unfortunately because of the UV-induced mortality it is difficult to carry out UV action spectra which would add to our understanding of the UV effect. In important experiments Okada *et al.*, (1974) have shown that UV-induced lesion to production of germ cells can be alleviated by the transplantation of unirradiated polar plasm. By utilizing this method of micro-injection, it should be possible to assay for the factors in the injected polar plasm which are prerequisite for restoring germ plasm activity.

1. *Autonomy of polar plasm.* Another approach to studying the biological

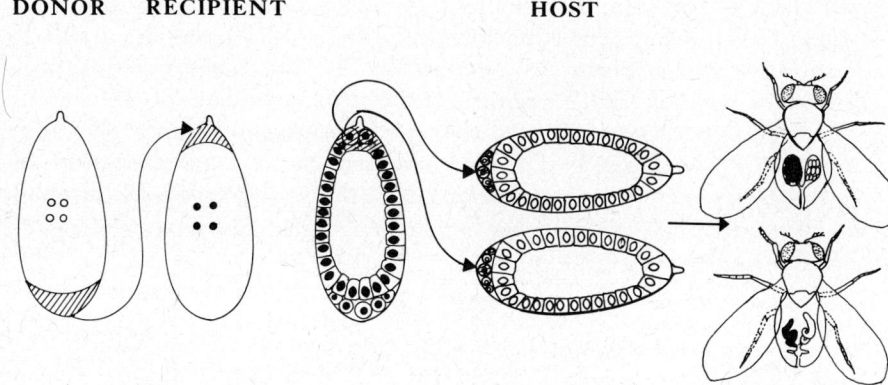

Fig. 1. Scheme for testing inductive properties of polar plasm. (after Illmensee and Mahowald, 1974).

properties of polar plasm is through transplantation of polar plasm to foreign locations in order to test for autonomy of function (Illmensee and Mahowald, 1974; 1976). The basic scheme for these transplantations is illustrated in Fig. 1. Between 5 and 100 pl of posterior polar plasm is transplanted to a foreign site which is subsequently analyzed for pole cell formation. At both the anterior tip and mid-ventral regions, posterior polar plasm is able to induce the formation of normal appearing pole cells which form prior to the appearance of the cellular blastoderm. Following transplantation to the posterior tip of a blastoderm stage embryo, these ectopically induced pole cells can function to produce gametes. These results clearly show that all the components required for pole cell formation are present at the posterior tip of the embryo shortly after fertilization and that they can function in a presumptive somatic region of the embryo. Moreover, it is clear that ectopic sites are excellent locations for testing the ability of components of the polar plasm to have inducing ability.

2. *Ontogeny.* The same transplantation scheme (Fig. 1) is readily adapted to testing the effectiveness of the posterior polar plasm prior to fertilization (Illmensee et al., 1976). The choice for donor polar plasm was suggested by previous ultrastructure studies which showed that the unique organelles of the germ plasm, the polar granules, have already appeared at mid-vitellogenic stages (Stage 10 of King, 1970). The results of these experiments are clear: polar plasm of unfertilized eggs, mature eggs (stage 14) and the chorionic filament stage (stage 13) is able to induce cells at the anterior tip which can function to produce gametes.

Posterior polar plasm from earlier stages of oogenesis is unable to function (results reviewed in Illmensee, 1976). These experiments show definitively that all the components required for germ cell determination are localized in the posterior polar plasm during oogenesis. The negative result with pre-stage 13 polar plasm came as a surprise since ultrastructurally the polar plasm does not change noticeably as the egg matures. However, the structure of the whole egg changes remarkably at this time in two obvious ways: firstly, glycogen yolk appears for the first time throughout the polar plasm; secondly, the volume of the oocyte decreases. Although the actual volume change has not been measured, the shrunken shape and the dense ultrastructure (Illmensee et al., 1976) clearly vouch for its existence. These changes may be signs of an ooplasmic maturation which is needed for proper functioning of the germ plasm.

A number of ultrastructural abnormalities were found in the embryos which received polar plasm from unfertilized and ovarian oocytes which should be mentioned. In the case of polar plasm from unfertilized oocytes, there was a clear increase in the amount of multivesicular bodies found at the anterior tip. These organelles, which contain acid phosphatase activity (Mahowald and Allis, in preparation), appear rapidly in the cortex after the oocyte leaves the ovary and remain in the cortex until the blastema stage when they move internally. However, in unfertilized eggs they increase in number during the first hours after leaving the ovary. Although pole cells induced at the anterior tip contained these lysosomes, this abnormality did not interfere with the induced pole cells differentiating as germ cells after transfer to the posterior tip.

The anterior tip which received polar plasm from ovarian eggs showed a clear inhibition of blastoderm formation (Fig. 2). A large number of blastoderm stage nuclei remained in a syncytial cortical cytoplasm throughout the time of cellular blastoderm formation. The extent of this inhibition of cell formation was approximately the region that contained transplanted plar plasm. The transplanted cytoplasm which contained polar granules, however, became segregated into cells prior to blastoderm formation. Thus, it appears that the processes leading to precocious pole cell formation are distinct from those leading to blastoderm formation.

Polar plasm from oocytes prior to stage 13 was ineffective in segregating pole cells or even polar granules. Instead, polar granules became isolated into cytolysosomal type structures. Thus, it is clear that important events in the differentiation of a functional polar plasm occur

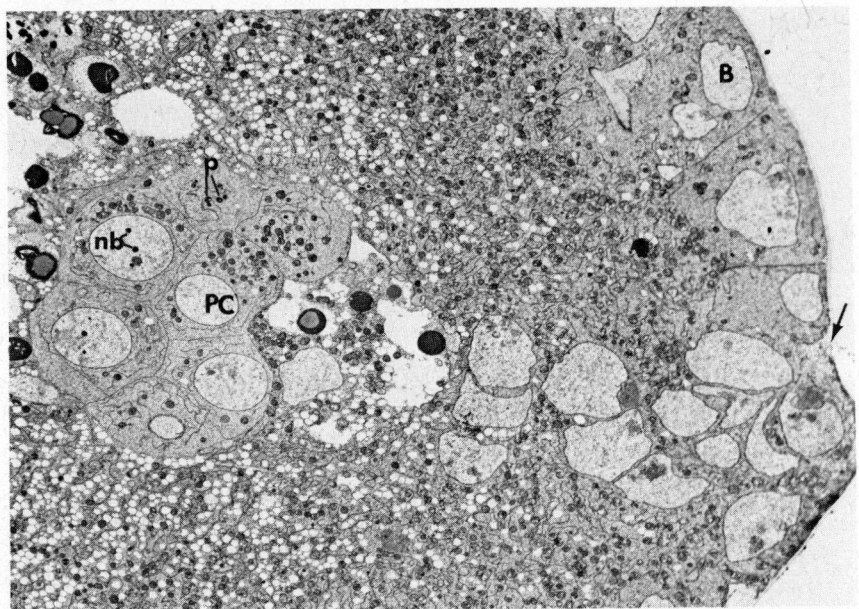

Fig. 2. Electron micrograph of a blastoderm-stage embryo which had received polar plasm from a stage 14 oocyte. Fourteen pole cells (PC) were found at the anterior tip. They contain both polar granules (p) and pole cell-specific nuclear bodies (nb). In addition, a cluster of nuclei remained outside the cellular blastoderm (B). The site of the injection is indicated by the arrow. (Illmensee et al., 1976).

between stages 12 and 13 which are needed for the proper segregation of pole cells. The possibility that at stage 12 the polar plasm is incomplete suggests that it will be possible to assay for a component that will make stage 12 polar plasm effective in pole cell formation.

3. *Interspecific transplantation.* The ability of polar plasm to function across species lines has been shown utilizing a similar transplantation scheme. Serial sections of 10 *D. melanogaster* embryos which had received germ plasm from *D. immigrans* were examined and in every instance induced pole cells were found (Mahowald et al., 1976). Thus, polar plasm of distantly related species within the same genus is clearly functional in producing pole cells, and these pole cells can produce germ cells. Preliminary results of interspecific transplantations have also been reported by Okada et al., (1974).

B. *Properties of Polar Granules*

Much attention has been given to the unique organelles of the germ plasm, the polar granules, because of the possibility that they might be

TABLE II

Properties of Polar Granules

1. Composed of a major basic protein of 95,000 daltons[1], which is synthesized prior to stage 10 of oogenesis[2]
2. First appear in posterior polar plasm at late stage 9 or early stage 10[3]
3. RNA positive in oocyte, negative in pole cells[4]
4. Associated with polysomes in pre-pole cell stage embryos[4]
5. Undergo fragmentation prior to pole cell formation and species-specific reaggregation in pole cells[5]
6. Transform into a "nuage-like" material during gastrulation[5]
7. Nuage continuously present in germ line cells[5]
8. Not inherited cytoplasmically[6]

[1]Waring et al., 1978
[2]Waring and Mahowald, 1978
[3]Mahowald, 1962
[4]Mahowald, 1971b
[5]Mahowald, 1971a
[6]Mahowald et al., 1976

the repository of the information for germ cell determination. There is no direct evidence that they are required for germ cell determination. However, because of their constant association with the polar plasm and pole cells, it is reasonable to postulate that they play a key role. Many of the characteristics of polar granules are summarized in Table II. On the basis of their association with polysome-like granules (Mahowald, 1968; 1977) and the cytochemically detectable presence of RNA prior to the appearance of polysomes and the absence of RNA after the polysomes disappeared from their periphery, Mahowald (1968) proposed that they might contain maternal mRNA which coded for specific protein(s) which in turn were responsible for cellular determination. In order to clarify any role in germ cell determination, it was obvious that these organelles would have to be chemically analyzed following their isolation. Considerable effort was made to isolate granules from mature eggs or pre-pole cell embryos since at these stages the organelles have RNA. Unfortunately this has not been possible (Mahowald, 1977). However, we have been able to obtain sufficient enrichment for polar granules from pole cells to enable us to identify the protein composition of the organelles and to study the time of synthesis during oogenesis.

The purification of mass quantities of pole cells from embryos at preblastoderm stages is possible with a sequence of isopycnic and sedimentation centrifugation (Allis et al., 1977). Between $1-2 \times 10^7$ cells can be prepared daily of which 80—90% are pole cells. In order to identify the proteins constituting the granule we compared similar subcellular

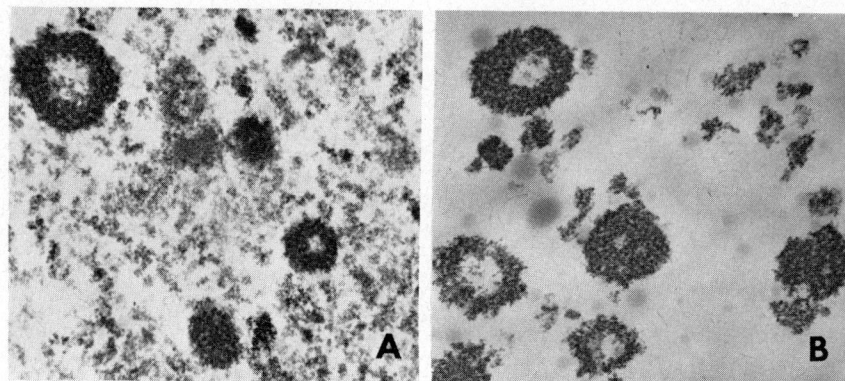

Fig. 3A. Electron micrograph of a particulate fraction from pole cells prior to sonication (A) and after sonication (B). (Waring et al., 1978).

fractions from cell populations enriched for pole cells and populations depleted of pole cells. Following differential centrifugation of detergent-treated cells a particulate fraction was obtained and analyzed by electron microscopy. Polar granules are a prominent component of the fraction from pole cells (Fig. 3A). The corresponding fraction from other embryonic cells consists of a fibrous matrix but lacks polar granules. We have prepared similar particulate fractions from cells distributed in an isopycnic gradient. Following SDS polyacrylamide gel analysis, one unique protein species is seen at 95,000 daltons for the particulate fractions from regions of the gradient containing pole cells (Fig. 4). Additional enrichment for polar granules can be achieved by sonication of a resuspended particle fraction (Fig. 3B). The only protein species that becomes enriched following sonication is the 95K dalton species. Based on this co-distribution with pole cells and the co-purification with polar granules, this 95K protein is probably a polar granule constituent. Inasmuch as no other protein species becomes similarly enriched, it appears to be the major constituent (Waring et al., 1978).

By utilizing two dimensional gels (O'Farrell, 1975) a number of further properties can be identified. This 95K dalton protein focuses in the pH range of 7.5 to 8.0. The basic character of this protein component of polar granules correlates well with its postulated role of localizing a specific maternal messenger RNA which codes for a protein needed for primordial germ cell determination. Greater resolution of basic proteins can be achieved with nonequilibrium pH gradient gel electrophoresis as

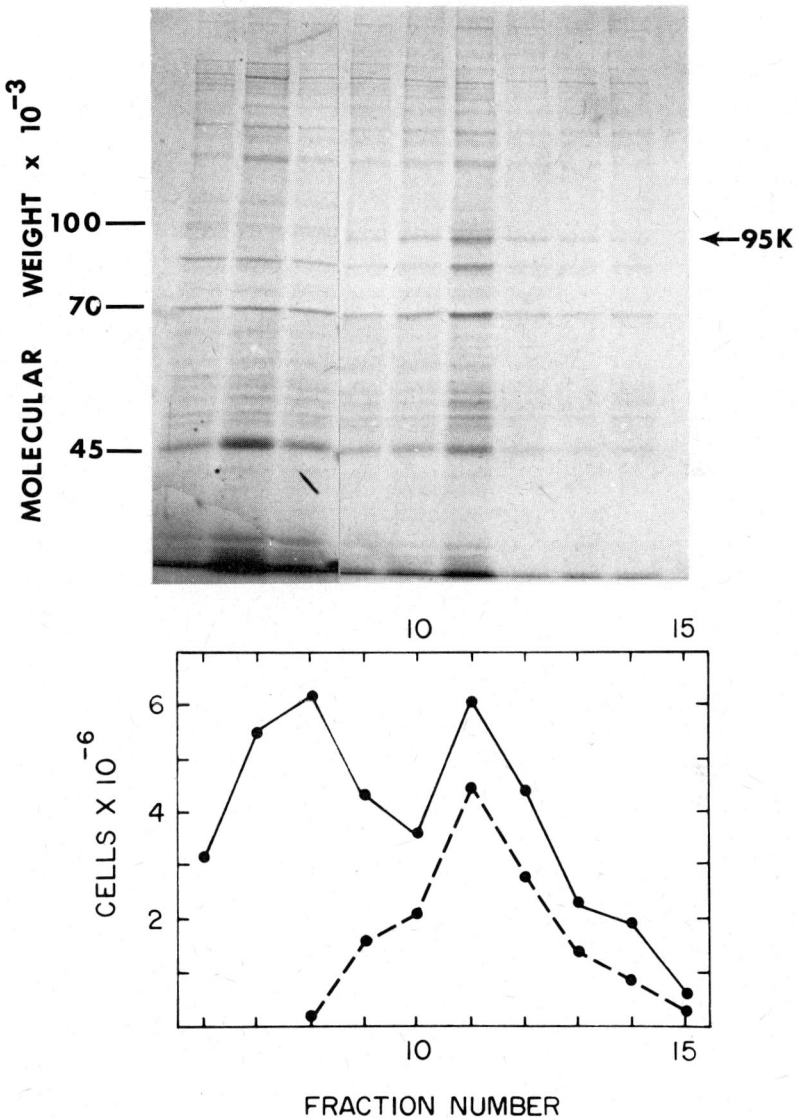

Fig. 4. SDS polyacrylamide gel electrophoretic analysis of proteins from particulate subcellular fractions (similar to those in Figure 3A). Cells from embryos in pole cell stages of development were fractionated on isopycnic density gradients. Proteins from particulate subcellular fractions were prepared from each cell fraction and electrophoresed in 7.5% acrylamide SDS gels. Total cells (———.———). Pole cells, determined by neutral lipid staining (---- ----). (Waring *et al.*, 1978).

the first dimension instead of isoelectric focusing (O'Farrell et al., 1977). With this procedure the polar granule protein can be identified in total pole cell lysates (Fig. 5). The fact that the polar granule protein is in a relatively uncluttered portion of the gel has facilitated the use of these gels to investigate the time of synthesis of this protein during oogenesis. Previously, Mahowald (1971a) had suggested that polar granules or a nuage-like derivative is always present in germ line cells of *Drosophila*. Although we have not yet completed a survey of the germ line during the complete life cycle of *Drosophila*, the following points are established. In ovaries of newly eclosed flies (containing only pre-vitellogenic stages) the synthesis of the polar granule protein can be detected but its synthesis is not observed in stages 10-14, the times at which the granule appears at the posterior tip. These results are consistent with the earlier suggestion (Mahowald, 1971a) that the nuage or dense bodies associated with the nuclear envelopes of nurse cells are actually precursors to the polar granules. This hypothesis also implies that this protein species should be synthesized at any time that nuage appears, e.g. during germ cell multiplication during larval and pupal stages. This result is being tested.

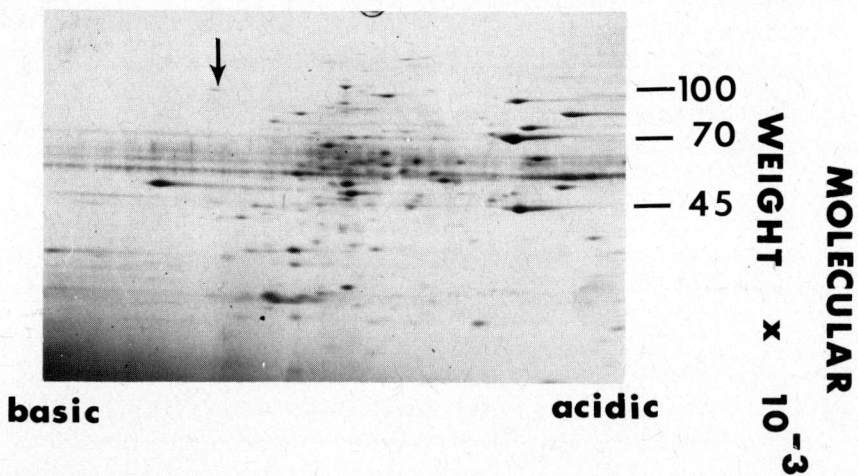

Fig. 5. Two-dimensional gel analysis of pole cell lysates utilizing nonequilibrium pH gradient in the first dimension and SDS polyacrylamide electrophoresis in the second dimension. The putative polar granule spot is marked with an arrow. (Waring et al., 1978).

We have not been able to detect any RNA species which are unique to the polar granule-rich fractions from pole cells. This result agrees with the cytochemical data (Mahowald, 1971a) which indicated that RNA was no longer present in the granules at the pole cell stage. From pre-pole cell stage embryos we have not obtained sufficiently pure preparations to identify any RNA species with the polar granules. Polysomes remain attached to polar granules during fractionation (Mahowald, 1977) so that this association with polar granules first seen in thin sections appears real. It will be important to determine whether or not any specific RNA species is associated with these organelles.

C. *Characteristics of Grandchildless (gs) Mutations*

1. *Gs of D. subobscura.* A number of maternal effect mutations exist which are called grandchildless because they affect the formation of germ cells. Since they have recently been reviewed (Mahowald, 1978), I will include only some recent results. The first *gs* gene found was by Spurway (1946) in *D. subobscura* and it is the best understood. The sterility of the F_1 progeny of a homozygous mother is clearly due to the failure of the pole cells to form (Fielding, 1967). The primary lesion of the mutation which leads to the failure of pole cell formation appears to happen during oogenesis (Mahowald *et al.*, 1979). First, although apparently normal numbers of polar granules appear during oogenesis, in mature eggs the granules at the posterior tip are selectively lost. A few granules remain at the periphery of the polar plasm. Secondly, following fertilization and early nuclear division, the migration of nuclei to both poles is delayed; in the case of the posterior tip, nuclei never move into the posterior polar plasm directly from the yolk. After the last periplasm nuclear division the lateral blastoderm nuclei spread across the posterior tip. At this time the original polar plasm is budded off from the posterior tip as a series of cytoplasmic vesicles containing only ribosomes, vesicles, and a fine granular ground substance. Thus, it appears that the failure in pole cell formation may not be associated with a failure in the initial formation of polar granules and polar plasm but with the maintenance and maturation of the polar plasm. After the polar granular protein has been identified in *D. subobscura* it should be possible to verify chemically the loss of polar granules in the mature egg.

2. *In D. melanogaster.* Thiery-Mieg (1976) has described a maternal effect mutation which causes considerable embryonic lethality. In addition, the few surviving embryos lack pole cells and are sterile. All embryos have a normal amount of polar granules by ultrastructural analysis. The *gs* phenotype of this mutant, called gs^{87}, shows a temperature sensitive period during vitellogenic stages of oogenesis, a period which is after the time of synthesis of the major polar granule protein. Since abundant polar granules are found in the mutant, the cause of the agametic effect must be different from that found in *D. subobscura*.

Two additional *gs* mutations have been identified on the X chromosome in *D. melanogaster* by Okada (personal commmunication). It is hoped that these mutations may assist in clarifying the functional components of the germ plasm.

III. POLE CELLS

A. *General Properties*

The general properties of pole cells are summarized in Table III (cf. Counce, 1973, for a detailed review). The number of preblastema nuclei that become segregated in pole cells has been estimated to be close to 10 both by morphology (Rabinowitz, 1941; Turner and Mahowald, 1976) and clonal analysis (Nissani, 1977; Wieschaus and Szabad, 1978). Following 1 to 3 additional divisions, the pole cells cease dividing in *D. melanogaster* until the time of gonadal divisions. In a number of other species pole cell divisions continue (Counce, 1963) but we do not have detailed knowledge concerning the rate of divisions or whether all cells divide.

TABLE III

Characteristics of Pole Cells in D. Melanogaster

1. Bud off from posterior tip prior to formation of cellular blastoderm;[1] contain nearly all polar granules and very little neutral lipid[2]
2. Clonally derive from 8 — 10 nuclei[3]
3. Divide 1 to 3 times before ceasing division until gonad stage[4]
4. Interdigitate with blastoderm cells prior to gastrulation; some reach boundary of yolk;[5] do not form secondary yolk nuclei[6]
5. Loosely attached to posterior midgut rudiment during gastrulation[7]
6. Some form midgut; some form primoridal germ cells[8]

[1]Huettner, 1923
[2]Allis et al., 1977
[3]Wieschaus and Szabad, 1978
[4]Counce, 1963

[5]Rabinowitz, 1941
[6]Mahowald et al., in preparation
[7]Turner and Mahowald, 1977
[8]Poulson, 1947; Poulson and Waterhouse, 1960

Not all pole cells in *Drosophila* become germ cells. Two additional fates have been described. First, between 10 and 20 pole cells have been thought to migrate back into the yolk during cellularization of the blastoderm to form secondary yolk nuclei (Rabinowitz, 1941). Huettner (1923) first noticed in histological preparations that the number of pole cells within the posterior midgut invagination during gastrulation is less than that at the posterior tip at the blastoderm stage. Rabinowitz (1941) described this putative migration into the yolk of some pole cells.

Mahowald (1962) described the ultrastructure of the posterior yolk nuclei, presumed to be derived from pole cells, and noted that they are typical yolk nuclei.

In a recent reexamination of embryos at this stage the basis for postulating this migration appears weak. Clearly, many of the pole cells interdigitate between the forming blastoderm cells and about 10 cells are found at the base of the furrows adjacent to the yolk. These pole cells have both polar granules and nuclear bodies. After serial section analysis with the EM of early stages of gastrulation, we find no interdigitation of pole cells at the surface of the blastoderm layer. At this time the pole cells form a plaque of cells in the forming posterior midgut invagination (Turner and Mahowald, 1976; 1977). A few cells are still present between or at the base of the posterior blastoderm cells. However, we find no accumulation of a new set of yolk nuclei in the posterior regions of the syncytial yolk mass that could have derived from pole cells. Consequently, from these observations (Mahowald et al., in preparation) we conclude that although most pole cells interdigitate with the blastoderm layer during cell membrane formation and thus may appear to be migrating through the layer, at the conclusion of blastoderm formation most of these cells are extruded from the layer. Some appear to remain between the blastoderm cells as though they were trapped. There is no evidence for the addition of new nuclei to the yolk mass.

A second fate of pole cells is to produce portions of the larval midgut. This fate is based upon histological studies of normal and UV-irradiated embryos (Poulson, 1947; Counce, 1963; Poulson and Waterhouse, 1960; Sonnenblick, 1950). In addition, a transplanted pole cell has produced both germ and gut derivatives (Illmensee et al., 1976). It is clear that not all pole cells reach the gonads. Although there are usually about 40 pole cells, the embryonic gonads average less than 10 cells/gonad (Poulson and Waterhouse, 1960; Wieschaus and Szabad, 1977). The remaining pole cells are thought to become integrated into portions of the midgut on the basis of histology (Poulson, 1947). Similar conclusions have been drawn from experiments in which the posterior tips of pre-blastoderm embryos were irradiated with UV light (Poulson and Waterhouse, 1960). Since both germ cells and calycocytes, a special cuprophilic cell of the midgut, were decreased by the irradition, both were concluded to have the same precursor. Unfortunately the UV-irradiation could have injured other cells of the presumptive endoderm in addition to the pole cells.

A number of points remain to be clarified concerning the fate of pole cells. Can any pole cell become a germ cell, or is the pole cell population at the blastoderm stage already differentiated into presumptive germ cells and gut cells? What is the fate of the pole cells which do not become germ cells? Is a gut cell type missing in an embryo lacking all pole cells as Poulson and Waterhouse (1960) suggest or is the gut normal as Fielding (1967) found for *D. subobscura*?

It will be difficult to follow the fate of pole cells within the posterior midgut invagination by means of electron microscopy because the cells lose both their polar granules and nuclear bodies during gastrulation (see below). Some other property of these cells will be needed, such as cell specific surface antigens, or other markers, before the diverse fates of pole cells can be finally answered.

B. *Pole Cell Specific Nuclear Body*

Recently, an unique nuclear organelle has been described in pole cells and its properties are summarized in Table IV. Nuclear bodies are first found in newly formed pole cells (Allis, 1978) as electron-dense structures distinct from nucleoli. They increase in size and acquire a characteristic hollow-spherical shape (Fig. 6a). Species specific characteristics are evident (Mahowald et al., 1976). In *D. melanogaster* the

TABLE IV

Properties of Nuclear Bodies

1. Present in pole cells only[1]
2. Form in young pole cells; between 1 and 6 per cell[2]
3. Dependent upon polar cytoplasm for their formation[3]
4. Stable to 2M NaCl and to pH between 5 and 9[4]
5. Disperse at time pole cells leave posterior midgut[2]
6. Composed of protein[1]

[1]Mahowald, 1977
[2]Allis, 1978
[3]Mahowald et al., 1976
[4]Karrer, unpublished observations

organelles have a solid electron dense hull (Fig. 6a) and they do not aggregate. In *D. immigrans*, the hull contains many electron lucid regions (Fig. 6b) and 2-3 nuclear bodies are regularly found associated with each other. In *D. subobscura* the hull has similar discontinuities as in *D. immigrans* but nuclear bodies do not aggregate (Fig. 6c). Nuclear bodies attain their distinctive shapes in pole cells of blastoderm stage embryos, and then they begin a sequence of fragmentation until they are nearly absent from pole cells in 6.5 hour embryos, a time when pole cells are leaving the lumen of the midgut.

A number of interesting properties of these organelles are known. Nuclear bodies appear in pole cells which have been induced ectopically

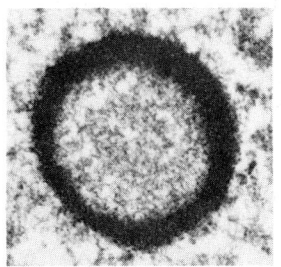 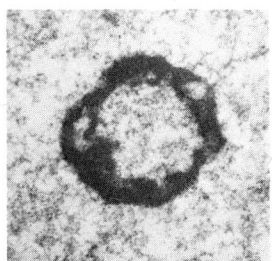 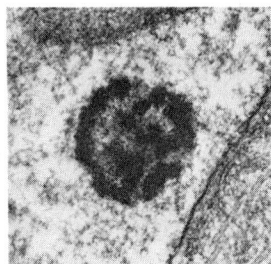

Fig. 6. Pole cell-specific nuclear body of (a) *D. melanogaster*, (b) *D. immigrans*, and (c) *D. subobscura*. x 40,000. (from Mahowald, 1977).

by polar plasm transplants (Illmensee and Mahowald, 1974; 1976). The morphology of the nuclear body in a pole cell induced in a *D. melanogaster* embryo by polar plasm from *D. immigrans* corresponds to that of *immigrans* (Mahowald *et al.*, 1976). Consequently, the nuclear body must depend upon the polar plasm for its formation and not the pole cell nucleus, even though it forms in the nucleus. Because of this dependence on the polar plasm, it is possible that the component proteins are either derived from polar granules or are synthesized from oogenetic mRNA localized in the polar plasm. It will be necessary to identify the protein constituents of nuclear bodies before we can determine their origin or function. We are currently attempting to purify these organelles.

C. *Pole Cells in Culture*

It is obvious that further understanding of the molecular events occurring during pole cell formation and germ cell determination will require the analysis of pole cells themselves. To this end we have developed procedures for obtaining large quantities of pole cells (Allis *et al.*, 1977). From these cells we have been able to identify the major polar granule protein (Waring *et al.*, 1978) and we have obtained fractions greatly enriched for nuclear bodies. However, if we wish to study the events leading to cellular determination, it is necessary to show that the isolated cells are duplicating cellular events *in vitro* that occur *in vivo*. Allis (1978) has studied the metabolic and growth properties of these cells during the first 6 hours in culture and found that they behave normally according to all criteria tested.

Somatic cells of the early gastrula have high levels of incorporation of precursors into both RNA and protein. Pole cells have similar levels of protein synthesis but were nearly devoid of RNA synthesis at the blastoderm stage although some RNA synthesis is detected a few hours later (Zalokar, 1976). Pole cells in culture duplicate these *in vivo* properties. Protein synthesis, as determined by quantitative autoradiography, is approximately the same for pole cells and the somatic cells which contaminate the pole cell culture. Autoradiography of ^{3}H-uridine incorporation indicates that initially in culture pole cells incorporate very little RNA relative to the somatic cells in the same culture but that by 6 hours in culture the pole cells have become active.

The third metabolic feature of pole cells concerns DNA replication and mitosis. In *D. melanogaster* pole cells cease mitosis by the blastoderm stage and do not begin again until they become incorporated into the embryonic gonad. In culture, we find that 30% of the pole cells incorporate ^{3}H-thymidine within the first hour in culture and that during the next five hours there is no further increase in incorporation. By utilizing colcemid-blocked mitosis we have found that the number of pole cells accumulating at metaphase rises rapidly for the first hour but after this time no further cells accumulate (Allis, 1978).

The fate of both polar granules and nuclear bodies were also studied in *in vitro* pole cells, and again the cells in culture remarkably paralleled the *in vivo* pole cells (Fig. 7). At 0 time 70% of the *in vitro* cells seen in thin section

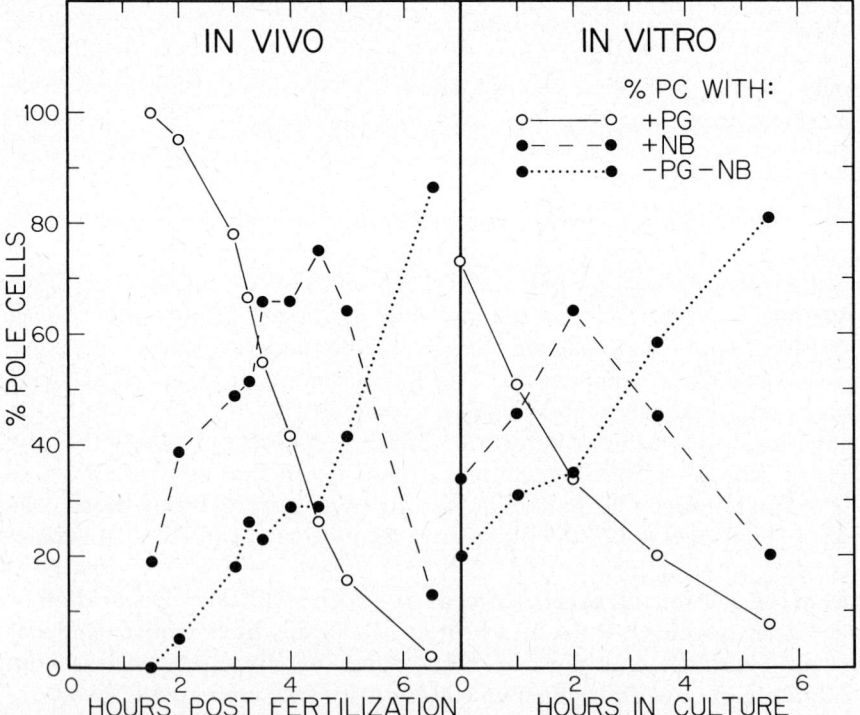

Fig. 7. Percentage of pole cells showing polar granules (PG) and/or nuclear bodies (NB) in random thin sections of embryos *(in vivo)* and of cells in culture *(in vitro)*. (From Allis, 1978).

had polar granules in the plane of section and this percentage dropped to less than 10% by six hours. Pole cell nuclei showing nuclear bodies rises from 32% at 0 time to 62% at 2 hours in culture and then drops to 20% at 6 hours. Analysis of the ultrastructural changes of pole cells *in vivo* remarkably parallel those seen *in vitro* (Allis, 1978). A comparison of this data with the similar data acquired for cells *in vivo* (Fig. 7) suggests two further points. Firstly, the cells in culture are at the 2.5 hour stage at 0 time in culture; secondly, the cells require 5.5 hours to reach the morphological state similar to that found *in vivo* at 6.5 hours.

The final test of normalcy in culture is biological function. Pole cells in culture (wild type genotype) have been transplanted back into $y\ sn^3mal$ embryos at the blastoderm stage utilizing the technique of van Deusen (1977). For control pole cells (i.e., cells taken directly from an embryo) the maximum efficiency for populating the germ line with pole cells is 50% because of the requirement that the germ cells be of the same sex as the host (van Deusen, 1977). Isolated populations of cells were always less efficient than controls (Table V) (Allis *et al.*, 1979) in spite of the fact that they were mixtures of both sexes. Nevertheless, the efficiency with which cultured pole cells could populate the germ line stayed constant in culture. Since pole cells in culture show the normal sequence of changes relating to age (e.g. Fig. 7), these data also suggest that pole cells from older embryos are still able to adapt to a blastoderm-stage embryo.

TABLE V

Capacity of Cultured Pole Cells to Function in vivo

Cells	# embryos receiving cells	% hatched	% enclosed	% fertile	% of fertile flies mosaic
Control[1]	106	77	53	46	40
Control[2]	98	60	38	35	0
0 time[3]	113	53	35	32	33
0.25 hr[3]	71	63	37	32	25
4 hr[3]	112	41	30	25	25

[1]Pole cells taken directly from another embryo.
[2]Portion of Renografin gradient containing less than 1% pole cells.
[3]Portion of Renografin gradient containing 60 - 70% pole cells.

In summary, Seecof and collaborators (1977) have previously shown that nerve and muscle precursor cells of the early gastrula faithfully continue their developmental program *in vitro*. Our data strongly suggest that pole cells also continue their *in vivo* program for the period studied. We have not yet extended our study to a later time, but we are now confident that a detailed analysis of both the synthesis of specific proteins, studied with 2-dimensional gel analysis (O'Farrell, 1975; O'Farrell et al., 1977), and specific RNAs studied after identifying specific DNA clones will make possible a molecular picture of these cells. This, in turn, will enable us to begin a study of the unique features occurring in pole cells which relate to germ cell determination.

IV. PERSPECTIVE

The understanding of the mechanisms involved in cellular determination will require not only an analysis of the ooplasmic components responsible for the cellular limitations but also the cellular changes consequent to the ooplasmic events. The germ plasm of *Drosophila* is clearly an excellent system for approaching these events. All of the required components for germ cell determination are produced during oogenesis and localized in the posterior polar plasm prior to fertilization. Techniques are available for manipulating the polar plasm by microinjection so that additional information will certainly be forthcoming. In addition, further screens for genetic mutations can readily be accomplished.

The important feature of the germ plasm for understanding determination is that we can isolate in sufficient quantities for molecular studies pole cells at the time when they are first determined. No other system offers both the cytoplasmic localization and the determined cells for analysis. We are currently testing the ability of pole cells which have been in culture for longer than six hours to function as germ cells. If pole cells do not lose this ability *in vitro*, then the molecular features of these cells should be relevant to understanding why these cells are germ cells. A major complexity, certainly, is the fact that probably 50% of the pole cells become portions of the gut. We have no idea at the present time how these two fates are determined. It is possible that pole cells are made in excess in normal situations and that only a limited number reach the gonads. The remainder respond to the endoderm environment and become midgut. If the number of pole cells has been diminished (by UV-irradiation or x-rays), then the surviving pole cells may preferentially reach the gonads.

Another feature of pole cells that is important in understanding determination is their surface properties. Whenever pole cells form, they are loosely associated with the surface. This suggests that these cells have low affinities for other cells. The development of pole cell-specific surface markers (e.g. antibodies) should make possible the analysis of the origin of these markers and their developmental fate. They may also provide a method of distinguishing between presumptive germ and gut cells.

We are at the point of being able to extend all aspects of this analysis to the molecular level. From such an analysis a detailed view of germ cell determination in *Drosophila* should be possible. The results obtained with this model system should facilitate the analysis of other examples of embryonic determination.

ACKNOWLEDGEMENTS

We wish to especially acknowledge the assistance of Dr. Allan Spradling in the RNA determination and the continued interest of Dr. Karl Illmensee in these studies. In addition, the faithful work of Sam Strait, Mary Lambton, Georgia Soltis, and Joan Caulton, in managing our mass cultures of flies is gratefully acknowledged. This work has been supported by NSF PCM 77-25427 and NIH HD-07983 to APM. C.D.A. is a predoctoral trainee (PHS-TOI-GM82), K.M.K. is a fellow of the Jane Coffin Childs Memorial Fund for Medical Research, E.M.U. is a predoctoral trainee (PHS T32 GM07227-04), and G.M.W. is an NIH postdoctoral fellow (HD-05302).

REFERENCES

Allis, C.D. (1978). "Isolation and Characterization of Pole Cells and Polar Granules from Embryos of *Drosophila melanogaster*". Ph.D. thesis, Indiana University. 132 pp.
Allis, C.D., Waring, G.L., and Mahowald, A.P. (1977). *Develop. Biol.* **56**, 372-381.
Allis, C.D., Underwood, E.M., and Mahowald, A.P. (1979). in press.
Beams, H.W. and Kessel, R.G. (1974). *Int. Rev. Cytol.* **39**, 413-479.
Bounoure, L. (1934). *Ann. Sci. Nat. 10e Ser* **17**, 67-248.
Bounoure, L. (1937). *C.R. Acad. Sci.* **204**, 1837-1839.
Childs, C.M. (1940). *Physiol. Zool.* **13**, 4-42.
Chung, H.-M., and Malacinski, G.M. (1975). *Proc. Nat. Acad. Sci. U.S.* **72**, 1235-1239.
Counce, S.J. (1963). *J. Morphol.* **112**, 129-145.
Curtis, A.S.G. (1962). *J. Embryol. Exp. Morphol.* **10**, 410-422.
Dohmen, M.R. and Verdonk, N.H. (1979). This volume.
Eddy, E.M. (1975). *Int. Rev. Cytol.* **43**, 229-280.
Fielding, C.F. (1967). *J. Embryol. Exp. Morphol.* **17**, 375-384.
Graziosi, G., and Marzari, R. (1976). *Wilhelm Roux' Arch.* **179**, 291-300.
Geigy, R. (1931). *Rev. Suisse Zool.* **38**, 187-288.
Hegner, R.W. (1908). *Biol. Bull,* **16**, 19-26.
Huettner, R.W. (1923). *J. Morphol.* **37**, 385-423.
Illmensee, K. (1976). In "Insect Development" (Lawrence, P., ed.), pp. 76-96. Symp. Roy. Entomol., Soc. London, No. 8. John Wiley & Sons, New York.
Illmensee, K., and Mahowald, A.P. (1974). *Proc. Nat. Acad. Sci. U.S.* **71**, 1016-1020.
Illmensee, K., and Mahowald, A.P. (1976). *Exp. Cell Res.* **97**, 127-140.

Illmensee, K., Mahowald, A.P., and Loomis, M.R. (1976). *Develop. Biol.* **49**, 40-65.
King, R.C. (1970). "Ovarian Development in *Drosophila melanogaster*". Academic Press, New York.
Mahowald, A.P. (1962). *J. Exp. Zool.* **151**, 210-215.
Mahowald, A.P. (1968). *J. Exp. Zool.* **167**, 237-262.
Mahowald, A.P. (1971a). *J. Exp. Zool.* **176**, 329-344.
Mahowald, A.P. (1971b). *J. Exp. Zool.* **176**, 345-352.
Mahowald, A.P. (1975). *Wilhelm Roux' Arch.* **176**, 223-240.
Mahowald, A.P. (1977). *Amer. Zool.* **17**, 551-563.
Mahowald, A.P. (1978). *In* "Mechanisms of Cell Change" (Ebert, J., ed.) John Wiley & Sons, N.Y., in press.
Mahowald, A.P. and Allis, C.D. (1978). in preparation.
Mahowald, A.P., Illmensee, K., and Turner, F.R. (1976). *J. Cell Biol.* **70**, 358-373.
Mahowald, A.P., Caulton, J.H., and Gehring, W.J. (1979). *Develop. Biol.*, in press.
Metschnikoff, E. (1866). *Z. Wiss. Zool.* **16**, 1-112.
Nissani, M. (1977). *Wilhelm Roux' Arch.* **182**, 203-211.
O'Farrell, P.H. (1975). *J. Biol. Chem.* **250**, 4007-4021.
O'Farrell, P.Z., Goodman, H.M., and O'Farrell, P.H. (1977). *Cell* **12**, 1133-1142.
Okada, M., Kleinman, I.A., and Schneiderman, H.A. (1974). *Develop. Biol.* **37**, 43-54.
Poulson, D.F. (1974). *Proc. Nat. Acad. Sci., U.S.* **33**, 182-184.
Poulson, D.F. (1950). *In* "Biology of Drosophila". (Demerec, M., ed.), pp. 168-274. John Wiley & Sons, New York.
Poulson, D.F., and Waterhouse, D.F. (1960). *Australian J. Biol. Sci.* **13**, 541-567.
Rabinowitz, M. (1941). *J. Morphol.* **69**, 1-49.
Sander, K. (1976). *Adv. Insect Physiol.* **12**, 125-238.
Schubiger, G., and Wood, W.J. (1977). *Amer. Zool.* **17**, 565-576.
Seecof, R.L. (1977). *Amer. Zool.* **17**, 577-584.
Smith, L.D. (1966). *Develop. Biol.* **14**, 330-347.
Smith, L.D., and Williams, M.A. (1975). *In* "Developmental Biology of Reproduction". (C.L. Market and J. Papaconstantinou, eds.), pp. 3-24. Academic Press, New York.
Sonnenblick, D.P. (1950). *In* "Biology of *Drosophila*". (Demerec, M. ed.), pp. 62-167. John Wiley & Sons, New York.
Spurway, H. (1946). *J. Genet.* **49**, 126-140.
Thierry-Mieg, D. (1976). *J. de Microscopie Biol. Cellulaire* **25**, 1-6.
Turner, F.R., and Mahowald, A.P. (1976). *Develop. Biol.* **50**, 95-108.
Turner, F.R., and Mahowald, A.P. (1977). *Develop. Biol.* **57**, 403-416.
Van Deusen, E.B. (1977). *J. Embryol. Exp. Morphol.* **37**, 173-185.
Waring, G.M., and Mahowald, A.P. (1978)., in preparation.
Waring, G.M., Allis, C.D., and Mahowald, A.P. (1978). *Develop. Biol.*, **66**, 197-206.
Wieschaus, E., and Szabad, J. (1978). *Develop. Biol.*,in press.
Wilson, E.B. (1928). "The Cell in Development and Heredity". 3rd ed. Macmillan Co. New York.
Zalokar, M. (1976). *Develop. Biol.* **49**, 425-437.

II. Maternal Effect Mutants of Development

Temperature Sensitive Maternal Effect Mutants of Early Development in *Caenorhabditis elegans*

David Hirsh

Department of Molecular, Cellular and Developmental Biology
University of Colorado
Boulder, Colorado 80309

I.   Introduction 149
II. Description of *C. elegans* 151
III. Isolation and Characterization of Zygote Defective Mutants 152
     A. Maternal Effects 153
     B. Critical Times of Temperature Sensitivity 155
IV. Morphology of Zygote Defective Mutants 160
V.   Conclusion 163
     References 165

I. INTRODUCTION

We have isolated temperature sensitive maternal effect mutants in the free-living nematode *Caenorhabditis elegans*. We use *C. elegans* for several basic reasons. It is easy to culture in the laboratory and it has a rapid life cycle. The genetics of *C. elegans* have been elucidated by Brenner (1974) and more recently have been refined by the lethal analysis studies of Herman *et al.* (1976, 1978). Both embryonic and postembryonic development can be observed directly and conveniently on the living worm with Nomarsky differential interference optics because egg shell and worm cuticle are transparent. The precise embryonic cell lineages of *C. elegans* are known from fertilization to the 200 blastomere stage (Deppe *et al.*, 1977; Nigon, 1949). All of the postembryonic somatic cell lineages are precisely known (Sulston and Horvitz, 1977; Kimble and Hirsh, 1978). It is very probable that soon the entire cell lineage tree for the somatic development from egg to adult will be known for *C. elegans.*

The use of *C. elegans* for studying development is also interesting in a historical context because many early studies on development were done with nematodes. While studying fertilization in nematodes, Van Beneden (1875) first observed that male and female gamete nuclei contribute equal chromosomal complements to zygote formation. Boveri (1899) traced embryonic lineages in *Ascaris* and described the stem cell pattern of the early divisions. Boveri also invoked the notion of determinants and provided evidence for germ line determinants in *Ascaris.* Considerable experimental evidence from *Drosophila* and amphibians, as well as *Ascaris,* enforces the belief that determinants are distinct entities that set the fate of the presumptive germ line cells (Smith, 1966; Illmensee and Mahowald, 1974). But what the molecular nature of those determinants is and whether they also exist for other embryonic lineages are still open questions. One way of probing the existence and nature of determinants is to ablate or modify them by mutation and try to analyze them genetically. This has been our approach.

If determinants exist that set the fates of embryonic cells exposed to them, it is reasonable that the determinants will be maternally donated to the egg. Therefore, a mutation that inactivates a determinant will be inherited as a maternal effect mutation. It is also reasonable to expect the maternally donated products to be proteins or mRNA's that would be translated later into proteins. If the determinants are proteins, then it should be possible to introduce temperature sensitive mutations into the genes coding for them. Temperature sensitive mutants have several advantages. They allow maintenance of lethal phenotypes because a clone can be kept at permissive temperature while a replica is tested for lethality at restrictive temperature. This is particularly advantageous for isolating maternal effect mutants. When a mutation arises, it is heterozygous in the first generation and homozygous in the second generation. The homozgotes survive because they come from a heterozygous mother. However, the adult homozygotes produce dead off-spring as the third generation. Therefore, the mutant phenotype is first recognized by the lethality. If the mutant is temperature sensitive, then a copy of the mutant exists at permissive temperature and can be propagated. Temperature sensitive phenotypes also are useful because time of gene function can be measured with temperature shift experiments (Suzuki, 1970; Hirsh and Vanderslice, 1976). We would expect determinants to function early in embryogenesis when basic lineages are being established.

Maternal effect lethal mutations that are not temperature sensitive can be isolated if genetic markers or balancer chromosomes are present so that the heterozygous sibling of the homozygous mutant can be maintained. These methods have been used in C. elegans by Wood and his co-workers (personal communication) and by Herman (1978).

II. DESCRIPTION OF C. ELEGANS

All of our studies, except where specifically noted, have used the Bristol strain of C. elegans. C. elegans is usually a self-fertilizing hermaphrodite but males occur spontaneously with a frequency of 1 in 700 animals (Hodgkin, 1974). Hermaphrodites are XX and contain 5 pairs of autosomes; males are XO. Males are used for performing genetic crosses. Hermaphrodites do not mate with each other. The genetics and the reproduction of C. elegans have been described by Brenner (1974), Herman et al. (1976), and Hirsh et al. (1976).

Adult C. elegans is about a millimeter long and contains both eggs and sperm. Oocytes bud off of the ovary and pass into the spermetheca where they are fertilized. Immediately after fertilization the vitelline membrane and chitinous shell form, sealing the embryo from the external world. The embryo is autonomous and can be dissected out of the uterus and will continue to develop. Embryos are impermeable to nearly all chemicals and the routine method for isolating large numbers of viable embryos is to dissolve the parents with 1% sodium hypochlorite — 0.5 N sodium hydroxide.

Fertilization to hatching takes approximately 12 hours and hatching to adulthood 45 hours at 25°. Embryonic cleavages begin almost immediately as the zygote passes down the uterus. At about the 50 blastomere stage, about two hours after fertilization, the embryo is laid. The remainder of embryogenesis occurs during the next 10 hours. A 250 μm first stage larva hatches from the egg shell. It grows in size and cell number during the next 45 hours before reaching adulthood. It undergoes four molts discarding its old cuticle at each molt. The majority of postembryonic cell proliferation is in gonadogenesis which occurs throughout juvenile growth. Upon hatching, the first stage larva contains 546 nongonadal somatic cells that increase to 808 cells in the adult. In addition, when the first stage larva hatches, it contains 4 primordial gonadal cells, two of which form the 143 somatic cells of the hermaphrodite reproductive system or the 56 somatic cells of the male reproductive system. The other two primordial gonadal cells form the

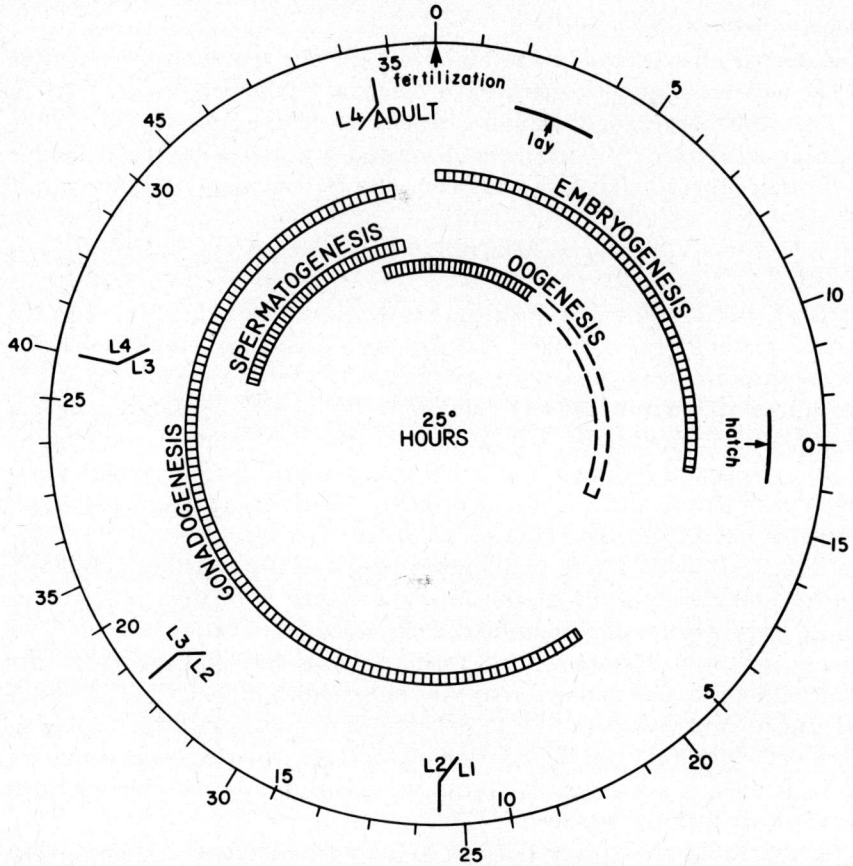

Fig. 1. Life cycle of *Caenorhabditis elegans*. The numbers on the outside of the circle indicate hours after fertilization at 25° and the numbers on the inside indicate hours after hatching. The larval stages L1, L2, L3 and L4 are separated by molts and the molts are designated at the times shown. Adults produce fertilized eggs for 4 days at 25° and oocytes for 8 days (Hirsh et al., 1976; Klass, 1977).

germ cells whose lineages are not known but would be interesting because *C. elegans* normally exists as a hermaphrodite which must segregate sperm from eggs during its development. The life cycle is shown in Fig. 1.

III. ISOLATION AND CHARACTERIZATION OF ZYGOTE DEFECTIVE MUTANTS

We isolated 223 temperature sensitive developmental mutants after screening approximately 7,500 F2 clones derived from EMS mutagenized parents. These mutants blocked at various places in the life cycle. Twenty-three of the mutants were classified as zygote defective mutants, abbreviated *zyg* mutants, because at the restrictive temperature, 25°, they never complete embryonic development and remain in the egg shell. Their permissive temperature is 16°. The naturally occuring Bergerac strain of *C. elegans* also displays a temperature sensitive *zyg* phenotype and was included in these studies.

All 24 zygote defective mutants are independent isolates. They are scattered among the six genetic linkage groups as determined by the standard genetic tests (Brenner, 1974). Genetic complementation tests between members of the same linkage groups show that one pair of mutants, B117 and B189, are allelic. All others complement each other. Most of the mutants are recessive, none are dominant but a few are semi-dominant.

A. *Maternal Effects*

We screened the 24 zygote defective mutants with three genetic tests to find the maternal effect mutants. These three tests are known as the self test (abbreviated S-test), the male rescue test (abbreviated R-test) and the heterozygous mutant male rescue test (abbreviated H-test). These tests have been described in greater detail elsewhere and they are discussed only briefly here (Hirsh et al., 1977; Wood et al., 1978). These tests are as follows:

S-test; $+/m\female \longrightarrow ++, +/m, m/m$
R-test; $+/+\male \times m/m\female \longrightarrow m/+$
H-test; $+/m\male \times m/m\female \longrightarrow m/+, m/m$

The S-test determines whether a + allele in a $+/m$ heterozygous mutant mother is sufficient to allow survival of the m/m homozygous mutant progeny. If the homozygous progeny survive, the mutant is classified as maternal or M in the S-test. If the homozygous mutant progeny do not survive, then the mutant is classified as non-maternal or N in the S-test. Twenty-one of the 24 independently isolated zygote defective mutants can be classified as maternal (M) in the S-test; that is, maternal gene expression is sufficient for zygote survival even if the gene is not expressed in the zygote.

The R-test asks whether sperm from a wild type male is sufficient to allow survival of $m/+$ heterozygous zygotes produced in an m/m homozygous mutant hermaphrodite. If the $m/+$ zygotes die, the mutant is classified as maternal or M in the R-test because the wild type sperm is not sufficient for survival. If the $m/+$ zygotes survive, the mutant is classified as nonmaternal or N in the R-test, indicating that either the paternally derived + allele or wild type sperm cytoplasm is sufficient for survival. The H-test is done to distinquish whether the wild type father contributed the + allele or a cytoplasmic factor (see below). In the R-test, half of the 24 *zyg* mutants were maternal (M) and half non-maternal (N). The combined results of the S and R-tests are shown in Fig. 2. Eleven mutants are strict maternal mutants (M,M). They strictly depend on maternal genome expression to survive. A mutation in a gene for an essential embryonic determinant would be in the M,M class.

Ten mutants are classified M,N; expression of either the maternal or the zygotic genome allows survival of the zygote. Two mutants are nonmaternal N,N. Only zygotic genome expression will allow survival of these two mutants. One mutant is N,M; it requires expression of both

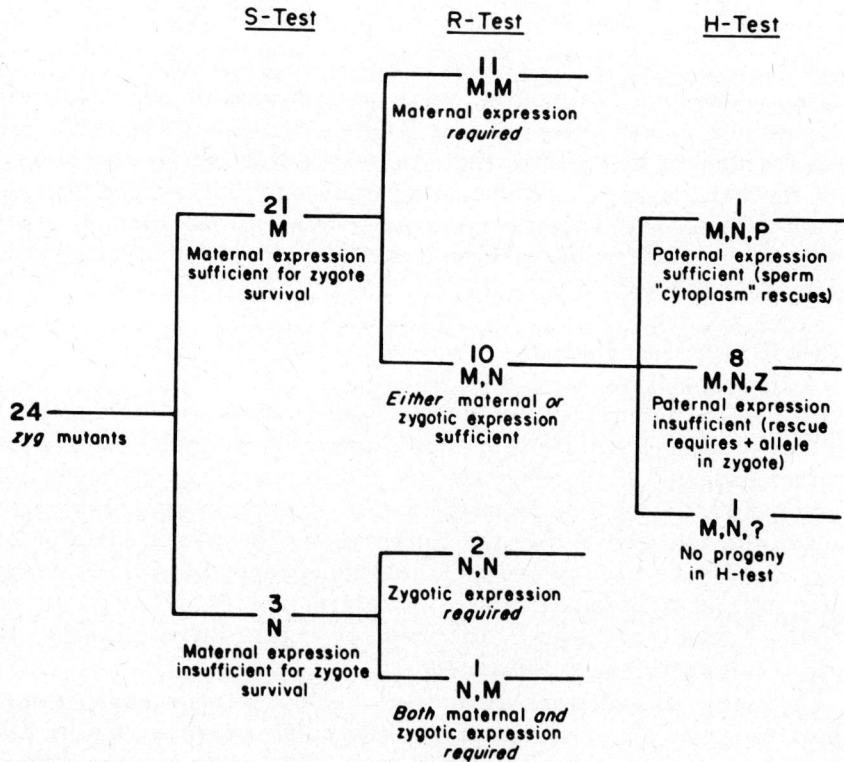

Fig. 2. Results of maternal effect tests on the zygote defective mutants.

the maternal and the zygotic genomes. Presumably a mutation in a gene product required continuously or in very large amounts for development would give a phenotype of this kind.

The H-test distinguishes whether the M,N mutants survived in the R-test because the wild type father contributed a + allele or wild type cytoplasm in the sperm. If only heterozygous progeny survive in the H-test, the + allele is responsible for survival. If both homozygous m/m mutant progeny and heterozygous $m/+$ progeny survive, a wild type cytoplasmic factor is responsible. Progeny of both genotypes survive in an H-test of the mutant, B235, indicating rescue by a paternal cytoplasmic factor. To know whether the egg also contains this cytoplasmic factor requires a reciprocal H-test cross (m/m ♂ x $m/+$ ♀) which is impossible because $b235/b235$ homozygous males are sterile at 25°. They make sperm but their tail morphology is abnormal. For eight M,N mutants only heterozygous mutant progeny survive in the H-test, indicating rescue by expression of the paternally derived + allele in the zygotes. One M, N mutant, B261, shows anomolous behavior in the H-test since it produces no progeny. It is being studied further.

B. *Critical Times of Temperature Sensitivity*

Temperature sensitive mutants can be analyzed with temperature shift experiments to define the critical time of temperature sensitivity, abbreviated *tcrit*. Different kinds of temperature shift data have been discussed previously (Hirsh and Vanderslice, 1976). Basically, a shift down experiment defines the latest time that the organism can be exposed to restrictive temperature and still be wild type. Or, conversely, it is the earliest time the organism can be exposed to restrictive temperature and show mutant phenotype. A shift up experiment defines the latest time the organism can be exposed to high temperature and display mutant phenotype. Or, conversely, it is the first time it can be shifted to restrictive temperature and give wild phenotype.

The *tcrit's* have been measured on 22 of the 24 *zyg* mutant strains. The *tcrit's* for embryos have been measured relative to the two cell stage because gravid hermaphrodites are reared at one temperature, two cell embryos are dissected from them, shifted at various times to the other temperature and then scored for survival. Alternatively, gravid hermaphrodites are reared at one temperature, shifted to another, two cell embryos dissected out and scored for survival. All of the times are converted to 25° hours by dividing the number of hours at 16° by two (Hirsh *et al.*, 1976; Byerly *et al.*, 1976). As shown below, it might not be correct to use a factor of two for some of the mutants.

Deppe *et al.* (1978) have described the embryonic cell lineages in *C. elegans* from fertilization to approximately 200 cells. We can approximate from their studies the stage of embryogenesis corresponding to the number of hours after 25° on our temperature shift graphs. Embryogenesis procedes at 20° approximately 75% as fast as it does at 25° (Byerly, *et al.* (1976); Hirsh, unpublished observations). We do not know which of several events to call gastrulation in *C. elegans* development; the major events that correspond to gastrulation in other organisms occur after the 24 cell stage in *C. elegans*. At the 24 cell stage which is approximately 70 mins. after the 2 cell stage, the two E cells move into the embryo and then divide. The P4 cell moves to fill the gap forming the blastopore. These initial migrations probably represent gastrulation in *C. elegans*. Other migrations occur at the 44 cell stage, approximately 85-90 mins. after the 2 cell stage, when the AB cell lineage descendents start to move inside the embryo. At this same time the MSt lineage cells start to move inside the embryonic cell mass. It is also worth noting that no nucleoli are visible at the 26 cell stage of the embryo indicating that zygotic ribosomal RNA synthesis has not begun yet and maternal ribosomes are probably being used.

Fig. 3 shows the results of temperature shift experiments on 22 of the 24 zygote defective mutants. All of the strict maternal mutants (M,M) change their responses to the temperature shifts during oogenesis and/or early embryogenesis. B101 is the only strict maternal mutant that remains temperature sensitive after the time of E cell migration but as discussed below, B101 appears to undergo embryonic cleavages more slowly than wild type.

The maternal effect mutants that were classified genetically as M,N,Z were interpreted as depending on either maternal or zygotic gene expression to survive. Most of these mutants change their responses to the temperature shifts after the time of E cell migration but there are exceptions. For example, mutants B185, B118, B220 and B188 display straight forward *tcrit's* but the mutant B185 is temperature sensitive for only 30 minutes one half hour after fertilization, indicating that the zygotic genome can be expressed before E cell migration.

The mutant B235, which is paternal (M,N,P), changes its reponse to temperature shifts early in its development and the mutant B261, which in the genetic tests shows anomalous behavior (M,N,?), is also anomalous in its *tcrit* because it is temperature sensitive before the completion of maturation of the parent and not afterwards. If the parent is reared at low temperature and embryos are shifted to high

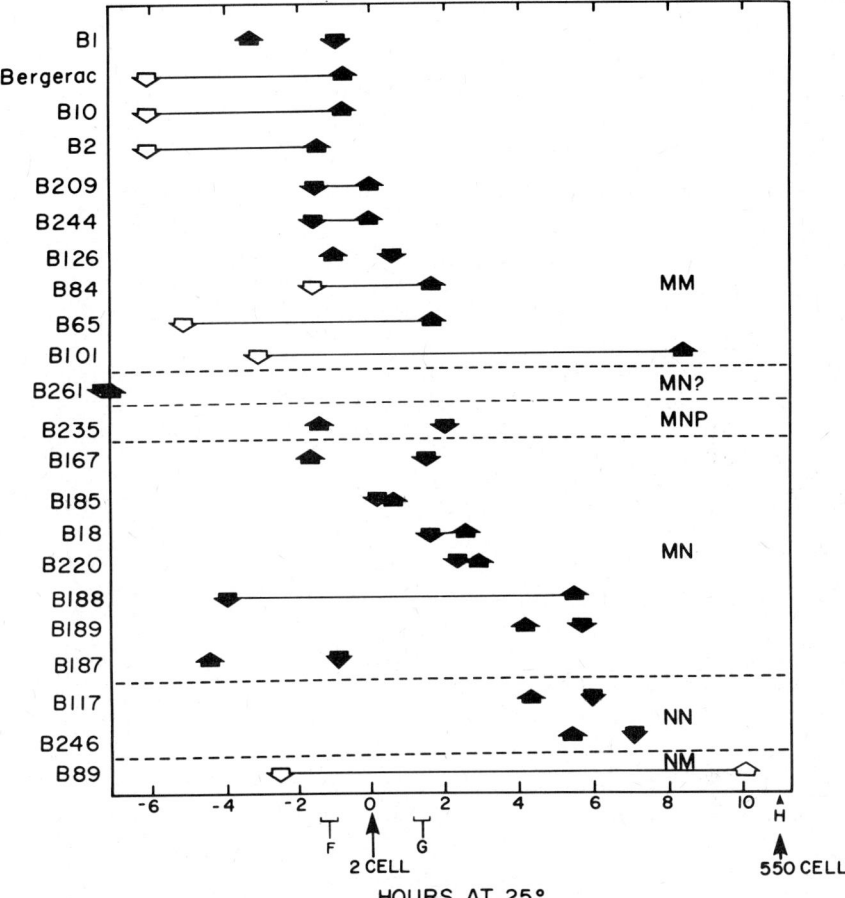

Fig. 3. The critical times of temperature sensitivity for zygote defective mutants. Arrows that point up designate the time when a change from mutant to wild phenotype occurs as a result of a shift up from permissive (16°) to restrictive (25°) temperatures. Arrows that point down designate the time when a shift down first causes a change from wild to mutant phenotype. The open arrows designate that the change in phenotype could begin earlier or end later than the times of the downward or upward arrows respectively. F designates the time of fertilization and G designates gastrulation which is taken as the time of the first blastomere migrations. H designates hatching of the larva which contains 550 cells. All the times are relative to normal embryogenesis at 25° with zero time taken as the time of the 2-cell embryo.

The dashed lines divide the mutant strains into the categories established from the maternal effect tests (see text and Fig. 2).

temperature, they are not temperature sensitive. The two non-maternal (N,N) mutants, B117 and B246, change their responses to temperature shifts late presumably reflecting the strict dependence on zygotic genome expression. Finally, the mutant, B89, which the genetic tests indicate requires both maternal and zygotic genome expression (N,M), displays a *tcrit* consistent with that classification because it is temperature sensitive throughout embryogenesis. Whenever it is shifted to restrictive temperature it stops development.

In all of our experiments, we assume that the temperature sensitive element is a protein. The *tcrit* measures when a temperature sensitive protein is required. Temperature sensitive proteins can either be temperature sensitive for synthesis (TSS) or thermal labile (TL) (Sadler and Novick, 1965; Jarvick and Botstein, 1975). A TSS protein made at low temperature is stable if exposed to high temperature after synthesis but a TL protein is temperature sensitive anytime.

In general, it is straightforward to interpret the *tcrit*'s when the shift down curve precedes the shift up curve; the shift down curve defines the beginning of the *tcrit* and the shift up curve defines the end. However, several of the mutants respond with a shift up curve that precedes a shift down curve. Three of the mutants classified as M,N,Z have *tcrit*'s when the shift up curve precedes the shift down curve. Two strict maternal mutants, B1 and B126, and both non-maternal (N,N) mutants also show this kind of *tcrit*.

There are three reasonable explanations for a shift up curve preceding a shift down curve. One is that the ratio of the rate of development at the restrictive temperature to the rate at the permissive temperature is not two in the mutants as it is in wild type. If the ratio is less than two, the curves can change their positions relative to each other. The ratio of developmental rates might be less than two because either the developmental rate at 16° is faster than wild type or, as is more likely with these mutants, the developmental rate at 25° is slower than wild type, particularly just before the lethal phenotype sets in.

A second possible explanation of this kind of *tcrit* is that excess TSS protein is synthesized in the interval between the shift up and the shift down (Hirsh and Vanderslice, 1976). An early shift up induces mutant phenotype. As the shifts up are progressively later, a point is reached where some protein synthesis has occurred at low temperature before the shift up. If the protein is TSS, then it is active because it was made at low temperature and if the protein is made in excess, then this small amount of synthesis is sufficient to produce wild phenotype. If the shift down occurs after the total period of protein synthesis, the animal will show mutant phenotype. However, if the shift down occurs just prior to the end of the protein synthetic period, then a small amount of new protein will be made at low temperature. Since the protein is normally made in excess, the small amount is enough to produce wild phenotype.

A third interpretation for a shift up curve preceding a shift down curve is particularly applicable to the interpretation of the M,N,Z mutants, which survive as a result of either maternal or zygotic genome expression. In these mutants, a TSS protein could be translated from maternal mRNA before fertilization and translated from zygotic mRNA after fertilization. Thus, each mutant would have two *tcrit's*, an early one corresponding to synthesis from maternal message, and a later one corresponding to the period of zygotic expression. In a heterozygous $m/+$ zygote derived from a homozygous mother, the zygotic genome expression of the wild type allele would provide wild type protein although maternally derived proteins would be denatured at high temperature.

It is possible to obtain further information about the *zyg* mutants by combining temperature shift responses with the known maternal effect classifications. By definition, the strict maternal genes represented by the M,M mutants must be transcribed maternally. There are a limited number of possible ways a maternal mRNA and the protein translated from it could act in development. The maternal mRNA could be translated before fertilization and the protein put into the egg or the mRNA could be translated in the oocyte or translation could be delayed until after fertilzation. A protein synthesized before fertilization could function before or after fertilization. If a protein is needed for oocyte maturation, it functions before fertilization; if a protein is an embryonic determinant, it functions after fertilization. Each of the possibilities for the appearance of a protein coded by a strictly maternal gene (M,M) can be assigned an expected *tcrit* for the protein being either temperature sensitive for synthesis or thermolabile. These expected *tcrit's* are shown in Fig. 4. The *tcrit's* of the M,M mutants correspond to some of these possible cases. The mutants, B1, B2, B10 and Bergerac, could represent TSS proteins needed for oocyte maturation or they could represent determinants put into the egg as TSS proteins. If the mutants B84 and B65 represent determinants, they must be put into the egg as proteins which are thermolabile. In contrast, B209, B224 and B126 could represent proteins that are synthesized after fertilization and therefore the mRNA's must be put into the egg by the mother. The proteins could either be TSS or TL. Therefore, it is possible to distinguish which mutants represent maternally derived proteins and which mutants represent maternally derived messages that are then translated in the zygote. The *tcrit* of B101 leaves many possible interpretations but it is clear that the protein is present during oogenesis because the mutant is temperature sensitive then as well as later. There is no case yet of a strict

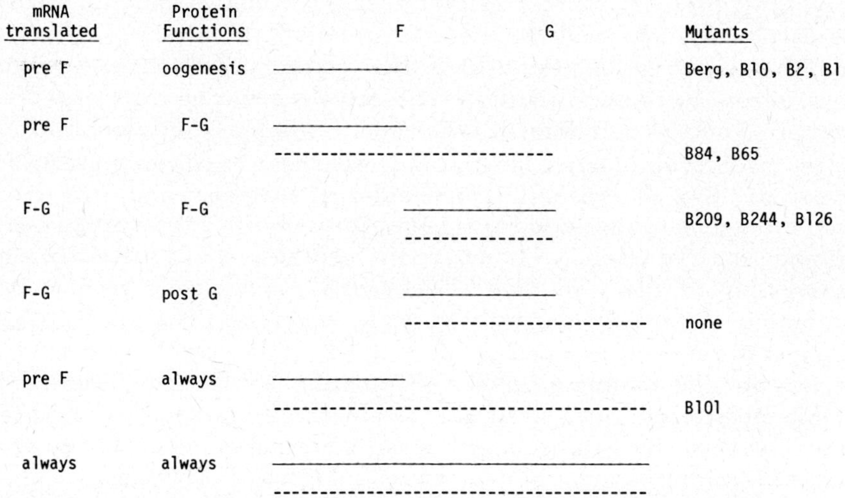

mRNA translated	Protein Functions	F	G	Mutants
pre F	oogenesis	——————		Berg, B10, B2, B1
pre F	F-G	———————		
		- - - - - - - - - - - - - - - - - - - -	- - - -	B84, B65
F-G	F-G		———————	
			- - - - - - - -	B209, B244, B126
F-G	post G		———————	
			- - - - - - - - - - - - - -	none
pre F	always	———————		
		- -	- - - -	B101
always	always	———————————————	———	
		- -	- - - -	

Fig. 4. The expected times of temperature sensitivity for strict maternal mutants as a function of when the maternal mRNA is translated and when the temperature sensitive protein functions. The expected $tcrit's$ are shown for proteins that are thermolabile (----) and proteins that are temperature sensitive for synthesis (—). Developmental time is from left to right with F designating the time of fertilization and G the time of gastrulation. The *zyg* mutants that have $tcrit's$ that correspond to the expected cases are listed on the right.

maternal mutant that is temperature sensitive from fertilization beyond gastrulation which would be the case of a TSS protein synthesized between fertilization and gastrulation and functioning after fertilization. These interpretations of the mutants make some simplifying assumptions, for example that the proteins do not exist as temperature insensitive zymogens. Even with these assumptions, these models are helpful for designing further tests with the maternal mutants.

IV. MORPHOLOGY OF ZYGOTE DEFECTIVE MUTANTS

We have examined the embryonic phenotypes of the *zyg* mutants in two ways. We have examined the early embryonic cleavages by direct observation with Nomarsky differential interference optics. Nematodes were shifted to restrictive temperature for 15 hours and then all of the zygotes were dissected out and observed microscopically at 25°. The morphology and timing of the first few cleavages were followed and compared to those of wild type embryos. We have also scored the number of nuclei to determine how far mutant embryos progress at restrictive temperature. Mutant worms were shifted to restrictive temperature for 15 hours, transferred to new plates and allowed to lay embryos for two hours. The adult worms were removed and dissolved with sodium hypochlorite-sodium hydroxide in order to isolate the *in utero* embryos which are in the one to fifty blastomere stages. Some of the embryos on the plates were harvested every two hours thereafter. Embryos were fixed and stained with Hoechst 33258 fluorescent dye which allows clear visualization of the nuclei. The nuclei in one hundred embryos were scored for each two hour interval of embryogenesis. A few mutants could not be scored this way because the adults hold onto the embryos at restrictive temperature. Therefore, these mutants were kept at 25° for 25 hours and batches of embryos were isolated, fixed, stained and examined and the distribution of embryonic stages scored within the total population.

The results of these examinations can be summarized by the following statements. All of the strict maternal mutants except the naturally occuring variant, Bergerac, have either abnormal early cleavages and/or stop dividing before the 100 nuclei stage. All mutants classified as M,N and N,N in the genetic tests proceed to advanced developmental stages before dying but two of the M,N mutants, the paternal mutant, B235, and the anomalous mutant, B261, also display abnormal early cleavages. No other M,N mutants produce abnormal early cleavages. It is interesting that the only mutants that are not strict maternal mutants that show abnormal early cleavages are these two mutants that also display non-standard behavior in the genetic tests and are also temperature sensitive early in development. The mutant B89, which is N,M and is continuously temperature sensitive, halts cleavage and degenerates rapidly whenever it is shifted to restrictive temperature.

The abnormal early cleavages are shown in Fig. 5 and a composite diagram of the different abnormal cleavages is shown in Fig. 6. The wild type first cleavage is transverse and asymmetrical. The larger cell is the AB blastomere. It undergoes a series of symmetrical divisions to give rise to cells that contribute to the hypodermis and the nervous system. The smaller cell is the P1 blastomere. It serves as a stem cell that gives rise to the five other basic embryonic lineages, E, MSt, C, D and the gonadal lineage (Deppe *et al.*, 1978). Some of the mutant zygotes shown in Figs. 5 and 6 undergo first cleavages that divide the egg into more than two parts. For example, the initial cleavage of B65 produces three blastomeres and the first cleavage of B235 gives three and sometimes four cells. The mutants B209 and B235 are polynuclear before cleavages start. The misoriented cleavage planes, such as the longitudinal cleavage plane of B244, are intriguing. Occasionally, one observes anuclear blastomeres that appear to be segregated from the others, as in B209. We have never

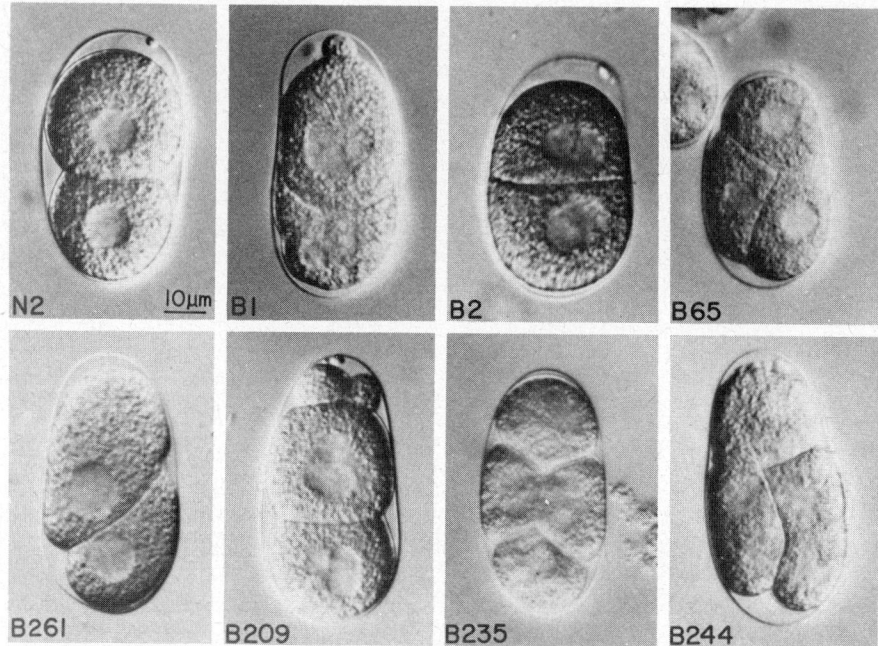

Fig. 5. Early cleavages of a wild type (N2) embryo and embryos from *zyg* mutants at 25°.

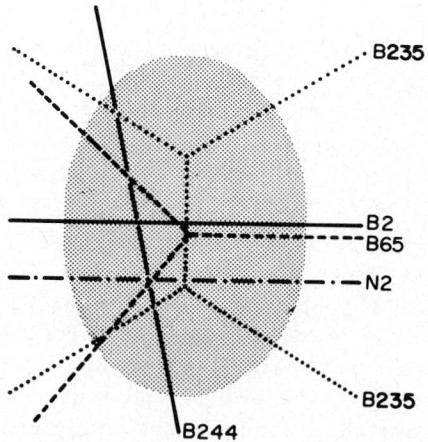

Fig. 6. Composite diagram showing the initial cleavage planes in embryos of several *zyg* mutants.

observed continued embryonic cleavages without karyokinesis, but there are mutants, such as B1, that continue karyokinesis in the absence of cytokinesis (Vanderslice and Hirsh, 1976). If an abnormal early cleavage occurs, it does not preclude continued nuclear division by that mutant. B244 is a good example because it cleaves initially longitudinally but proceeds to have more than 200 nuclei.

V. CONCLUSION

We now have isolated strict maternal mutants that are temperature sensitive early in development, display consistently abnormal early cleavages, and die before completing embryogenesis. We can make educated guesses as to which strict maternal mutants depend on receiving maternal mRNA's and which depend on receiving maternal proteins. These criteria fit what might be expected for a mutation in an embryonic determinant. However, as yet, there is no evidence that these mutations or the abnormal cleavages that result from the mutations disrupt a specific cell lineage or the differentiated fates of a subset of cells while all of the other cells continue their normal lineages and differentiation. The embryonic lineages are known in *C. elegans*, but assays are needed for the differentiated states of the different cells. In one instance, the nematode itself has provided a visible indicator of differentiation. When the embryo consists of 100 cells, the gut or E cell lineage has 8 cells that are fluorescent and birefringent. In the adult nematode, the 32 intestinal cells also are fluorescent and birefringent. John Laufer and Paolo Bazzicalupo (personal communication) have done experiments similar to those of Whittaker's (1973) to show that this fluorescent and birefringent signal of gut cell differentiation is confined to blastomeres of the appropriate lineage even in the absence of cell division. It will be particularly interesting to see what is the distribution of gut cell fluorescence and birefringence in the *ts* maternal effect mutants that have abnormal cleavages. We will want to examine the patterns of the differentiated states in several cell lineages after a few divisions in the mutants and compare them with wild type. Perhaps if we had more assays of differentiated states, we could detect if only one lineage is disrupted or if cells take on differentiated phenotypes that are not normally seen in them but are normally found in other cells in the embryo, implying that there had been some mutationally induced rearrangement of determinants. The only case described in experimental developmental studies thus far that seems to fit this description of a mutation that specifically affects determinants is the grandchildless mutant of *Drosophila* (Fielding, 1967). Grandchildless might be the only nonlethal determinant mutant because it disrupts only the final step in the developmental hierarchy, namely the germ line formation, so that other aspects of development remain unaffected. Many other known mutants in other organisms fit our strict maternal (M,M) classification. Zalokar *et al.* (1975) and Gans *et al.* (1975) surveyed a large group of female sterile mutants in *Drosophila* as did Bakken (1973). Rice and Garen (1975) reported on three strict maternal effect lethal blastoderm mutants. The ova deficient mutant in Axolotl blocks embryogenesis at the onset of gastrulation and is a strict maternal mutant (Humphrey, 1966). The stability of the o^+ state has raised hopes for it being an embryonic determinant in the classical sense (Brothers, 1976). But none of these numerous mutations has been shown to affect differentiation or cell lineages in a specific way that would be expected for a mutation in an embryonic determinant. The embryonic lethal mutant in *Drosophila*, rudimentary, that would be M,N in our scheme, provides an important lesson. The rudimentary mutation is in aspartic transcarbamylase which is a common anabolic enzyme (Norby, 1973). Mutations in metabolic enzymes may be what all of our maternal effect *zyg* mutants, as well as non-maternal *zyg* mutants, represent. In this regard, it is noteworthy that nearly all of the *zyg* mutants described here also have postembryonic temperature sensitive phenotypes, some with discrete periods of temperature sensitivity. This suggests that most of the mutants represent proteins that function in postembryonic development as well as embryonic development. This behavior would not be expected for proteins acting solely as embryonic determinants. However, the strict maternal mutants B2 and B244, the paternal mutant B235, and the anomalous mutant B261 have no phenotypes other than the embryonic lethality. Perhaps further studies on these mutants will resolve whether any of them contains a mutated determinant.

ACKNOWLEDGEMENTS

I am indebted to Drs. William B. Wood and Ralph Hecht for help with some of the experiments described here and for many stimulating discussions. I also am grateful to Steven Carr, Kimberly Johnson and Rebecca Vanderslice for help with the experiments. This work was supported by grants GM 19851 and GM 70465 from the US Public Health Service.

REFERENCES

Bakken, A.H. (1973). *Develop. Biol.* **33**, 100-122.
Boveri, T. (1899). Die Entwicklung von Ascaris megalocephala mit besonderer Rucksicht auf die Kernverhaltnisse. Gustav Fischer, Jena.
Brenner, S. (1974). *Genetics* **77**, 71-94.
Brothers, A.J. (1976). *Nature* **260**, 112-115.
Byerly, L., Cassada, R.C., and Russell, R.L. (1976). *Develop. Biol.* **51**, 23-33.
Deppe, U., Schierenberg, E., Cole, T., Krieg, C., Schmitt, D., Yoder, B., and von Ehrenstein, G. (1978). *Proc. Nat. Acad. Sci. U.S.* **75**, 376-380.
Fielding, C. (1967). *J. Embryol. Exp. Morphol.* **17**, 375-384.
Gans, M., Audit, C. and Masson, M. (1975). *Genetics* **81**, 683-704.
Herman, R.K., Albertson, D.G. and Brenner, S. (1976). *Genetics* **83**, 91-105.
Herman, R.K. (1978). *Genetics* **88**, 49-65.
Hirsh, D. and Vanderslice, R. (1976). *Develop. Biol.* **49**, 220-235.
Hirsh, D., Oppenheim, D., and Klass, M. (1976). *Develop. Biol.* **49**, 200-219.
Hirsh, D., Wood, W.B., Hecht, R., Carr, S. and Vanderslice, R. (1977). In "Molecular Biology of Eukaryotic Systems" (Abelson, J.N., Wilcox, G. and Fox, C.F., eds.), pp. 347-356. Academic Press, New York.
Hodgkin, J.A. (1974). Genetic and anatomical aspects of the *Caenorhabditis elegans* male, University of Cambridge Thesis.
Humphrey, R.R. (1966). *Develop. Biol.* **13**, 57-76.
Illmensee, K. and Mahowald, A.P. (1974). *Proc. Nat. Acad. Sci. U.S.* **71**, 1016-1020.
Jarvik, J. and Botstein, D. (1973). *Proc. Nat. Acad. Sci. U.S.* **70**, 2046-2050.
Klass, M.R. (1977). *Mech. Aging and Develop.* **6**, 413-429.
Kimble, J. and Hirsh, D. (1979). *Develop. Biol.*, in press.
Nigon, V. (1949). *Ann. Sci. Nat.* 11e serie, II.
Norby, S. (1973). *Hereditas* **73**, 11-16.
Rice, T.B. and Garen, A. (1975). *Develop. Biol.* **43**, 277-286.
Sadler, J.R. and Novick, A. (1965). *J. Mol. Biol.* **12**, 305-327.
Smith, L.D. (1966). *Develop. Biol.* **14**, 330-343.
Sulston, J.E. and Horvitz, H.R. (1977). *Develop. Biol.* **56**, 110-156.
Suzuki, D.T. (1970). *Science* **170**, 695-706.
Van Beneden, E. (1883). *Arch. Biol.* **4**, 265.
Vanderslice, R. and Hirsh, D. (1976). *Develop. Biol.* **49**, 236-249.
Whittaker, J.R. (1973). *Proc. Nat. Acad. Sci. U.S.* **70**, 2096-2100.
Wood, W.B., Hecht, R., Carr, S., Wolf, N., Vanderslice, R., and Hirsh, D., in preparation.
Zalokar, M., Audit, C., and Erk. I. (1975). *Develop. Biol.* **47**, 419-432.

A Specific Case of Genetic Control of Early Development: the *o* Maternal Effect Mutation of the Mexican Axolotl

Ann Janice Brothers
Department of Zoology
University of California
Berkeley, California 94720

I. Introduction 167
II. Description of the Phenotype of the *o* (for Ova Deficient) Maternal Effect Mutation...............................168
  A. Correction of the Gastrular Arrest 169
  B. Characteristics of the o+ Substance Synthesis 170
  C. Biochemical and Cytological Characterization of the Mutant Eggs 171
III. Stability of Nuclear Activation 173
IV. Characterization of the Correction of the Mutant Phenotype 177
V. Retention of the Capacity to Interact with the o+ Substance 178
VI. Discussion of the Possible Mode of Action of the o+ Substance 178
  References 182

I. INTRODUCTION

The nature of the control of gene action during development is one of the central problems in Developmental Biology. Morphogenetic substances apparently are synthesized during oogenesis, stored in the egg and are eventually arranged in a pattern which acts after fertilization to control the development of the zygote into the differentiated multicellular organism (Morgan, 1927; Illmensee and Mahowald, 1974; Davidson, 1976). This control must involve an interaction between the zygote nucleus and the components of the egg cytoplasm. The nature of these morphogenetic substances and the manner in which they function are unknown. Genes which exert maternal effects through modifications of the egg cytoplasm are therefore of special interest, since they provide a means of approaching the problem of how the egg cytoplasm acts to control early development (Briggs, 1973).

II. DESCRIPTION OF THE PHENOTYPE OF THE o (FOR OVA DEFICIENT) MATERNAL EFFECT MUTATION

A simple recessive maternal effect mutant (o, for ova deficient) was discovered in the Mexican axolotl (Humphrey, 1966). Females homo-

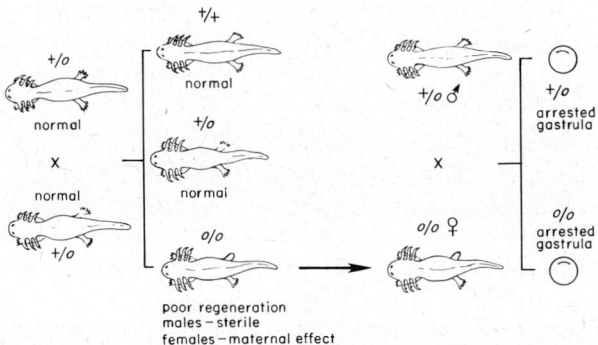

Fig. 1. A maternal effect mutation (o, for ova deficient). Matings between heterozygotes (+/o) produce offspring of the expected genotypes. Of these, the +/+ and +/o genotypes are completely normal. The homozygous recessives (o/o) also are normal through embryonic and larval life. During juvenile development the homozygous recessives generally exhibit a reduced ability to regenerate limbs, and may grow more slowly than their normal sibs. Adult homozygous mutant males remain more or less juvenile with respect to secondary sexual characteristics and the testes remain small, with the majority of the germ cells in the spermatogonial stage. Germ cells display degeneration, usually in spermatogonial stage. Adult homozygous mutant females have normal appearing ovaries and produce mature eggs indistinguishable in appearance from eggs produced by a normal female. Eggs spawned by a homozygous female can be fertilized with sperm from either a normal (+/+) or heterozygous (+/o) male. The cleavage of the mutant eggs appears completely normal until late blastula stage, approximately 1 day after fertilization. Then the cleavage rate drops and the embryos enter gastrulation with cells considerably larger than those of normal embryos at the same developmental stage. Mutant embryos may proceed part way through gastrulation, but usually not beyond the crescentric blastopore stage, where they arrest. In the spawnings from one female the mutant embryos proceeded through gastrulation before arresting. In no instance have mutant embryos formed neural folds (Humphrey, personal communication; Briggs, 1972). This gastrular arrest of mutant eggs occurs whether or not the normal allele was introduced by the sperm at fertilization.

zygous for *o* produce eggs which exhibit normal response to fertilization and are indistinguishable from normal fertilized eggs during cleavage and early blastula stages (Fig. 1). At mid- to late-blastula stages the mutant eggs (eggs spawned by an *o/o* female) show a slowing of cell division (Carroll, 1974). The eggs will form a dorsal lip but always stop developing at some point during gastrulation. All eggs spawned by one female will arrest at the same point, but eggs spawned by another female may arrest at a slightly different point in gastrulation. The mutant eggs never differentiate neural folds or other axial organs, regardless of whether or not the normal allele is introduced by the sperm at fertilization. Ovary transplants have demonstrated that the genetic lesion occurs within the ovary itself and its expression is not affected by the maternal environment (Humphrey, 1966).

Animal hemisphere cells from mutant blastulae grafted onto normal recipient blastulae show no correction of the mutant phenotype. The host embryo undergoes gastrulation and organogenesis but the grafted mutant cells do not participate in the normal development of the recipient embryo (Cassens, 1968).

A. *Correction of the Gastrular Arrest*

These data suggest a gene-controlled modification of the egg cytoplasm which acts to affect fundamental events in early morphogenesis. Mutant eggs exhibit a cytoplasmic deficiency which can be corrected by the injection of plasm from normal oocytes or eggs

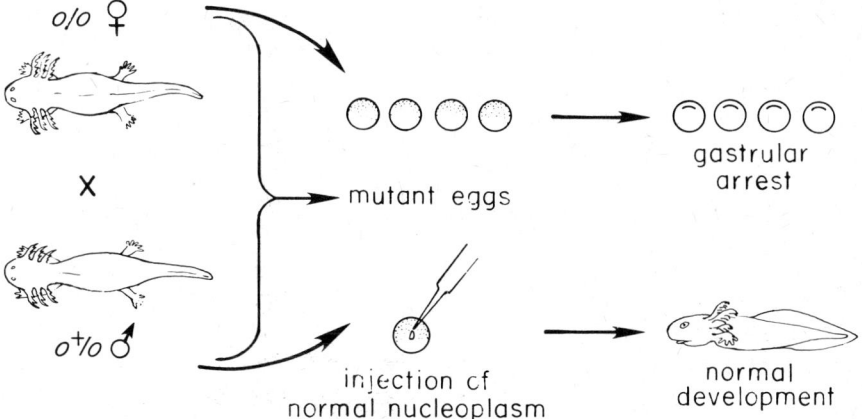

Fig. 2. Females heterozygous for *o* produce eggs which always exhibit gastrular arrest. This arrest can be corrected by the injection of normal mature egg cytoplasm or the germinal vesicle nuclear sap from normal ovarian oocytes.

(Briggs and Cassens, 1966). Injection of nucleoplasm from a normal oocyte nucleus (germinal vesicle) into a mutant egg before first cleavage completely corrects the arrest at gastrulation (Fig. 2). Injection of normal egg cytoplasm into only one blastomere of a mutant egg after first cleavage results in correction of the injected side only. The uninjected side shows the mutant phenotype, gastrular arrest, while the injected side forms neural folds and axial structures. This shows that the active factor cannot diffuse between the blastomeres.

The injection of cytoplasm from mutant eggs (Cassens, 1968) or the nucleoplasm from mutant germinal vesicles (Brothers, 1976) into normal fertilized eggs produces no affect on normal development, demonstrating that the genetic lesion does not involve an inhibitory action of some component present in mutant eggs or oocytes. The results of the germinal vesicle and cytoplasmic injections are summarized in Table I.

TABLE I

Injection of Cytoplasm or Germinal Vesicle Nucleoplasm

		developmental stage		
		blastula	neurula	larva
a.	mutant eggs	+	−	−
b.	normal eggs	+	+	+
c.	mutant eggs injected with cytoplasm from normal eggs	+	+	+
d.	mutant eggs injected with nucleoplasm from normal oocyte germinal vesicles	+	+	+
e.	normal eggs injected with cytoplasm from mutant eggs	+	+	+
f.	normal eggs injected with nucleoplasm from mutant oocyte germinal vesicles	+	+	+

B. *Characteristics of the o+ Substance Synthesis*

During oogenesis the normal allele of the *o* gene seems to direct the synthesis of an *o*+ substance and this substance apparently is concentrated in the germinal vesicle. This *o*+ substance is present in the germinal vesicle by early "lampbrush" stage and its synthesis continues through most of oocyte growth. It can be recovered from the cytoplasm of a normal mature egg after the germinal vesicle breaks down, dispersing its contents into the oocyte cytoplasm. The active component is found in the germinal vesicles of oocytes of widely separated amphibian species (Table II). In all instances the substance appears to be functionally equivalent to the active substance present in axolotl germinal vesicles in the regulation of early embryonic development, since it acts to correct the arrest at gastrulation when injected into mutant eggs (Briggs, 1972).

TABLE II

Correction of o Maternal Effect by Nuclear Sap from Oocytes of Various Amphibian Species

Source of nuclear sap	Number of recipient mutant eggs	Complete Cleavage	Correction (neurulation)
Controls (no injection)	612	550	0
Rana pipiens	51	42	40
Rana catesbieana	20	17	17
Xenopus laevis	67	56	40
Hyla crucifer	23	17	10
Pseudacris triseriata	16	14	10

Nuclear sap was obtained from the germinal vesicles of fully grown ovarian oocytes and injected into fertilized eggs of *o* females. Volumes ranged from about 0.001 cu. mm to 0.005 cu. mm. Corrections to advanced embryonic stages were obtained at all dose levels. (From Briggs, 1972.)

The active factor can be recovered from axolotl germinal vesicles, mature oocyte cytoplasm, the cytoplasm of fertilized eggs and early cleavage stages. It is no longer present in detectable amounts after the late blastula stage, suggesting that it has become fixed to some cell structure, rendering it unextractable, or that it has been degraded. The active component also cannot be recovered from adult testis, liver or spleen (Briggs and Justus, 1968). Preliminary characterization of the $o+$ substance indicates that it may be of high molecular weight and that it depends upon a protein or proteins for its activity, since the active factor is sensitive to heat and to trypsin digestion, but insensitive to digestion with DNase (Briggs and Justus, 1968).

C. *Biochemical and Cytological Characterization of the Mutant Eggs*

The first detectable differences between the mutant and normal eggs appear at mid-blastula stage. During cleavage stages and early blastula stage the mutant and normal eggs exhibit similar patterns of protein synthesis, as compared by double isotope acrylamide gel electrophoresis experiments. By the late-blastula stage mutant eggs show a difference in their incorporation of the isotope, apparently retaining the pattern seen during earlier stages of development, whereas the normal late-blastula shows an increased incorporation of the isotope into some proteins (Malacinski, 1971). Mechanically enucleated amphibian eggs also show this failure to alter or modulate the pattern of isotope incorporation into newly synthesized proteins (Malacinski, 1972).

During blastulation, there is a sharp reduction in DNA synthesis in mutant embryos. By the mid-to-late-blastula stages the mutant embryos also show a drastic drop in the mitotic index and the incorporation of (^{3}H)-thymidine, as compared to normal embryos. Karyotypes of mutant and normal blastulae are similar and the pattern of cold induced secondary constrictions is essentially the same for mutant and normal blastulae. In mutant embryos the nucleolus may show a slight precociousness in the assumption of a spherical form, appearing as such at mid-blastulation, whereas the spherical nucleolar form is not observed in normal embryos until late blastulation. In mutant embryos there is also a tendency for the nucleolus to be retained at the nucleolar organizer locus through prophase of the mitotic cycle, whereas in normal embryos the nucleolus usually disappears in early prophase (Table III). Mutant embryos which are heterozygous (o+/o) for the o mutation exhibit one abnormally long nucleolar organizer. Since the nucleolar organizer is located near the end of the short arm of chromosome 4 this suggests that the o gene may be located on chromosome 4. It would be expected that if the gene were located on another chromosome, interchromosomal effects would act to cause both nucleolar organizers of the heterozygote to be abnormally long (Carroll, 1974).

TABLE III

Metaphase Nucleolar Retention[a]

Parental genotype (Female) (male)	Number of Spawnings	Number of embryos		
		No nucleoli	1 nucleolus	2 nucleoli
o+/o+ × o+/o+	28	84	7	0
o/o × o+/o+	22	6	153	0
o+/o × o+/o	5	8	17	9
o+/o × o+/o+	2	6	6	0

[a]From Carroll (1974)

Mutant and normal embryos show no autoradiographically detectable incorporation of ^{3}H-uridine during cleavage and early blastula stages. Normal late-blastulae exhibit intense incorporation of ^{3}H-uridine, whereas mutant embryos do not (Carroll, 1974). These results suggest that the mutant blastula may lack a functional genome. It is well established that development can proceed through cleavage and blastulation without a functional genome. Maternally inherited transcripts can provide sufficient information to support development up to gastrular stages; however, gastrulation and organogenesis do require the presence of a functional nucleus (Davidson, 1976; Gurdon, 1974). Apparently, by gastrulation, a majority of the transcripts on the polysomes are not of maternal origin, but are newly synthesized messenger RNAs (Nemer, 1975; Galau et al., 1976). Treatment of embryos during early to mid-blastulation with inhibitors of RNA synthesis (Guidice et al., 1968; Brachet and Denis, 1963), heat (Gilchrist, 1933) or by X-irradiation (Neyfakh, 1964) inhibits organogenesis. These experimental results have been interpreted to suggest that new transcripts required for later differentiation may be synthesized during blastulation.

The pattern of RNA synthesis in normal axolotl eggs is currently under examination (Malacinski and Brothers, 1978). It is expected that an analysis of the pattern of RNA synthesis in the mutant eggs will provide information about the effect of the o+ substance upon gene activity during development.

The data indicate that the o+ substance may be a protein(s) synthesized during oogenesis, stored in the germinal vesicle and then dispersed into the cytoplasm of the mature egg upon germinal vesicle breakdown. The o+ substance is no longer extractable from eggs after late-blastula stages, indicating that it may have become fixed to some subcellular particle or degraded. While the normal allele of the o gene is functional during oogenesis, it apparently does not function during early development. The presence of the o+ substance in the egg cytoplasm seems to be absolutely essential for the normal activation, during blastulation, of the nuclear genes required for gastrulation and organogenesis.

III. STABILITY OF NUCLEAR ACTIVATION

Functional analysis of how the o+ substance acts to control gene expression and whether this nucleo-cytoplasmic interaction acts to produce a stable alteration in the capacity of the nucleus to support development can be most rigorously tested by nuclear transplantation.

A test of the heritability of the nuclear activation is constructed by transplanting nuclei (which are exposed to the $o+$ substance in the normal egg cytoplasm) from various stages of normal blastulae into enucleated mutant eggs (which lack the $o+$ substance). If the activation is not heritable, then the recipient mutant should in all cases develop like typical mutant eggs and arrest at gastrulation. If the activation is stable and heritable then it should persist in the absence of the $o+$ substance. Therefore the activated nuclei but not the unactivated ones should promote normal development of the recipient mutant eggs (Fig. 3).

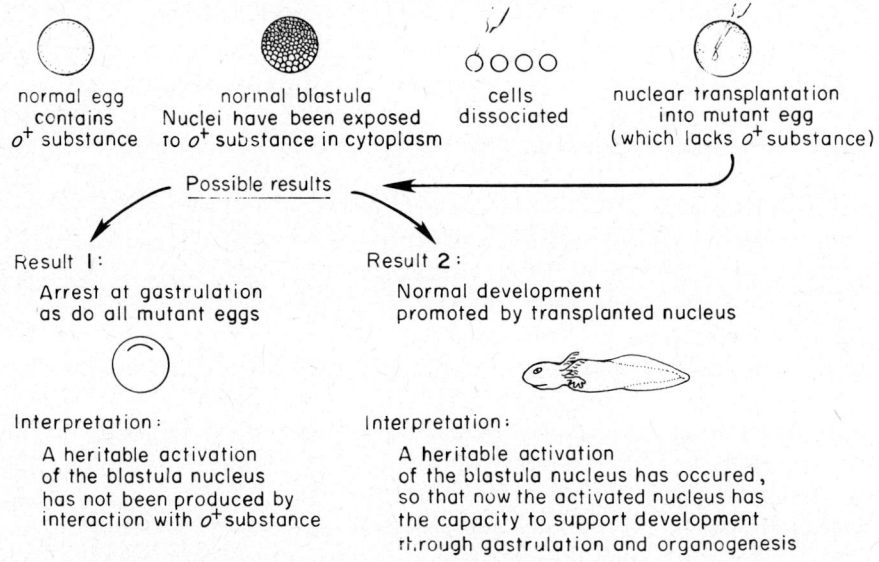

Fig. 3. Diagram of the nuclear transplantation experiments to test the heritability of nuclear activation.

The results (Table IV) of such a series of nuclear transplants demonstrate that nuclei of normal early blastulae, stages 8 to 8 1/4, cannot improve the development of the recipient enucleated mutant eggs — they all arrest at gastrulation. However, nuclei taken from normal mid-blastulae, stage 8 1/2, and from late-blastulae and gastrulae, are capable of promoting the normal development of the recipient enucleated mutant eggs — those nuclear transplants develop into fertile adults.

TABLE IV

Capacity of Normal Blastula Nuclei to Promote Development of Recipient Mutant Eggs[a]

Developmental stage of the nuclear donor	Total (100%)	Complete blastula	Number (%) of embryos surviving to each stage				Feeding larva (1 month post-hatching)
			Early-gastrula (stage 10)	Neurula	Post-neurula	Larva	
Fertile mutant eggs — no transplant	888	883 (99%)	883 (99%)	0			
Early blastula (stages 8-8¼)	234	130 (56%)	127 (54%)	0			
Mid-blastula (stage 8½)	87	41 (47%)	37 (42%)	25 (29%)	21 (24%)	18 (20%)	16 (18%)
Late mid-blastula (stage 8¾)	36	27 (48%)	24 (43%)	19 (34%)	16 (28%)	15 (27%)	13 (23%)
Late blastula (stage 9)	141	85 (60%)	69 (49%)	49 (35%)	46 (33%)	37 (25%)	28 (20%)
Late gastrula (stages 11½-12)	218	122 (56%)	75 (34%)	58 (27%)	43 (20%)	39 (18%)	25 (11%)
Control — transplants of nuclei from a normal blastula into an enucleated normal egg							
Early blastula (stage 8)	157	80 (51%)	75 (48%)	63 (40%)	54 (34%)	49 (31%)	37 (23%)
Late blastula (stage 9)	108	67 (62%)	66 (61%)	57 (53%)	48 (44%)	33 (30%)	29 (28%)
Mid-gastrula (stages 10½-11½)	98	56 (57%)	50 (51%)	36 (37%)	24 (24%)	22 (22%)	17 (17%)

[a] From Brothers (1976)

[b] The staging series used is a modification of the stage series described by Harrison[21] and Carroll[18]. The modifications used in this paper are based on the developmental stage, measured by the time after first cleavage, at 18°C. Stage 8=15-18 h. Stage 8¼=18-20 h. Stage 8½=20-22 h. Stage 8¾=22-24 h. Stage 9=24-29 h. The genetic markers used for the nuclear transplants were those governing pigmentation (d), (m) and in some instances recessive mutations affecting early development (f), (g) and a cell autonomous lethal (ut) (Briggs, 1973; Humphrey, 1975).

A definitive test of the mitotic heritability of the nuclear activation is provided by doing serial clonal nuclear transplants of activated nuclei. Enucleated mutant eggs are used for recipients at each transplant generation. Activated nuclei (taken from normal late-blastulae, stage 9) are transplanted into enucleated mutant eggs. Those nuclear transplants are allowed to develop to early-blastulae, stage 8. (Previous experiments had shown that stage 8 normal blastula nuclei were not capable of altering the mutant phenotype of the recipient mutant eggs; those stage 8, early-blastulae, had not yet shown a stable alteration in their capacity to promote normal development.) However, when the nuclei from the nuclear transplants of stage 9 donor normal blastulae are sacrificed at stage 8, early-blastula, and are again transplanted into a second generation of recipient enucleated mutant eggs, they can support the normal development of those mutant eggs (Fig. 4). Members of each of those serial transplant clones develop normally and some are raised to sexually mature adults. The spawnings from those animals develop normally and the genetic markers of the donor nuclei are expressed. These experiments demonstrate that the nucleus, once it has passed through the activation period, retains the capacity to support normal development through gastrulation and neurulation. This activation is a stable condition and is mitotically heritable for at least 30 mitotic generations (Brothers, 1976).

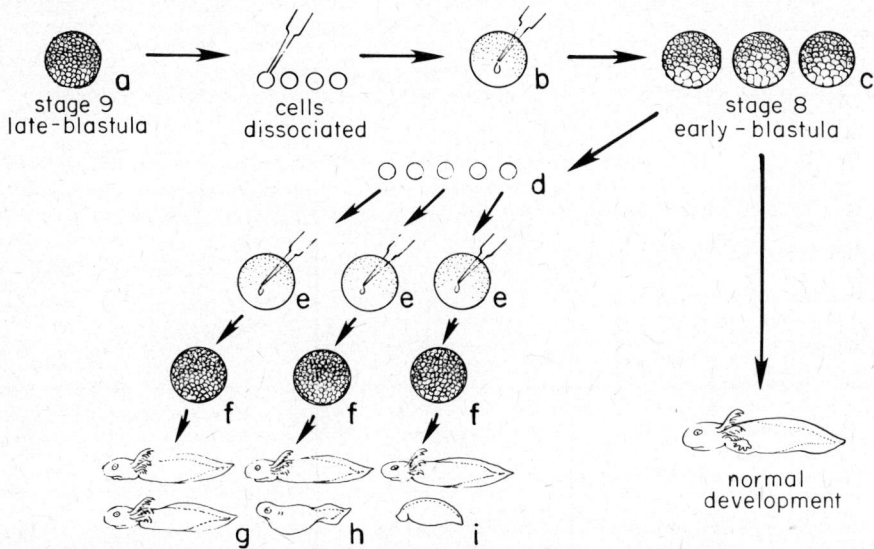

Fig. 4. Serial clonal nuclear transplants of activated nuclei using enucleated mutant eggs for recipients at each transplant generation. a, The nuclear donor is a normal stage 9, late blastula which shows intense ^{3}H-uridine incorporation in autoradiographs; b, the dissociated cells are used for nuclear transplants into an enucleated mutant egg; c, the nuclear transplants are allowed to develop to stage 8, early blastula stage. A few are sacrificed to serve as nuclear donors for the serial clonal transplants, and a portion of each nuclear donor blastula is taken for autoradiographic analysis. There is no detectable ^{3}H-uridine incorporation. d, The stage 8 nuclear donor blastula is dissociated in Ca^{2+}- and Mg^{2+}-free Steinberg's medium plus 0.005 M Na citrate, pH =7.4; e, serial clonal nuclear transplants are done, again using enucleated mutant recipient eggs; f, the serial clonal transplants are allowed to develop to stage 9, late blastula stage, where a few blastulae from each serial clone are used for autoradiographic analysis. They show intense ^{3}H-uridine incorporation. (Brothers, 1976).

These experiments indicate that the capacity to express the nuclear genes necessary for gastrulation and neurulation is stably altered and can be maintained apparently independently of the continuation of the pattern of RNA synthesis exhibited by mid- to late-blastula nuclei. In *Xenopus*, ribosomal RNA synthesis is not detectable until gastrulation.

Nuclei which are synthesizing rRNA, when transplanted into enucleated recipient eggs, do not continue synthesizing rRNA in detectable amounts, but show a reactivation of rRNA synthesis at gastrulation, as do normal embryos (Gurdon and Brown, 1965).

Further experiments to test whether or not the activation can be maintained independently of the continuation of RNA synthesis have been suggested by Professor A. A. Neyfakh, Leningrad University (personal communication) and are currently in progress.

IV. CHARACTERIZATION OF THE CORRECTION OF THE MUTANT PHENOTYPE

The nature of the correction that is obtained when $o+$ substance is injected into mutant eggs before first cleavage can provide some information about the nature of the nuclear activation. The question of whether the observed correction shows a quantitative response in proportion to the amount of $o+$ substance injected or whether the response is an all or none phenomenon is important. Injection of various volumes of nucleoplasm for germinal vesicles into mutant eggs showed that as little as 2-6% of the volume of a germinal vesicle was sufficient to produce full correction of the recipient mutant eggs (Table II and Table V). Sexually mature animals were obtained from rescued mutant eggs injected with the small volumes of germinal vesicle nuclear sap (Briggs, 1972).

TABLE V[a]

Correction of Mutant Eggs Obtained with Different Volumes of Germinal Vesicle Nuclear Sap

Volume of normal oocyte nuclear sap injected into mutant eggs		Development of recipient eggs	
Volume cu. mm	% of volume of one axolotl germinal vesicle	larvae	sexually mature adults
0.002 to 0.004	3 to 6	+	+
0.008 to 0.016	12 to 24	+	+
0.021 to 0.036	33 to 55	+	+

[a]From Briggs (1972)

V. RETENTION OF THE CAPACITY TO INTERACT WITH THE $o+$ SUBSTANCE

Transplantation of mutant blastula nuclei into normal enucleated recipient eggs tests the question of whether or not nuclei retain the capacity to interact with the $o+$ substance throughout early development. Nuclei of mutant eggs are not exposed to the $o+$ substance since those eggs lack it. However injection of nucleoplasm containing $o+$ substance into mutant eggs before first cleavage enables them to develop normally, showing that those nuclei can respond to the presence of the $o+$ substance. Transplantation of nuclei from mutant blastulae of various stages into enucleated normal eggs presents the first contact that those nuclei have with the $o+$ substance. This provides a test of the competence of those nuclei to respond to the $o+$ substance present in the recipient egg cytoplasm. The results show that though late-cleavage and early-blastula stage nuclei can respond and promote the normal development of the recipient eggs, nuclei from late mid-blastula, stage 8 3/4, and late-blastula, stage 9, cannot respond. Late-blastula nuclei are incapable of supporting normal cleavage when transplanted into recipient normal eggs. This suggests that late-blastula nuclei which have not been activated by an exposure to the $o+$ substance are injured and are unable to function properly, even when exposed to the active factor in the recipient normal egg cytoplasm (Brothers, 1976).

VI. DISCUSSION OF THE POSSIBLE MODE OF ACTION OF THE $o+$ SUBSTANCE

During oogenesis the genome apparently directs the synthesis of substances which are stored in the egg cytoplasm and act after fertilization to affect early development. Normal development involves the interaction of these substances and the genetically identical nuclei of the early embryo: these nucleo-cytoplasmic interactions are believed to be involved in the production of the critical event whereby cells become determined to different developmental pathways (Briggs and King, 1959; Rossi et al., 1975). Thus as a consequence different genes will be turned on and turned off in the different cell types. The mechanisms which control this cellular determination and whether or not this regulation of gene expression involves stable alterations of the genome are fundamental questions of Developmental Biology (King and Briggs, 1956; Illmensee, 1976).

Cellular determination is a highly stable and heritable capacity of individual cells (Hadorn, 1966; Wieschaus and Gehring, 1976). This determined state is stably maintained even though the differentiated functions are not expressed (Konigsberg, 1963; Coon, 1966; Cahn and Cahn, 1966; Davidson, 1974). The pattern of gene expression can be controlled and modified by extracellular or cytoplasmic factors, and by gene dosage, but once it is established, the capacity to express a specific state of determination is inherited over many cellular generations (Ephrussi, 1972; Brown and Weiss, 1975).

Cytoplasmic control of nuclear synthesis activity has been well documented, and the migration of proteins between the cytoplasm and the nucleus has been reported for a variety of systems (Legname and Goldstein, 1972; Kroeger et al., 1963). Heterokaryons between chicken erythrocytes and HeLa cells show that RNA synthesis is resumed in the erythrocyte nuclei (Harris, 1974) and that chicken specific proteins are synthesized (Appels et al., 1974). This activation of RNA synthesis is accompanied by the appearance of human nuclear antigens in the chicken erythrocyte nuclei (Ringertz and Savage, 1977).

Somatic cell nuclei injected into amphibian oocytes swell, their chromatin becomes dispersed, protein exchange between the injected nuclei and the oocyte cytoplasm occurs, and RNA synthesis is very active. The expression of genes can be detected by a coupled transcriptional and translational system (Gurdon et al., 1976a, b).

Nuclei from somatic cells exhibiting a characteristic pattern of protein synthesis, when injected into oocytes exhibit a cessation of synthesis of those proteins (Gurdon et al., 1976a; Etkin, 1976). Recently evidence has been presented for selective gene expression: apparently the transcriptional pattern of the injected somatic cell nuclei is being modified, or reprogrammed, by components of the egg cytoplasm. There is not only a turning off of previously synthesized somatic cell proteins but, in addition, there is an activation of genes normally expressed in oocytes (DeRobertis and Gurdon, 1977; DeRobertis et al., 1977).

Brain cell nuclei which do not normally undergo cell division show active DNA synthesis after injection into *Xenopus* egg cytoplasm (Graham et al., 1966), and it has been shown that there is a movement of cytoplasmic proteins into these nuclei (Merriam, 1969). Recently the active component has been partially isolated and characterized. Apparently it is a protein and it has been demonstrated to have an effect upon DNA synthesis and the formation of replication eyes *in vitro* (Benbow and Ford, 1975). To date, this constitutes one of the most complete characterizations of a cytoplasmic component which interacts with the DNA to affect very early developmental events.

The association of proteins with DNA or the chromosomes and the subsequent modification of DNA or chromosome behavior has been described for both procaryotes (Alberts and Sternglanz, 1977; Bourgeois and Pfahl, 1976) and eucaryotes (Frenster, 1965; Gilmour and Paul, 1973). Steroid hormones become associated with cytoplasmic receptors; this complex is transported to the nucleus where it becomes associated with the chromatin and acts to mediate gene transcription (Yamamoto and Alberts, 1975). It would appear that the germinal vesicle of amphibians serves as a selective storage receptacle for proteins synthesized during oogenesis and which are dispersed to the egg cytoplasm upon germinal vesicle breakdown at oocyte maturation (Ecker and Smith, 1971; Briggs, 1972). The differential accumulation of certain classes of proteins in the germinal vesicle has been demonstrated (Paine and Feldherr, 1972; Bonner, 1975; Clark and Merriam, 1977). Recent experiments have demonstrated that the germinal vesicle contains components which can transcribe injected DNA fragments (Mertz and Gurdon, 1977) with fidelity equivalent to that observed *in vivo* (Brown and Gurdon, 1977). Also there is a component of the germinal vesicle nucleoplasm which acts to stabilize injected DNA fragments in supercoiled form protecting them against degradation by deoxyribonucleases (Wylie *et al.*, 1978).

Proteins synthesized in the oocyte cytoplasm are later found in the embryonic nuclei during cleavage and blastulation (Ecker and Smith, 1971). During embryogenesis there is a migration of acidic proteins, or their sub-units, from the cytoplasm into the nucleus of determined cells. When these determined (late-gastrula endodermal cell) nuclei are transplanted into enucleated recipient eggs, there is a selective loss of acidic proteins (not histones) from the transplanted nucleus during the first cleavage division in the egg cytoplasm. Subsequently there is an appearance of acidic proteins in the transplanted nucleus which parallels the same pattern observed for embryonic nuclei in the course of normal development (DiBerardino and Hoffner, 1975). This is interpreted to reflect the reprogramming of the transplanted nucleus by the egg cytoplasm.

Treatment of *Drosophila melanogaster* embryos with ether produces a clonally inherited phenocopy of *bithroax* due to an interference with the activation of the *bithorax* complex. The phenocopy is apparently not due to an interference with the mechanism of the expression of the gene products of this gene complex; apparently the phenocopy is produced solely by an interference with the activation. Recent experiments have indicated that a regulator mutation *(Rb-pbx)* acts during this period of embryonic activation of the *bithorax* complex and that the result of the action of this "super-repressor" mutation is a mitotically heritable alteration of determination (Garcia-Bellido, 1977).

Elegant demonstrations of a cytoplasmically controlled stable nuclear determination which is established at one point in the nuclear "life cycle" has been shown for *Paramecium* (Sonneborn, 1954) and *Tetrahymena* (Nanney, 1956). In *Paramecium* exposure of the micronucleus just prior to its differentiation into a macronucleus (but not after the macronuclear differentiation) to a specific cytoplasmic factor acts to cause a stable alteration in the expression of the mating type. This is a heritable alteration of nuclear determination through all subsequent clonal generations and it cannot be changed even with later exposure to the cytoplasm of the opposite mating type.

The $o+$ substance is synthesized during oogenesis and stored in the germinal vesicle of ovarian oocytes. Upon maturation the contents of the germinal vesicle are dispersed into the egg cytoplasm. The presence of the $o+$ substance in the egg cytoplasm is essential for normal development through gastrulation and organogenesis. The nucleo-cytoplasmic interaction between the $o+$ substance and the nucleus acts to produce a stable nuclear activation during blastulation. The sensitive period for this activation is restricted to one point during embryonic development, mid-blastula stage. If the embryonic nucleus has not been exposed to the $o+$ substance by mid-blastulation the nucleus is irreversibly damaged and unable to subsequently respond to the presence of the active factor in normal egg cytoplasm.

Apparently this alteration of nuclear capacity to support normal development is the result of the activation of the nuclear genes required for gastrulation and neurulation.

The nuclear activation is dependent upon the presence of the $o+$ substance within the egg cytoplasm, but the experiments do not demonstrate that there is a direct interaction between the $o+$ substance and the embryonic genome. Hopefully, further experiments will yield some insight into this question.

It is indeed tempting to speculate that the $o+$ substance might be a regulatory molecule, protein, such as has been suggested to be involved in eucaryotic gene regulation (Frenster, 1965; Davidson and Britten, 1973; Yamamoto and Alberts, 1976).

ACKNOWLEDGEMENTS

I am indebted to Robert W. Briggs and George M. Malacinski for discussions concerning this paper. The research presented in this manuscript was possible because of Dr. R.R. Humphrey to whom all who have worked on the o mutation were indebted. He was a rigorous scientist and a modest, warm, courageous human being.

This research has been supported by grants from N.S.F. and N.I.H. The axolotl colony at Indiana University supplied the eggs and animals and is supported by N.S.F. grant DEB 75-17664 A01.

REFERENCES

Apples, R., Bolund, L., Goto, S. and Ringertz, N.R. (1974). *Exp. Cell Res.* **85**, 182-190.
Alberts, B. and Sternglanz, R. (1977). *Nature* **269**, 655-661.
Benbow, R.M. and Ford, C.C. (1975). *Proc. Nat. Acad. Sci. U.S.* **72**, 2437-2441.
Bonner, W.M. (1975). *J. Cell Biol.* **64**, 431-437.
Bourgeois, S. and Pfahl, M. (1976) Repressors. *In* Advances in Protein Chemistry (C.B. Anfinsen, J.T. Edsall and F.M. Richards, eds.), Vol., 30, pp. 1-90. Academic Press, N.Y.
Brachet, J. and Denis, H. (1963). *Nature* **198**, 205-206.
Briggs, R. (1972). *J. Exp. Zool.* **181**, 271-280.
Briggs, R. (1973). *In* "Genetic Mechanisms of Development" (F.R. Ruddle, ed.), 31st Symp. of the Soc. for Develop. Biol. pp. 169-199. Academic Press, New York.
Briggs, R. and Cassens, G. (1966). *Proc. Nat. Acad. Sci. U.S.* **55**, 1103-1109.
Briggs, R. and King, T.J. (1959). *In* "The Cell" (J. Brachet and A.E. Mirsky, eds.), Vol. 1, pp. 538-617. Academic Press, New York.
Briggs, R. and Justus, J.T. (1968). *J. Exp. Zool.* **147**, 105-116.
Brothers, A.J. (1976). *Nature* **260**, 112-115.
Brown, D.D. and Gurdon, J.B. (1977). *Proc. Nat. Acad. Sci. U.S.* **74**, 2064-2068.
Brown, J.E. and Weiss, M.C. (1975). *Cell* **6**, 481-494.
Cahn, R.D. and Cahn, M.B. (1966). *Proc. Nat. Acad. Sci. U.S.* **55**, 106-114.
Carroll, C.R. (1974). *J. Exp. Zool.* **187**, 409-422.
Cassens, G. (1968). An analysis of the maternal effect of gene o in *Ambystoma mexicanum*. Ph.D. Thesis, Indiana University, Bloomington.
Clark, T.G. and Merriam, R.W. (1977). *Cell* **12**, 883-891.
Cohen, L.H., Newrock, K.M. and Zweidler, A. (1975). *Science* **190**, 994-996.
Coon, H.G. (1966). *Proc., Nat. Acad. Sci. U.S.* **55**, 66-73.
Davidson, E.H. (1976). Gene Activity in Early Development. Second ed., Academic Press.
Davidson, E.H. and Britten, R.J. (1973). *Quart. Rev. Biol.* **48**, 565-613.
Davidson, R.L. (1974). *In* "Somatic Cell Hybridization" (R.L. Davidson and F. de la Cruz, eds.), pp. 131-150. Raven Press, New York.
Davis, F.M. and Adelberg, E.A. (1973). *Bact. Rev.* **37**, 197-214.
De Robertis, E.M. and Gurdon, J.B. (1977). *Proc. Nat. Acad. Sci. U.S.* **74**, 2470-2474.
De Robertis, E.M., Partington, G.A., Longthorne, R.F. and Gurdon, J.B. (1977). *J. Embryol. Exp. Morphol.* **40**, 199-214.
Di Berardino, M.A. and Hoffner, N.J. (1975). *Exp. Cell. Res.* **94**, 235-252.

Ecker, R.E. and Smith, L.D. (1971). *Develop. Biol.* **24,** 559-576.
Ephrussi, B. (1972). Hybridization of Somatic Cells. Princeton University Press.
Etkin, L.D. (1976). *Develop. Biol.* **52,** 201-209.
Frenster, J.H. (1965). *Nature* **206,** 680-683.
Galau, G.A., Klein, W.H., Davis, M.M., Wold, B.J., Britten, R.J. and Davidson, E.H. (1976). *Cell* **7,** 487-505.
Garcia-Bellido, A. (1977). *Amer. Zool.* **17,** 613-629.
Gilchrist, F.G. (1933). *J. Exp. Zool.* **66,** 15-51.
Gilmour, R.S. and Paul, J. (1973). *Proc. Nat. Acad. Sci. U.S.* **70,** 3440-3445.
Guidice, G., Mutolo, V., and Donatuti, G. (1968). *Wilhelm Roux' Arch.* **161,** 118-129.
Graham, C.F., Arms, K. and Gurdon, J.B. (1966). *Develop. Biol.* **14,** 349-381.
Gurdon, J.B. (1974). The Control of Gene Expression in Animal Development. Harvard Univ. Press., Cambridge.
Gurdon, J.B. and Brown, D.D. (1965). *J. Mol. Biol.* **12,** 27-35.
Gurdon, J.B., DeRobertis, E.M. and Partington, G. (1976a). *Nature* **260,** 116-120.
Gurdon, J.B., Partington, G.A. and DeRobertis, E.M. (1976b). *J. Embryol. Exp. Morphol.* **36,** 541-553.
Hadorn, E. (1967). *In* "Major Problems in Developmental Biology." (M. Locke, ed.), pp. 84-104, Academic Press, New York.
Harris, H. (1974) Nucleus and Cytoplasm. Oxford University Press.
Humphrey, R.R. (1966). *Develop. Biol.* **13,** 57-76.
Humphrey, R.R. (1975). *In* "Handbook of Genetics" (R.C. King, ed.), Vol. 4, pp. 1-23. Plenum Publ. Corp., New York.
Illmensee, K. (1976). *In* "Insect Development" (P.A. Lawrence, ed.), 8th Symp. of the Royal Entomological Soc. of London, pp. 76-96. Blackwell Scient. Pub., London.
Illmensee, K. and Mahowald, A.P. (1974). *Proc. Nat. Acad. Sci. U.S.* **71,** 1016-1021.
King. T.J. and Briggs, R. (1956). *Cold Spring. Harb. Symp. Quant. Biol.* **21,** 270-290.
Konigsberg, J. (1963). *Science* **140,** 1273-1284.
Kroger, H., Jacob, J. and Sirlin, J.L. (1963). *Exp. Cell Res.* **31,** 416-423.
Legname, C. and Goldstein, L. (1972). *Exp. Cell Res.* **75,** 111-112.
Malacinski, G.M. (1971). *Develop Biol.* **26,** 442-451.
Malacinski, G.M. (1972). *J. Exp. Zool.* **181,** 409-420.
Malacinski, G.M. and Brothers, A.J. (1978)., unpublished.
Merriam, R.W. (1969). *J. Cell Sci.* **5,** 333-349.
Mertz, J.E. and Gurdon, J.B. (1977). *Proc. Nat. Acad. Sci. U.S.* **74,** 1502-1506.
Morgan, T.H. (1927). Experimental Embryology. Columbia University Press, New York.
Nanney, D.L. (1956). *Amer. Nat.* **90,** 291-307.
Nemer, M. (1975). *Cell* **6,** 559-570.
Neyfakh, A.A. (1964). *Nature* **201:** 880-884.
Paine, P.L. and Feldherr, C.M. (1972). *Exp. Cell Res.* **74,** 81-98.
Ringertz, N.R. and Savage, R. (1977). Cell Hybrids. Academic Press, New York.
Rossi, M., Augusti-Tocco, G. and Monroy, A. (1975). *Quart. Rev. Biophys.* **8,** 43-119.
Sonneborn, T.M. (1954). Proc. of the 9th Intern. Congress of Genetics. *Carylogia,* Suppl., pp. 307-325.
Wieschaus, E. and Gehring. W. (1976). *Develop. Biol.* **50,** 249-265.
Wylie, A.H., Gurdon, J.B. and Price, J. (1977). *Nature* **268:** 150-152.
Yamamoto, K.R. and Alberts, B.M. (1976). *Ann. Rev. Biochem.* **45,** 722-746.

Maternal Effect Mutations that Alter the Spatial Coordinates of the Embryo of *Drosophila melanogaster*

Christiane Nüsslein-Volhard
*European Molecular Biology Laboratory**
6900 Heidelberg / F.R.G.

I. Introduction 185
II. The Genetic Approach to Embryonic Pattern Formation 187
    A. Embryogenesis in *Drosophila melanogaster* 187
    B. Models of Embryonic Pattern Formation 189
III. The Antero-Posterior Pattern: *Bicaudal* 191
    A. The Phenotypes of the *Bicaudal* Mutation 191
    B. Testing of Models 197
IV. The Dorso-Ventral Pattern: *Dorsal* 200
    A. The Recessive *Dorsal* Phenotype 200
    B. The Dominant *Dorsal* Phenotype 206
    C. Interpretation of the *Dorsal* Phenotypes 208
V. Concluding Remarks 209
    References 211

I. INTRODUCTION

This report is divided into three sections. In the first section, the properties of the system — embryogenesis in *Drosophila melanogaster* — are treated in some detail. This is necessary for understanding the deviations from normal development observed in mutant embryos. It further includes a discussion of current ideas about embryonic pattern formation and a possible way to test them with maternal effect mutants. In the second and third sections the phenotypes of two maternal effect mutations — *bicaudal* and *dorsal* — will be described. They suggest that

*Much of the previously unpublished work presented here has been done in the Biologisches Institut I (Zoologie), Freiburg, F.R.G.

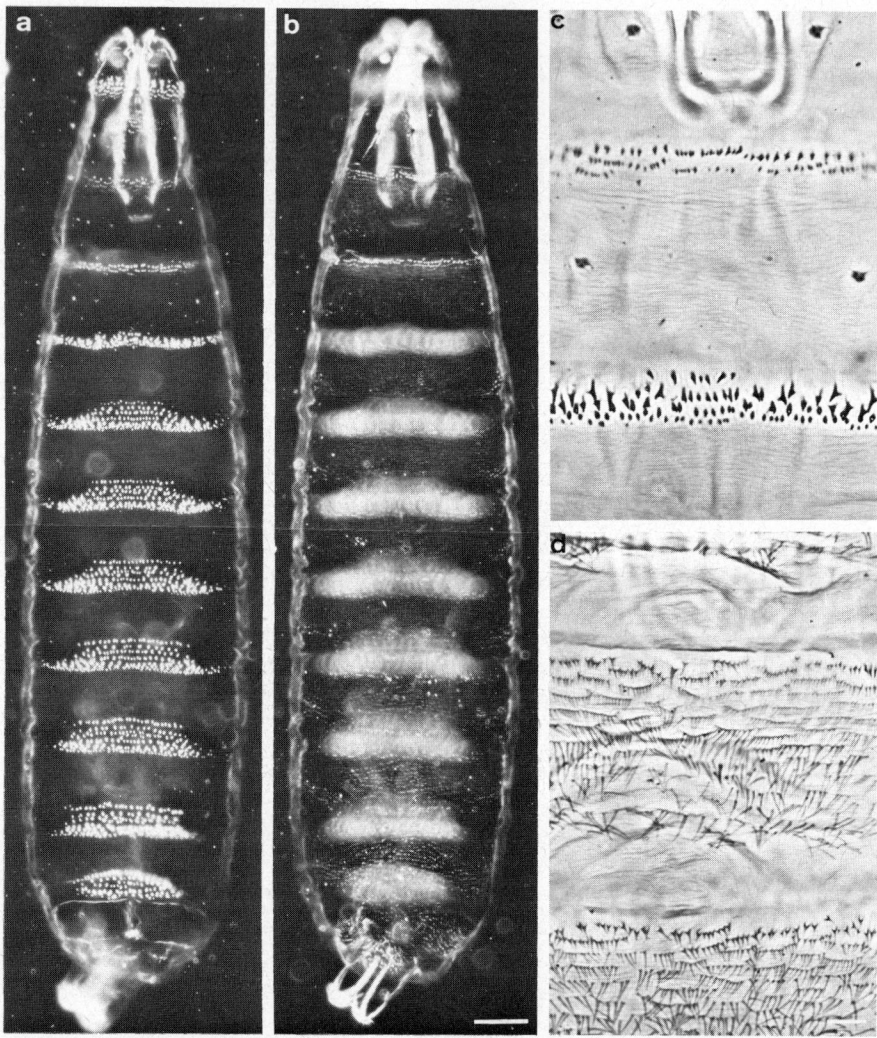

Fig. 1. The integument of a *Drosophila* first instar larva. (a) ventral pattern, (b) dorsal pattern, (c) and (d) details from the third thoracic segment, (c) ventral (d) dorsal. Anterior at top in all figures. The pattern (from top to bottom) shows a continuous array of 3 thoracic and 8 abdominal segments. Each segment, on the ventral side, is marked anteriorly by a belt of denticle hook rows, which are narrow and fine in the thorax and more prominent in the abdomen. The dorsal side is covered with a segmentally repeated pattern of fine hairs in all segments. The sclerotinized parts of the head are involuted and come to lie inside the first and second thoracic segment. The protrusions at the posterior end of the larva are the posterior spiracles. The larvae were fixed and cleared according to the procedure of Van der Meer (1977a); (a) and (b) are dark field pictures; the scale is 50 μm. (c) and (d) are phase contrast pictures; the scale is 10 μm.

embryonic pattern formation is achieved by two gradients, in antero-posterior and dorso-ventral direction, defining the spatial coordinates of the developing embryo.

II. THE GENETIC APPROACH TO EMBRYONIC PATTERN FORMATION

A. *Embryogenesis in Drosophila melanogaster*

1. Drosophila embryogenesis results in a simple metameric pattern. For studying the principles and molecular mechanisms of embryonic pattern formation, mutations which lead to an alteration of a pattern are valuable tools. Embryogenesis in *Drosophila melanogaster* is an obvious system not only because *Drosophila* is genetically the best known higher eucaryote, but also because embryonic development results in a simple metameric pattern of 11 repeating units, the 3 thoracic and 8 abdominal segments of the hatching larva. This cuticular pattern of the larval integument, the hypoderm (Fig. 1), is well marked with structures indicating position and polarity. As will be shown in this report, the pattern may be dramatically altered by mutations, the changes indicating overall changes in the spatial organization of almost the entire embryonic material.

Embryogenesis in *Drosophila* occurs outside the maternal organism in a relatively large egg (500 μm x 180 μm) and is thus accessible to other experimental approaches like cell transplantation and injection of substances. Embryogenesis is fast and initially simple. Development of the *Drosophila* embryo begins with a number of rapid synchronous nuclear divisions. The nuclei migrate to the surface of the egg and are enclosed in cell membranes to form, at 3 h of development, the first cellular state, the blastoderm. This uniform single cell layer is divided up into the three germ layers in the rapid process of gastrulation (see Fig. 8). The mesoderm derives from a longitudinal infolding at the ventral side of the egg, the ventral furrow. Additional invaginations occur at the anterior ventral side and the posterior tip of the egg, giving rise to the anlagen of the endodermal gut (see Fig. 5a,c). Gastrulation is accomplished by a stretching of the ventral cell layers along the dorsal side of the egg (germ band extension). This process is followed by primary organogenesis, little of which can be observed directly as it occurs in the interior of the embryo. By 10 h the germ band unrolls (germ band shortening), the head portion involutes and the future organs come to lie at approximately their final position. By 22 h the final differentiation is achieved and the larva hatches from the egg.

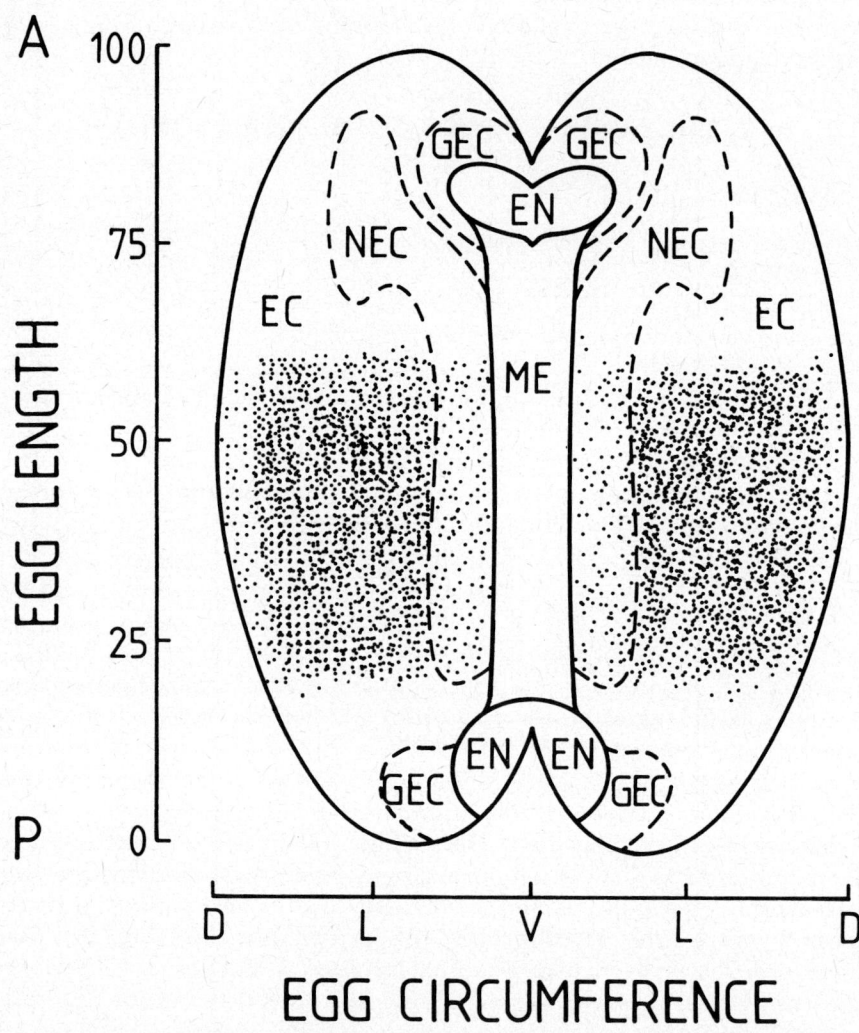

Fig. 2. A fate map of the *Drosophila* embryo at the blastoderm stage, slightly modified from Poulson (1950). The egg is cut open along the dorsal midline and the left and right halves spread out to approximate a flat sheet of cells. Anterior at top. Ventral at the middle of the Figure. The vertical scale (egg length) indicates the antero-posterior axis at the longest extension of the egg; the horizontal scale (egg circumference) is twice the dorso-ventral axis at the widest point of the egg. The egg length is about 500 μm, the circumference (πx diameter) is approximately the same. 500 μm corresponds to roughly 100 blastoderm cell diameters. Me = Mesoderm anlage, En = Endoderm anlage, Ec = Ectoderm, Gec = gut ectoderm, Nec = neurogenic ectoderm. A = anterior, P = Posterior, D = dorsal, L = Lateral, V = ventral. The shaded area represents the anlage for the entire larval hypoderm (see Fig. 1) as deduced from the frequency and position of hypodermal defects following UV-laser microbeam irradiations (Lohs-Schardin and Nüsslein-Volhard, 1979).

2. *A fate map for the larval hypoderm.* On the basis of histological observations Poulson (1950) has constructed a fate map of the blastoderm, a simplified version of which is shown in Fig. 2. Except for a longitudinal strip at the ventral midline, forming the mesodermal and endodermal anlagen, the larger fraction of the egg forms ectodermal derivatives. Within the ectoderm anlage, the ventral most cells flanking the mesoderm anlage give rise to the nervous system, while anteriorly and posteriorly a portion invaginates to form the ectodermal fore- and hindgut respectively. In order to locate the anlage for the larval hypoderm more precisely, we have used a UV-laser microbeam to disable discrete portions of the blastoderm (M. Lohs-Schardin and C. Nüsslein-Volhard, 1979). Irradiation within a well defined region of the blastoderm (Fig. 2, shaded area) yielded defects in the hypoderm in up to 90% of the irradiated individuals. Defective larvae lacked a small region of cuticle in their otherwise normal integument. The position of the defects within the hypodermal pattern correlated closely to the site of irradiation on the blastoderm. Irradiation at 50% egg length yielded defects mainly in the metathorax, while at 20% the last abdominal segment frequently was damaged. The correlation between site of irradiation and position of the defect was used to construct a fate map of the larval hypoderm. Fig. 2 shows that the anlage for the larval hypoderm occupies two separate regions on the left and right side of the blastoderm, extending in anterior direction from 20% (8th abdominal segment) to 60% (prothorax) the length of the egg. All but the shaded region eventually comes to lie inside the larva. It is important to note that the anlage for the right and left halves of the integument are separated by the future mesoderm and nervous system and come together only at gastrulation. Within these regions, the anlagen for the individual thoracic and abdominal segments are evenly spaced, each occupying a narrow transverse strip on the blastoderm, the antero-posterior extent of which corresponds to not more than 3-4 cell diameters.

B. *Models of Embryonic Pattern Formation*

1. *"Localized determinants" versus "gradients".* How do the hypoderm anlage cells and indeed all the cells in different regions of the egg "learn" which part of the larva they should produce? What kind of information is contained in the egg structure to achieve the spatial arrangement of organs and segments, to create and maintain polarity? These are old questions of classical embryology and, although extensively studied in many organisms, no convincing decision between various proposed

concepts has been achieved. (For review see e.g. Graham, 1976; for pattern formation in insect embryogenesis, Sander, 1976). Embryological observations lead to the formulation of models, two extremes of which may be called the concept of "localized determinants" and the concept of "gradients". The former postulates many different morphogenetic substances laid down in the egg at specific regions, each substance determining a particular organ or tissue to be formed at the site of its presence. It is conceptually very simple; however, it leaves unexplained how a complex spatial arrangement of the "determinants" is achieved during oogenesis. The concept of gradients postulates that the position within the egg is specified by the concentration rather than the nature of a morphogen. The concentration of the morphogen is assumed to change continuously along the axis of the egg in a gradient. Thus given two gradients of two morphogens oriented perpendicular to each other, each cell within a two-dimensional sheet of cells is provided with a unique set of positional values defining its spatial coordinates. In a second step of "gradient interpretation", cells within a range of morphogen concentrations are determined for a single sort of cell differentiation. The initially continuous gradient thus may lead to a discontinuous pattern; e.g., below a certain threshold concentration cells will form a pattern element A, while above it they will form element B.

Localized determinant and gradient hypotheses may not necessarily be contradictory since a gradient may lead at some stage of development to an unequal distribution of several determinants and thus become mosaic with age. It is also possible that some developmental decisions in an organism are reached by the one principle and other decisions by the other. For example, it is conceivable that the three germ layers would be determined by localized determinants, while the further subdivisions would depend on gradients, or vice versa. As will be discussed below, evidence suggests that in *Drosophila* the patterning of the entire somatic tissue is achieved by gradients, whereas the germ line precursor cells appear to be determined by a localized determinant. (Illmensee *et al.*, 1976; see also Mahowald, this volume).

2. Maternal effect mutants suggest gradients. In order to test these two and other possible models, mutations that alter the embryonic pattern are of great value because mutations, in contrast to many other experimental approaches are expected to affect primarily only a single component of the developing system, the primary gene product. Since the egg and thus the structural information it contains for patterning of the embryo is made from the genome of the mother during oogenesis, the type of mutants which are informative are maternal effect mutants: an altered embryonic pattern depends upon the genome of the mother which laid the egg, not on that of the embryo itself (review by Wright, 1970). In general, if the concept of localized determinants were correct, one should be able to find a class of maternal effect mutations each of which causes the embryo to fail to form a particular organ or tissue, while the remainder of the body pattern is normal. In contrast, the gradient hypothesis would predict that a single mutation may affect the entire embryonic pattern, since one morphogen gradient defines many different tissues.

In this paper the phenotypes of two maternal effect mutations will be described. In both, the entire embryonic organization is dramatically altered. The types of alterations suggest that embryonic pattern formation in *Drosophila* is achieved by two gradients, in antero-posterior and dorso-ventral direction, defining the spatial coordinates of the developing embryo.

III. THE ANTERO-POSTERIOR PATTERN: *BICAUDAL*

A. *The Phenotypes of the Bicaudal Mutation*

1. Bicaudal is a maternal effect mutation mapping at a single locus. Bicaudal is a recessive maternal effect mutation causing some of the embryos produced by mutant females to develop two abdominal ends arranged in mirror image symmetry rather than the normal sequence of head, thorax and abdomen (Fig. 3a-c). The mutant was isolated and many of its interesting features already described by Bull (1966). Owing to the very low penetrance observed and the variable expression of the phenotypes, however, a detailed genetic and phenotypic analysis only became possible after the penetrance had been raised considerably (Nüsslein-Volhard, 1977). Recombination and deletion mapping revealed that *bicaudal* maps as a single point mutant on the right arm of the second chromosome (2-67). Since hemizygous females, in particular those with a particular chromosome carrying a deletion for *vestigial* and adjacent loci (vg^B), produce mutant embryos with much higher frequency than homozygotes, it appears that the mutation is hypomorphic, i.e. the phenotypes are produced by a lesser than normal amount of the *bicaudal* gene product. The expression of the mutant phenotype furthermore depends strongly on other factors than genetic, such as temperature and age of the female. Best producers are very young females at high temperature.

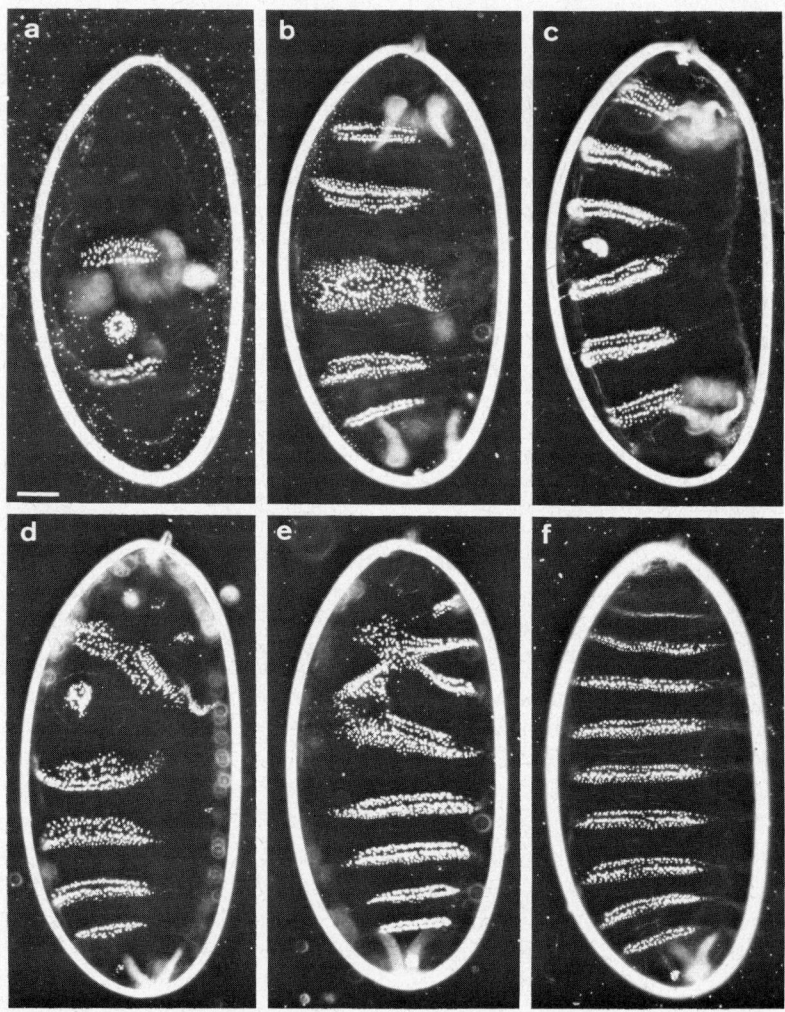

Fig. 3. The ventral hypoderm pattern of *bicaudal* pheontypes. (a-c) symmetrical bicaudal embryos with 1 1/2 (a), 3 (b) and 3 1/2 (c) posterior segments duplicated. (d) asymmetrical bicaudal embryo with 4 1/2 segments with normal polarity and 1 1/2 segments with reversed polarity. The duplicated spiracles (out of focus) were rudimentary. (e) unilateral bicaudal. The left lateral half has a plane of symmetry in the third abdominal segment whereas the right half contains all 8 abdominal segments, with normal polarity. This embryo had two rudimentary spiracles at the anterior dorsal part. (f) headless embryo, showing a complete sequence of 8 abdominal and 3 thoracic segments; the head is missing entirely, as evident from the lack of the prominent mouth parts (see Fig. 1). Anterior at top in all the figures. The scale is 50 μm.

2. Bicaudal induces a range of different antero-posterior pattern abnormalities. Analysis of the segment pattern of late lethal embryos produced by mutant females revealed that the *bicaudal* phenotype is highly variable and that a number of different pattern abnormalities occur. Genetic analysis showed that this variation is not caused by several independent mutations but rather has to be regarded as reflecting various degrees of expression of the mutant character. The *bicaudal* phenotypes have been described and discussed in detail (Nüsslein-Volhard, 1977) and much of this needs not to be repeated here. Some observations, however, require special mention. In all abnormal embryos, the antero-posterior arrangement of segments, their total number, and often their polarity is affected. Embryos from mutant females belong to either of three large categories, (a) symmetrical bicaudal embryos, (b) normal, hatching larvae, (c) intermediates between the two. Symmetrical bicaudal embryos have two abdominal ends arranged in mirror image symmetry, each with exactly the same number of segments. The number of segments thus duplicated varies between 1 and 5, the most frequent being three. The third category is more heterogeneous ranging from asymmetrical bicaudal embryos in which the posterior part with normal polarity always contains more segments than the anterior reversed part, to embryos with normal polarity throughout, but missing anterior-most pattern elements ("headless" embryos) to various extents. Examples displaying some of the patterns found are shown in Fig. 3.

3. The frequency distribution of different phenotypes is bimodal. Although there is a continuous range of patterns, with increasing numbers of posterior segments with normal polarity, the symmetrical bicaudal embryos on the one end of the spectrum, and the normal, polar, larvae at the other are the most frequent patterns, while any type of intermediate pattern is comparatively rare (Fig. 4). This observation holds also for collections from females with lower penetrance (e.g. old females): the frequency of normal larvae increases at the expense of *both* the bicaudal and the intermediate phenotypes, while the size distribution (segment number) of symmetrical bicaudal embryos does not change significantly. The finding of a bimodal distribution of phenotypes suggests that the *bicaudal* mutation interferes with a pattern forming mechanism which has a polar and a symmetrical structure as alternative stable states.

4. Bicaudal affects the organization of the entire somatic tissue. Although no extensive histological studies of mutant embryos have been made, it may be stated safely that the location of the internal organs corresponds to

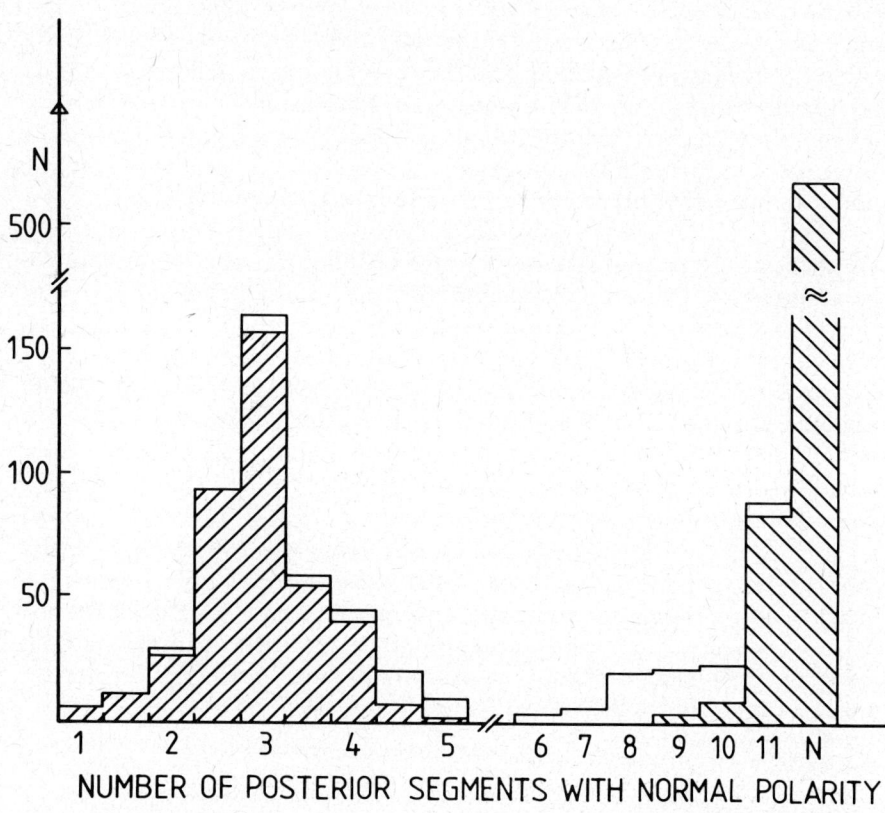

Fig. 4. Frequency distribution of bicaudal phenotypes depending on the size of the part with normal polarity. The data are taken from a collection of 2300 eggs from bic $L^2spl vgB$ females, 1300 of which developed to a larval stage. 12% of those showed phenotypes other than the ones recorded here (Nüsslein-Volhard, 1977). The shaded areas represent the frequency of symmetrical bicaudal embryos (up to 5 posterior segments with normal polarity) and embryos with normal polarity (headless embryo and normal larvae, 9 segments and larger). Added on top (not shaded) are the frequencies of asymmetrical bicaudal embryos (up to 6 segments) and headless-bicaudal embryos (6-11 segments). Mixed class embryos (3%) were counted twice. Note the change of scale in abscissa and ordinate.

the pattern abnormalities observed in the hypoderm (Bull, 1966; our own observations. In bicaudal embryos at gastrulation, "posterior" midgut invaginations occur at both the *posterior* and the *anterior* parts of the embryos (Fig. 5b). Bicaudal embryos regularly have malpighian tubules (specialization of the ectodermal hindgut) duplicated in the anterior end. Culturing anterior and posterior parts of bicaudal embryos in adult females and testing the differentiation capacities of the imaginal discs in larval transplants revealed that the anterior part frequently forms

COORDINATE MUTANTS IN *DROSOPHILA*   195

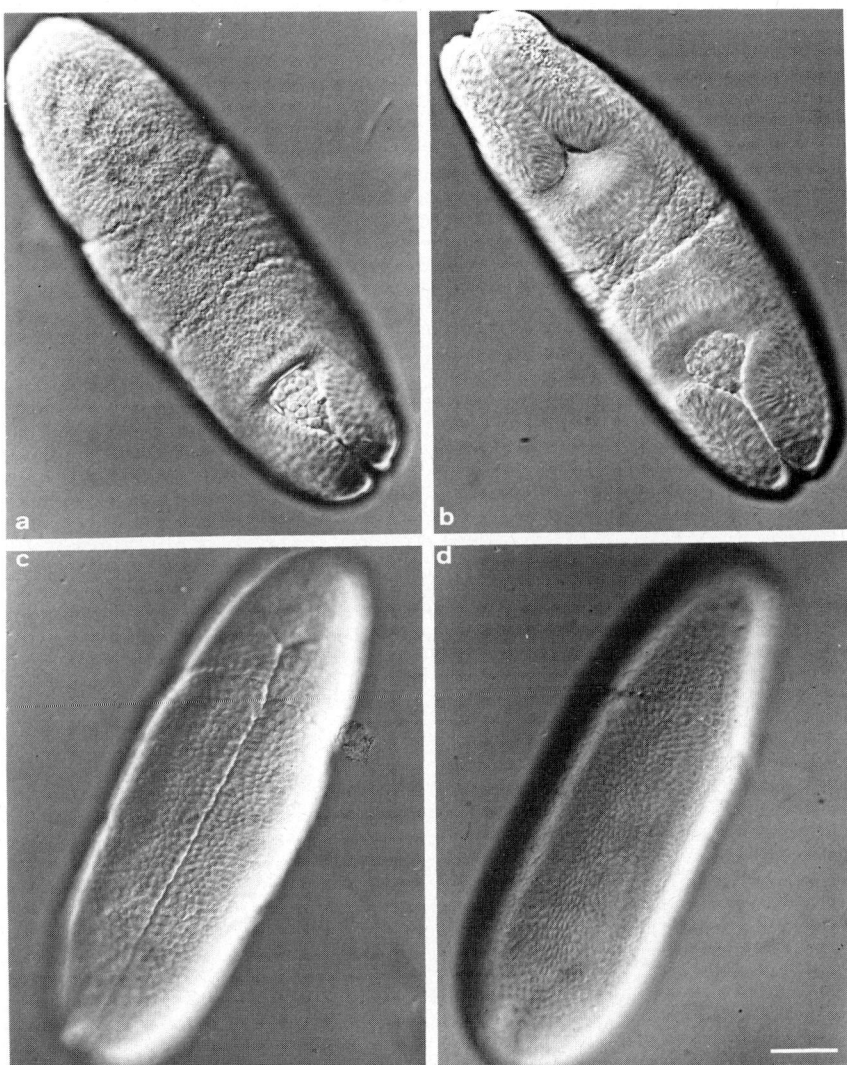

Fig. 5. Gastrulation in normal and mutant embryos. (a) a wildtype embryo in the process of posterior midgut invagination from the dorsal side (b) a bicaudal embryo of about the same age. Note the absence of pole cells in the anterior invagination. (c) the same wildtype embryo as in (a), but from the ventral side. The longitudinal line marks the ventral furrow; at the anterior end of it is the anterior midgut invagination. (d) *dorsal*-dominant embryo at about the same age. Note the absence of both ventral furrow and anterior midgut invagination. Anterior end up in all figures. All embryos were fixed in Heptane/Glutaraldehyde and, after removal of the vitelline membrane, mounted in Faure's medium. Nomarski interference optics. The scale is 50 μm.

genital disc structures, indicating that the symmetrization extends to the anlagen of the imaginal structures (Gehring and Nüsslein-Volhard, unpublished). From the first stages of embryogenesis where antero-posterior polarity in a normal *Drosophila* embryo is observed, bicaudal embryos are distinguishable from their normal siblings. The only exceptions to the perfect symmetry of gastrulating bicaudal embryos are the pole cells, precursors of the germ line, which are, as in normal embryos, found only at the original posterior end (Bull, 1966, Fig. 5b). Thus the location of pole cells seems to be a process independent of the patterning of the somatic tissue.

5. *Bicaudal affects the posterior as well as the anterior part of the embryo.* As shown earlier, the posterior half of the blastoderm in normal development forms the entire abdomen of 8 segments (Fig. 2). In bicaudal embryos, the posterior part with normal polarity never contains more than 5 abdominal segments. It frequently contains much less, the average being 3 (Fig. 4). Already at gastrulation, normal and duplicated parts of the embryos are of equal size, or the anterior part is smaller, never the posterior (Fig. 3a-c, 5b). This indicates that the posterior half of the egg gives rise to much fewer segments than in a normal egg. Thus, in bicaudal embryos the entire fate map is changed, not only that in the anterior portion of the egg. Similar considerations applied to the intermediate phenotypes (i.e. asymmetrical bicaudal and headless embryos) reveal that this statement also holds for embryos showing only a weak expression of the mutant phenotype.

6. *In the antero-posterior direction, different parts of the embryo form an integrated pattern.* In general, any particular segment within a mutant pattern is flanked by segments which are also its nearest neighbors in the normal segment pattern. Segments at the plane of symmetry often are not complete, in that the plane of symmetry may cut through any position within a segment, through the denticle belt at the anterior part of a segment or the naked cuticle forming the posterior part of a segment. In addition, although there is a great variation in segment numbers in bicaudal embryos, in the vast majority of the cases, the number of segments of the anterior part is identical to its posterior counterpart. A possible exception to this normal neighborhood relation is provided in asymmetrical bicaudal embryos, the anterior part being smaller than the posterior part and usually displaying an irregular pattern, as if some segments were skipped (Fig. 3d). In asymmetrical bicaudal embryos, the plane of symmetry is shifted anteriorly, the posterior part with normal polarity being much

larger than the anterior, reversed part. This suggests that the skipping of segments results from an anlage which was too small to form the complete sequence of segments. Since a segment in normal embryos stems from a strip of only 3-4 cells wide, this interpretation seems to be reasonable.

7. *A few cases display dorso-ventral or left-right discontinuities.* Abrupt rather than continuous changes in the hypoderm pattern are observed in two phenotypes which have been termed "headless-bicaudal" and "mixed-class". In headless-bicaudal embryos, the ventral hypoderm pattern displays normal polarity and a normal sequence of segments throughout, while in the dorsal pattern, posterior-most structures are found at the anterior end. Such embryos consist of a complete abdomen terminated at the anterior end with a pair of rudimentary spiracles. The mixed class consists of embryos in which the right and left halves display antero-posterior patterns belonging to two different classes of phenotypes. Frequently, they are mixtures of asymmetrical bicaudal and headless bicaudal embryos, i.e. patterns ranging very close together in the spectrum of phenotypes. The pattern discontinuity is restricted to the anterior portion of the egg, where in the ventral midline parts of segments with different positional values and opposing polarity meet (Fig. 3e). In this context it should be remembered that the anlage for the left and right parts of the ventral hypoderm lie quite far apart on the fate map, separated by anlagen for internal organs, the patterning of which is difficult to assess (Fig. 2). Thus, what appears to be a discontinuity in the hypoderm may actually develop from a less abrupt change of positional values in the fate map. An independence of the patterning of right and left sides has also been found by Van der Meer (1977b) in double abdomens produced in eggs of the beetle *Callosobruchus maculatus* after temporary constriction of the anterior portion of the egg.

B. *Testing of Models*

1. *The concept of an anterior determinant does not explain the bicaudal phenotype.* It is very difficult, if not impossible, to explain the various features of bicaudal embryos in terms of localized determinants, if one takes into account that the primary cause of the abnormal patterns is the less than normal amount of a single component, the *bicaudal* gene product. Double abdomen formation in *Smittia* following various treatments of the anterior egg pole has been interpreted by Kalthoff (1978, see also this volume) to result from the inactivation of an "anterior determinant". In

Smittia, unlike *Drosophila*, the duplication includes almost the entire abdomen; little variation in segment number is reported and intermediate phenotypes do not seem to occur. The anterior determinant concept, in our opinion, does not explain any of the *bicaudal* phenotypes. The spectrum of phenotypes, the smallness and variability in segment number in symmetrical bicaudal embryos and the coordinate change of pattern in anterior *and* posterior egg part are not accounted for by this and possibly any "localized determinant" concept. On the other hand, the striking similarity in the phenomenon of double abdomen formation in so closely related species strongly suggests an underlying mechanism common to both systems.

2. A gradient defining the antero-posterior coordinate of the Drosophila embryo may best explain the bicaudal phenotypes. Double abdomen formation in both *Smittia* (Meinhardt, 1977) and *Drosophila* are well compatible with a gradient concept, in which a normal pattern is produced by a monotonic gradient and double abdomens by a symmetrical gradient. The curves in Fig. 6 describe the spectrum of *bicaudal* phenotypes by the change of positional values along the antero-posterior coordinate; the transverse lines indicate threshold levels for head, thorax and abdomen. In this Figure, the normal pattern is represented by a monotonic gradient with a highpoint at the posterior pole of the egg, and a symmetrical bicaudal embryo by a U-shaped gradient with highpoints at both the posterior and the anterior poles. While in the normal pattern any positional value appears only once along the antero-posterior axis, in the symmetrical gradient the upper values appear twice while lower values do not appear at all. This shape readily describes the symmetry, the reversal of polarity and the much lower and variable segment number, features regularly observed in bicaudal embryos. Intermediate phenotypes are described as asymmetrical gradients with a steep slope close to the anterior pole. This reflects the finding that in large symmetrical bicaudals and asymmetrical bicaudals the anterior, reversed part is smaller and less well developed than its posterior counterpart (see above).

3. A molecular mechanism creating polar and symmetrical gradients: the Meinhardt-model. A model for the establishment of stable concentration gradients has been proposed by Gierer and Meinhardt (1972; Gierer, 1977) and recently has been adapted by Meinhardt (1977) for pattern formation in insect embryogenesis. The model is based on the kinetics of molecular interaction and diffusion and involves two substances. Both substances — the activator and the inhibitor in terms of the model — are produced at

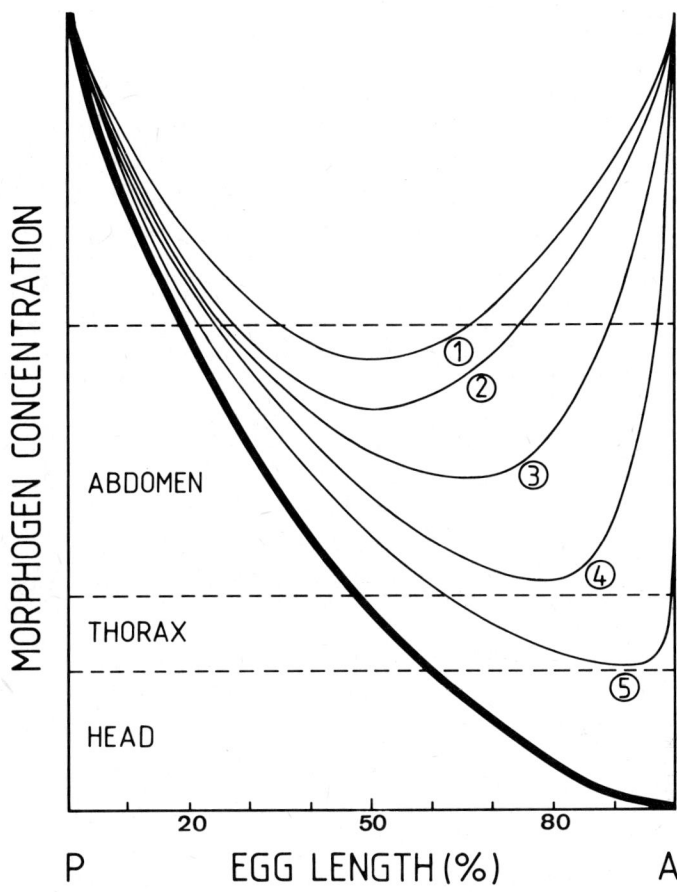

Fig. 6. Hypothetical gradients of morphogen concentration defining the antero-posterior coordinate in normal and bicaudal embryos. The horizontal scale represents the antero-posterior egg axis; posterior pole, left. The vertical scale refers to the concentration of a morphogen which is presumed to define the positional value at any point along the antero-posterior axis. The dotted horizontal lines indicate the threshold levels for abdomen, thorax and head. They are chosen such that the gradient for the normal pattern (heavy, monotonic line) defines the antero-posterior extent of the anlagen for abdomen and thorax on the blastoderm fate map: the position of the anterior and posterior rims of the abdomen anlage are defined by the values on the horizontal scale where the gradient crosses the lower and upper thresholds: 48% and 20% egg length respectively; the corresponding values for thorax formation are 60% and 48%. The U-shaped curves specify the patterns of various *bicaudal* phenotypes. (1) and (2) correspond to symmetrical bicaudal embryos as presented in the Fig. 3a-c. Only a small portion of the abdomen is specified at the middle of the egg. (3) describes an asymmetrical bicaudal embryo (e.g. Fig. 3d). (4) a headless-bicaudal embryo; almost the entire abdomen is specified. (5) corresponds to a headless embryo (Fig. 3f), consisting of the entire thorax and abdomen. The curves do not represent computer simulation of a quantitative model but indicate the approximate shape such a model should yield in order to explain the *bicaudal* phenotypes.

the posterior egg pole and spread by diffusion. The activator stimulates its own production and that of the inhibitor. The inhibitor diffuses faster than the activator and inhibits production of the activator outside the activated area. This interaction, if certain kinetic criteria are met (Gierer and Meinhardt, 1972) creates a graded distribution of activator and inhibitor along the antero-posterior axis of the egg, with a highpoint at the posterior pole. The concentration gradients of inhibitor and activator are stable with time and are quickly re-established following local disturbances. If, however, the range of inhibition is reduced, a new activator peak may form at the anterior pole and a symmetrical, U-shaped gradient will develop. This can be achieved by various means, e.g. by partly destroying the inhibitor or its source, or a leakage at the anterior pole, or increased activator production. Thus according to this model it is relatively easy to conceive how a single mutation may lead to the complex and dramatic alteration of the pattern observed. Since, however, the range of inhibition is dependent on a number of parameters of the system (Meinhardt, 1977) it is more difficult to assess the role the *bicaudal* gene product plays in this process. On the other hand, the model predicts that mutations in a number of different genes would give essentially the same phenotype. Recently we have found a maternal effect mutant producing the same spectrum of phenotypes as *bicaudal*. It appears to be dominant and *not* allelic to *bicaudal*.

IV. THE DORSO-VENTRAL PATTERN: *DORSAL*

A. *The Recessive Dorsal Phenotype*

1. Dorsal has a recessive and a dominant maternal effect phenotype. In this section the phenotypes of a newly isolated maternal effect mutant, *mat (2) dorsal* (abbreviated as *dorsal* or *dl*), will be discussed. This mutation affects the dorso-ventral (or transverse) pattern of the *Drosophila* embryo in a similarly dramatic way as *bicaudal* affects the antero-posterior (or longitudinal) pattern. As illustrated in the fate map (Fig. 2), the dorso-ventral pattern contains the dorsal and ventral hypoderm, the neurogenic ectoderm and the mesoderm as pattern elements. It is a bilateral pattern, each longitudinal half of the embryo containing all the pattern elements, arranged in mirror image symmetry, the plane of symmetry cutting through the mesoderm anlage at the ventral midline and the dorsal hypoderm at the dorsal midline.

Depending on the genetic constitution of the female, *mat (2) dorsal* may exert two distinct alterations of this pattern. The first phenotype, the recessive phenotype, is expressed in embryos from homozygous *dl/dl* females. It consists of a dorsalization of the entire embryo such that structures characteristic for the dorsal side are formed in all regions of the egg. The second phenotype is dominant and may be described as an intermediate between the recessive phenotype and wildtype in that structures are formed in ventral egg regions which are normally derived more laterally. In contrast to the recessive phenotype, which is fully penetrant and not temperature sensitive, the expression of the dominant phenotype depends upon temperature and the genetic "background" of the female, as will be discussed below.

2. The differentiated dorsal embryo consists of a yolk-filled tube of dorsal hypoderm. At the time when control embryos hatch, embryos from *dl* females have developed a long tube of larval hypoderm with many constrictions, wound up irregularly in the egg case. The tube is covered with a cuticle marked with a hypodermal hair pattern characteristic for the dorsal side of a normal *Drosophila* embryo (Fig. 7a, b). Apart from large masses of unconsumed yolk, the hypodermal tube appears empty and internal organs have never been found in *dorsal* embryos.

Although appearing at first sight rather irregular, the cuticular pattern in *dorsal* embryos, in some regions at least, reveals an orderly arrangement of hairs. Frequently, transverse rows are found, reminiscent of the rows marking the segment borders in wildtype larvae. The hair pattern changes its character along the antero-posterior axis: anteriorly, the pattern is similar to dorsal thorax, while posteriorly it resembles the dorsal pattern of the abdominal segments. Normal antero-posterior polarity in *dorsal* embryos is furthermore indicated by the orientation of the hairs and by the presence of terminal markers: in the anterior portion of the embryo, a brownish sclerotised structure is found which may be the labrum, part of the mouth armature, while at the posterior end tiny spiracles without tracheae frequently develop. Cuticular patterns typical for ventral or lateral hypoderm are never found in *dorsal* embryos. In particular, the dense prominent denticle rows marking the segment borders on the ventral side of a normal larva are always lacking (Fig. 7).

3. Gastrulation in dorsal embryos is symmetrical along the dorso-ventral axis. The failure in *dorsal* embryos to form tissue normally derived from ventral and lateral regions of the egg may have either of three reasons: a) cells are not formed at the ventral egg parts, b) cells at the ventral egg parts do not develop or differentiate, c) cells at the ventral egg parts are programmed

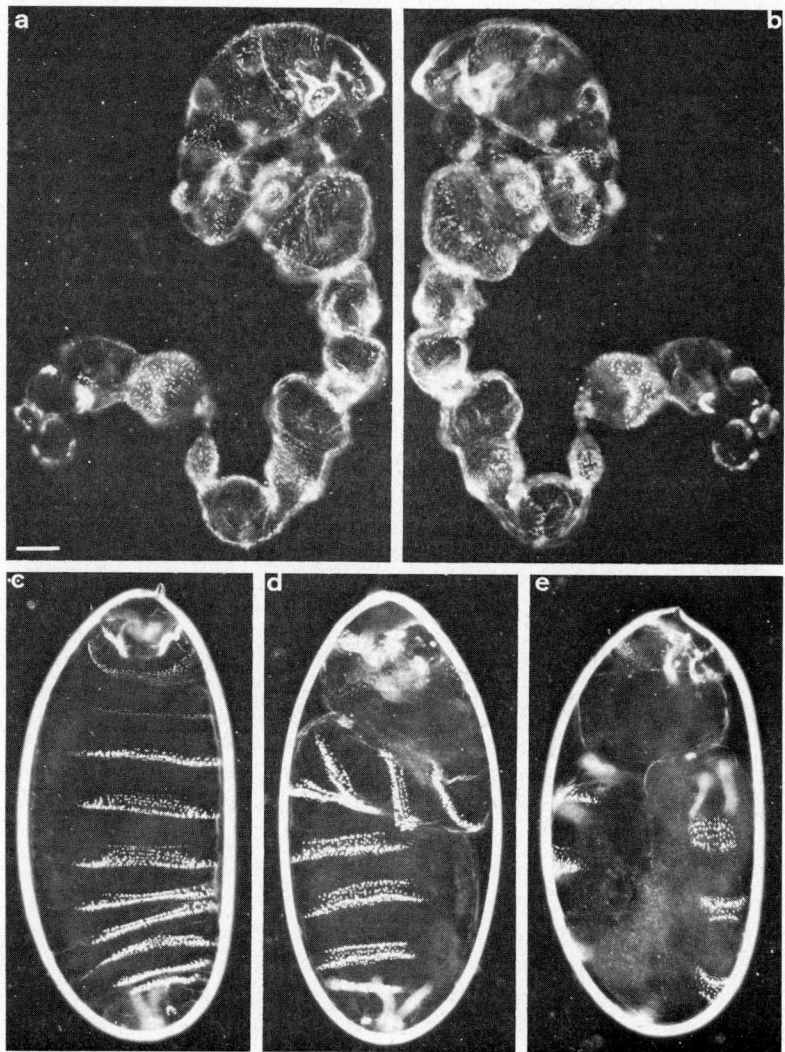

Fig. 7. The hypoderm pattern of dorsal phenotypes. a) and b) show the cuticle of an embryo released from the egg case revealing the recessive *dorsal* phenotype. a) and b) show pictures taken from the two sides of the preparation. Note the absence of ventral hook rows (see Fig. 1a) and the similarity of the pattern of fine hairs at both sides of the embryo. c) - e) *dorsal*-dominant embryos, showing various degrees of expression of the mutant phenotype. c) a "straight" embryo; it is almost normal, except for a disarrangement of the mouth parts and an irregular spacing of the ventral hook rows. d) a "knot" embryo; the hypoderm bag is wound up in an S-shape within the egg case. e) a "tail up" embryo; the hypoderm is wound up in a U-shape, such that the posterior tip of the embryo comes to lie very close to the thorax. Note the narrowness of the ventral denticle hook rows in this embryo. Anterior up in all figures. Darkfield optics. The scale is 50 μm.

to form dorsal instead of ventral structures. Observations on earlier stages of living and sectioned material indicate that the third possibility is realized. *Dorsal* embryos form a normal-appearing cellular blastoderm. Gastrulation, however, is highly abnormal in that it shows no sign of dorso-ventral polarity (Fig. 8). At the ventral egg side, neither a ventral furrow nor the anterior midgut invagination is formed. Instead, the blastoderm cell layer thickens and develops transverse folds at both the ventral and the dorsal side of the embryo. At the posterior tip of the egg, the cells which would normally form the posterior midgut pocket flatten out, presumably being pulled anteriorly by the remainder of the blastoderm. An indentation appears between vitelline membrane and the posterior part of the embryo at both the dorsal and the ventral sides of the egg. The pole cells stay at the tip. With increasing developmental time, deep folds appear successively in all regions of the egg. Longitudinal sections indicate that the *dorsal* embryo has remained a continuous single cell layer also at later stages, although with many infoldings. Thus the lack of ventrally or laterally derived structures in *dorsal* embryos does not stem from a failure of cells to form or to differentiate in these egg regions.

4. Dorsal affects the entire embryonic organization. It appears that normal cells are formed in all egg regions in *dorsal* embryos, but these cells are all channeled into a differentiation pathway typical for dorsally derived cells: cells in the ventral midline form dorsal hypoderm instead of mesoderm, ventro-lateral cells form dorsal hypoderm instead of the nervous system, lateral cells form dorsal hypoderm instead of ventral and lateral hypoderm. Thus like in *bicaudal* embryos, the entire fate map is changed.

At present it is not possible to decide which of the two alternative interpretations of the *dorsal* pattern is correct: it may be a mirror image duplication of the normal dorsal pattern, or it may result from a complete abolition of dorso-ventral polarity, which would yield a rotationally symmetric structure. The difficulty of interpretation stems from the lack of positional markers within the dorsal hair pattern and the rather disorganized patterns formed by *dorsal* embryos. The regular folding pattern observed in gastrulating *dorsal* embryos looks much the same if viewed from either the lateral or dorsal side of the embryo, suggesting rotational symmetry. On the other hand, if a duplication would include only a small fraction of the dorsal half of the embryo (analogous to the small fraction of the posterior pattern duplicated in *bicaudal* embryos), a mirror image duplication would be very hard to detect.

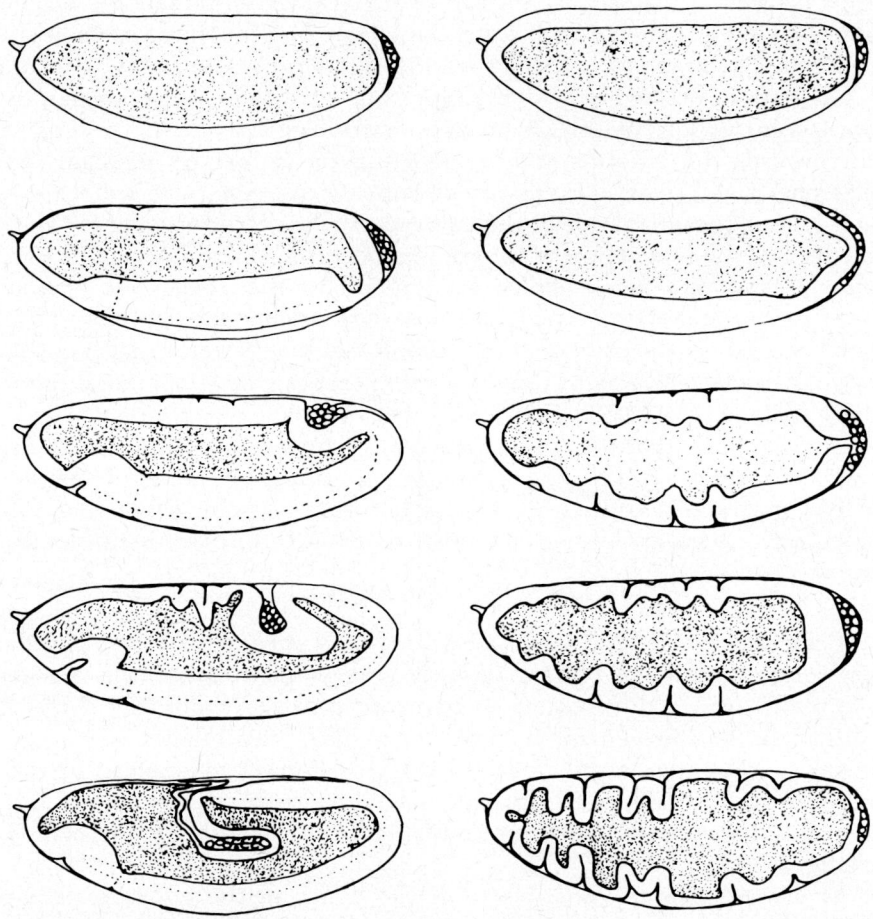

Fig. 8. Gastrulation in normal and dorsal embryos. Semi-schematic drawings of photographs from living and fixed material. Left, a gastrulating normal embryo (3-4h). Right, a dorsal embryo of corresponding age. Anterior, left: Ventral side of egg at bottom in all figures. For further explanation see text.

5. *Dorsal is a single point mutation and specially affects embryonic pattern formation.* For the interpretation of the *dorsal* phenotype it is of prime importance to know whether mutations at more than one locus are involved. Segregation and recombination analysis positioned *dorsal* to the left arm of the second chromosome at map position 52.9. Complementation tests with a set of overlapping deficiencies (Wright *et al.*, 1976) revealed that the *dorsal* locus lies within a small region of approximately 10 bands (36C2-4; 36C11) on the giant chromosome. Thus it is very likely that the *dorsal*

phenotype is caused by a mutation at only a single locus. Since heterozygotes of dorsal and a deficiency covering the dorsal locus, DF(2)137, show a phenotype which is indistinguisable from the homozygous phenotype, it is suggested that dorsal is an amorphic mutation, i.e. the phenotype caused by the lack of a functional dorsal gene product. The fact that the dorsal phenotype shows complete penetrance and is not temperature-sensitive supports this notion.

Although dorsal exerts a dramatic effect on embryonic organization, it does not seem to affect other vital functions during the development of the fly. The viability of both homozygotes and hemizygotes is normal at low as well as high temperature. Mutant females lay eggs with a rate similar to that of control females and mutant males have normal fertility (Table I). Moreover, the development of the dorsal phenotype is independent of the embryonic genotype, and is not rescued by the paternal genotype. These data suggest that the expression of the dorsal gene is restricted to oogenesis.

TABLE I

Maternal Effect of Dorsal

Parental genotype[a]	Number of females	Number of eggs[b]	Eggs per day per female	Eggs hatched[c] (%)
dl/dl X dl/+	8	797	38.6	0
dl/+ X dl/dl	7	669	37.0	92
dl/dl X +/+	8	820	39.7	0

a) The exact genotypes were: $dl/dl = dl\ cn\ sca\ /\ al\ dp\ b\ dl\ pr\ cn\ sca$; $dl/+ = dl\ cn\ sca\ /\ al\ dp\ b\ pr\ cn\ sca$. +/+ males were from an Oregon R Stock.

b) Eggs were collected from individual females at 22° in three successive intervals (total period 62 h), starting 4-6 days after eclosion of females.

c) At least 90% of the eggs from dl/dl females showed the dorsal phenotype, while the others were probably unfertilized. The unhatched eggs from $dl/+$ females were mostly undeveloped.

B. The Dominant Dorsal Phenotype

1. The dorsal-dominant phenotype is temperature sensitive and dependent on the genetic background of the female. While at low temperature the eggs from heterozygous $dl/+$ females develop normally, at high temperature, a small fraction of the embryos develop to the larval stage but fail to hatch. This dominant temperature-sensitive maternal effect lethality is variable and futhermore depends strongly on the genetic background of the female. Preliminary evidence suggests that deficiencies for *vg* and adjacent loci, including the *bicaudal* locus, have a particularly strong enhancing effect. Because $dl\ bic/+vg^D$ females show a more extreme expression of the dominant *dorsal* phenotype than $dl\ +/+\ vg^D$ females, it seems that the enhancing effect of the *vg* deficiencies is due to hemizygosity of the *bic* locus. However, more experiments are needed to clarify this point. The description and interpretation to follow must therefore be regarded as preliminary. The dominant *dorsal* phenotype which will be discussed below, is observed in eggs from $dl\ bic/+\ vg^D$ females.

2. Dorsal-dominant embryos lack organs derived from the ventral region of the egg. *Dorsal*-dominant embryos consist of a bag of well developed larval hypoderm arranged more or less regularly in the egg case (Fig. 7c-e), filled with masses of unconsumed yolk. Few, if any, internal organs can be distinguished. The embryos are inert and lack any signs of muscular movements. Analysis of the pattern in embryos released from the egg case and stretched out reveal that segmentation in these embyos is quite normal: they have the normal number and sequence of segments in the antero-posterior direction and the antero-posterior distance between adjacent segments is constant, indicating that the irregular distance of segment borders as observed in unhatched embryos is caused by the lack of interior organs, in particular muscles, rather than defects in the hypoderm itself. The dorso-ventral hypoderm pattern, however, is frequently abnormal in that the ventral-most pattern elements are reduced as is evident from the often much smaller transverse width of the ventral denticle rows. This effect is variable in *dorsal*-dominant embryos and strongest in the most extreme type, the "tail up" embryos (Fig. 7e). These observations suggest that in *dorsal*-dominant embryos ventral pattern elements, including ventral parts of the hypoderm, are missing to various extents.

3. Dorsal-dominant embryos do not form the ventral furrow. The failure of *dorsal*-dominant embryos to form ventrally derived organs is not caused by a failure of cells to form in ventral egg parts. The cellular blastoderm is perfectly normal. During gastrulation the first deviation from normal development is observed: *dorsal*-dominant embryos do not form the ventral furrow nor the anterior midgut invagination (Fig. 5d). As mentioned earlier, these invaginations separate the mesoderm and part of the endoderm anlage from the remainder of the embryo. In *dorsal*-dominant embryos, the ventral part remains a single cell layer, while the other morphogenetic movements at gastrulation, like the cephalic furrow, the posterior midgut invagination and the extension of the "germ band" appear quite normal. Observations on late stages indicate that *dorsal*-dominant embryos frequently lack the nervous system, which in normal embryos forms a very prominent mass of cells within the germ band. Some of the embryos fail to undergo the process of shortening of germ band and presumably give rise to the "tail up" embryos (Fig. 7e). In summary, the lack of internal organs as observed in differentiated *dorsal*-dominant embryos is explained by a failure of the respective anlagen to form early in development.

4. *In dorsal-dominant embryos, the fate of ventral cells is changed.* What is the fate of the cells in the ventral region of *dorsal*-dominant embryos? Our observations suggest that they develop structures derived from the more lateral regions in normal eggs. Cells which normally would form mesoderm, now form nervous system or ventral hypoderm instead, depending on the expression of the mutant phenotype. One major argument for this interpretation is that the hypoderm in *dorsal*-dominant embryos forms a closed tube. If a large portion of the ventral cells would die or degenerate some time in development (which we do not observe) one would expect the hypoderm to form a sheet, rather than a tube, because the anlage for the right and left side of the ventral hypoderm is separated on the fate map. A more direct evidence for a different programming of the ventral cells stems from laser irradiation experiments: as mentioned earlier, irradiation of the normal blastoderm in the ventral midline never yields defects in the larval hypoderm, while more lateral irradiations frequently give rise to such defects. When *dorsal*-dominant blastoderms, however, are irradiated at the ventral midline, the resulting larvae frequently show defects in the ventral hypoderm pattern; furthermore, the defect frequency is highest in the larvae which show the strongest expression of the phenotype, the "tail up" larvae. This experiment strongly suggests that cells at the ventral midline, which form mesoderm in normal embryos, are included in the anlage for the ventral hypoderm in *dorsal*-dominant embryos.

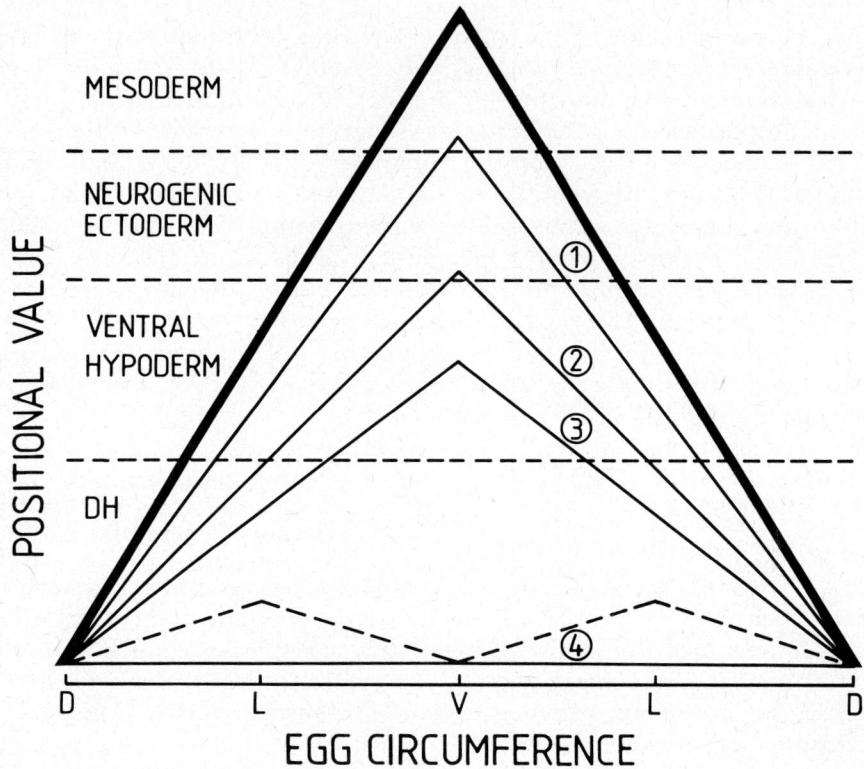

Fig. 9. Hypothetical gradients defining the dorso-ventral coordinate in normal and dorsal embryos. The horizontal scale gives the position along the egg circumference, from dorsal (left) to ventral (middle) to dorsal (right). The vertical scale refers to a positional value which is thought to in some way correspond to a morphogen concentration. The horizontal dotted lines indicate the threshold levels for mesoderm, neurogenic ectoderm, ventral hypoderm and dorsal hypoderm. They are chosen such that the upper, heavy line specifies the position of these anlagen in the blastoderm fate maps of a normal embryo. (1), (2) and (3) describe various phenotypes observed in *dorsal*-dominant embryos, as shown in Fig. 7c, d and e. In (3) only dorsal hypoderm and part of the ventral hypoderm is specified, which accounts for the lack of parts of ventral cuticle in the most extreme *dorsal*-dominant phenotype (Fig. 7e). (4) describes the recessive *dorsal* phenotype, the base line if *dorsal* leads to a complete abolition of dorso-ventral polarity, the zig-zag-dotted line if it would result in a shallow mirror-image duplication of the dorsal-most egg part. For further information see text.

C. *Interpretation of the Dorsal Phenotypes*

1. Arguments against localized determinants. The lack of internal organs in *dorsal*-dominant embryos cannot be simply explained by a failure, for structural reasons, of invaginations to form because the only invaginations which are affected are the ventral furrow and the anterior midgut anlage, whereas the cephalic furrow and the posterior midgut invagination appear normal. Moreover, structures which do not derive from invaginations, such as the nervous system and the ventral hypoderm, are affected as well. For similar reasons the absence of a "mesoderm determinant" in the recessive phenotype, or its presence in subnormal amount in the dominant phenotype, fails to explain the effect *dorsal* has on the entire spectrum of ventral tissues. A more general "ventral determinant" concept also does not adequately describe the dorsal phenotype because it fails to explain why structures typical for more lateral (in *dorsal*-dominant embryos) or dorsal (in the recessive *dorsal* phenotype) positions would be formed ventrally.

2. *A gradient defining the dorso-ventral coordinate may best explain the dorsal phenotypes.* Analogous to the explanation of the *bicaudal* phenotypes by an antero-posterior gradient, the spectrum of *dorsal* phenotypes may be described by assuming a dorso-ventral gradient. This gradient would normally be symmetrical, given the bilaterality of the egg. Such a gradient with a highpoint at the ventral midline is schematically decribed in Fig. 9. The recessive *dorsal* phenotype would be explained by the absence of the gradient, resulting in the abolition of dorso-ventral polarity and the formation of dorsal hypoderm and only dorsal hypoderm at all regions along the dorso-ventral coordinate. The *dorsal*-dominant phenotypes would arise if the height the gradient reaches at the ventral midline is reduced. This leads to a change in positional values in all cells along the dorso-ventral axis, and ventral cells form more lateral structures.

V. CONCLUDING REMARKS

The two maternal effect mutants described above share a number of common features, the most striking of which is that the entire embryonic organization is affected. Some structures are lacking, while others are formed at atypic regions of the egg. The transformations, however, do not result in juxtapositions of normally separated elements, but are coordinated such that an integrated pattern is formed. The types of patterns observed suggest that the spatial organization in normal development of the *Drosophila* embryo is achieved by a set of at least two gradients, oriented perpendicular to each other and defining an antero-posterior and a dorso-ventral coordinate.

In *Drosophila*, and insects in general, the embryonic axes normally correspond to the axes of the outer egg shell, suggesting that the orientation of the embryonic gradients is triggered by the same principle governing the polarity of the egg shells, presumably the polarity of the oocyte. The polarity of the oocyte may be transmitted to the seemingly homogeneous egg cell by means of the orientation of the gradients. In *bicaudal* and *dorsal,* the egg shell polarity (and presumably the oocyte polarity) is, however, not affected and an apolar or symmetrical embryo can develop from an initially polar environment. A possible interpretation of the pattern in *bicaudal* and *dorsal*-dominant embryos is that in the former the gradient in the embryo is initiated properly, but is not stable and in the latter it does not reach its normal height. This interpretation is supported by the finding that in both *bicaudal* and *dorsal*-dominant embryos the mutant phenotype may still be influenced by temperature shifts after the egg is laid (unpublished). In the recessive *dorsal* embryos the dorso-ventral gradient may not be established at all.

In this context it would be interesting to find mutants which affect the egg shell polarity and study their effects on embryonic organization. A promising case was recently reported by Lohs-Schardin and Sander (1976) who found a double-headed embryo in an egg shell which clearly showed an anterior duplication, having a micropyle at both anterior and posterior ends. Further, female sterile mutants developing bipolar follicles, but unfortunately no mature eggs, have been described by Gill (1963).

A gradient mechanism implies at least two steps in pattern specification. In the first step, the establishment of a stable gradient defines the spatial coordinates, while in a second step the initially continuous slope is interpreted in discrete portions. Sander (1976) has proposed the term *coordinate mutants* for mutants affecting the first step, in order to distinguish them from mutants which affect the second, interpretation, step. By this definition, both *bicaudal* and *dorsal* are coordinate mutants. The phenotypes associated with the *bithorax* pseudoallele series (Lewis, 1964) may result from a misinterpretation of the antero-posterior gradient, leading to the wrong type of segment in an otherwise normal pattern. In a similar manner, the *Notch* embryonic lethals (Poulson, reviewed in Wright, 1970) may be explained as misinterpretation of the dorso-ventral gradient, in that a much larger fraction of the ectoderm forms nervous tissue. These two mutants are zygotic, the transformation depending on the genotype of the individual affected, rather than on that of the mother. The coordinate mutants described in this paper, on the other hand, are strictly maternal. Since the sample of well characterized mutants in both categories is still small and to some extent non-random, generalizations are difficult. It remains to be seen whether the sole contribution of the mother to pattern formation in the embryo is the establishment of a set of spatial coordinates, while all further subdivisions rest upon the genome of the embryo.

ACKNOWLEDGEMENTS

I thank Dr. Margit Lohs-Schardin for her contribution of part of the previously unpublished work, Claus Christensen for the photographic prints and Eric Wieschaus for drawing Fig. 8. I am much indebted to Prof. K. Sander and Dr. E. Wieschaus for stimulating discussions and critical reading of the manuscript. This work was supported by long term fellowships from the European Molecular Biology Organization and the Deutsche Forschungsgemeinschaft.

REFERENCES

Bull, A.L. (1966). *J. Exp. Zool.* **161**, 221-241.
Gierer, A. (1977). *Curr. Top. Develop. Biol.* **11**, 17-59.
Gierer, A. and Meinhardt, H. (1972). *Kybernetik* **12**, 30-39.
Gill, K. (1963). *J. Exp. Zool.* **152**, 251-277.
Graham, C.F. (1976). In "The Developmental Biology of Plants and Animals" (C.F. Graham and P.F. Wareing, eds.), pp. 14-27. Blackwell, Oxford.
Illmensee, K., Mahowald, A.P. and Loomis, M.R. (1976). *Develop. Biol.* **46**, 40-65.
Kalthoff, K. (1978). *J. Cell Sci.* **29**, 1-15.
Lewis, E.B. (1964). In "The Role of Chromosomes in Development" (M. Locke, ed.), pp. 231-252. Academic Press, New York.
Lohs-Schardin, M. and Sander, K. (1976). *Roux' Archives* **179**, 159-162.
Lohs-Schardin, M. and Nusslein-Volhard, C. (1979). in preparation
Meinhardt, H. (1977). *J. Cell Sci.* **23**, 117-139.
Nüsslein-Volhard, C. (1977). *Roux' Archives* **183**, 249-268.
Poulson, D.F. (1950). " "Biology of *Drosophila*" (M. Demerec, ed.), pp. 168-270. New York, Hafner.
Sander, K. (1976). *Adv. Insect Physiol.* **12**, 125-238.
Van der Meer, J. (1977a). *Dros. Inf. Serv.* **52**, 160.
Van der Meer, J. (1977b). Thesis, Nijmegen.
Wright, T.R.F. (1970). *Adv. Genetics* **15**, 262-396.
Wright, T.R.F., Hodgetts, R.B. and Sherald, A.F. (1976). *Genetics* **84**, 267-285.

III. Pattern Formation in Developing Systems

An Analysis of Cell-Surface Patterning in *Tetrahymena*

Joseph Frankel

Department of Zoology
University of Iowa
Iowa City, Iowa 52242

I.	Introduction	215
II.	Description of the System	218
III.	Short-Range Positioning	222
IV.	Relational Systems of Long-Range Positioning	225
	A. Control of Latitudes	226
	B. Control of Longitudes	230
V.	What are the Underlying Mechanisms?	233
	A. The Duality of Positional Systems	233
	B. Do Pre-existing Ciliary Structures Serve as Reference Points?	235
	C. What Measuring System(s) Does the Cell Use?	238
VI.	Conclusions	242
	References	244

I. INTRODUCTION

Ciliated protozoa are unicellular, yet manifest an extraordinarily complex cell surface organization. In the light of our rapidly expanding knowledge of the mosaic nature of cell membranes generally (e.g., Nicolson *et al.*, 1977), this complexity of the ciliate surface may not be exceptional. What is exceptional, however, is that manifestations of this complexity are readily visible at the light microscopical level, and are amenable to counting and measurement and also to microsurgical and genetic manipulation.

Abbreviations: OA: Oral area; CV: Contractile vacuole; CVP: Contractile vacuole pore; LM: Longitudinal microtubule band.

The classic method for experimental analysis of regulation of pattern has been the same in ciliates as in metazoa: microsurgery. The most extensively operated ciliate is the large, trumpet-shaped *Stentor coeruleus*. T. H. Morgan (1901) was operating on this organism at about the same time as he and his students were carrying out pioneering microsurgical studies on hydroids. A *Stentor* renascence began in the late 1940's, especially after Vance Tartar worked out grafting procedures for this organism. The major insights emerging from Tartar's work on *Stentor* relate to the pre-eminent role of the cell surface layer (cortex) in the regulation of pattern: nucleated cell fragments deprived of the cortex could survive for a long time but could not form any of the ciliary organelles of the cell surface; such cells, if provided with a small patch of cortex, could reconstitute a small but structurally complete stentor (Tartar, 1961, p. 45). Tartar further made the major discovery, first reported in the 14th symposium of this series, that the site of origin and the spatial orientation of the most prominent cell-surface organelle system, the feeding apparatus, are determined by a specific pattern-juxtaposition within the cell surface itself (Tartar, 1956). Furthermore, this pattern-juxtaposition (the "zone of stripe-width contrast") was shown to be vegetatively self-propagating through longitudinal extension over what could be viewed as a lengthening cylindrical surface which sequentially becomes cut up into segments by transverse division furrows. (e.g. see Tartar, 1962, p. 23).

Tartar's observations and insights have been extended in two different conceptual directions. First, Tartar noticed that when two of the pattern juxtapositions determining the site of development of feeding structures are grafted close to one another, they tend to interact in a manner suggestive of mutual inhibition. On the basis of these and other observations, Tartar in 1956 wrote of ". . an intimate relatedness of pattern components . . ." (Tartar, 1956, p. 84). Gotram Uhlig pursued this topic in an exhaustive microsurgical analysis that led him to conclude that cortical development in *Stentor* is controlled by two interacting orthogonally disposed gradients, one horizontally encircling the cell and the other vertically disposed along the apical-basal axis (Uhlig, 1959, 1960). Uhlig's two-gradient hypothesis was expressed in terms similar to those applied to regeneration in hydroids by Huxley and De Beer (1934, chapter 8) and to development of amphibian embryos by Kühn (1955, lecture 16), and could rather easily be reformulated in Wolpert's (1969) newer and more precise terminology of positional information (Frankel, 1974). Similar formulations could also be applied to results of microsurgical studies in the structurally very different hypotrich ciliates (Jerka-Dziadosz, 1974).

The major paradigm-organism for the second conceptual approach to ciliate patterns is actually not *Stentor*, but rather *Paramecium*. Tartar had worked with this genus before he turned to *Stentor*, and inferred from microsurgical experiments that the formation of new feeding structures is restricted to a highly localized region (Tartar, 1941, 1954), a conclusion that was definitively established by Hanson's analysis of doublet paramecia by localized ultraviolet microirradiation (Hanson, 1962; Hanson and Ungerlieder, 1973). Sonneborn then took advantage of the great genetic opportunities of the *"Paramecium aurelia"* species group and exhaustively proved that cells of identical genotype and common internal fluid cytoplasm could serially propagate differences in the multiplicity of feeding structures and associated organelles (Sonneborn, 1963). Subsequently and most spectacularly, Beisson and Sonneborn (1965) demonstrated the autonomy of the polarity and asymmetry of a longitudinally propagated organelle set, the ciliary meridian. Tartar's early conclusion, that in *Paramecium* "All the new structures present in the daughter cells, excepting the new contractile vacuoles, arise in direct genetic continuity with previously existing parts, a striking case of cytoplasmic inheritance" (Tartar, 1941) was confirmed and successively elaborated into the concept of "cytotaxis" (Sonneborn, 1964) and later into the idea of positioning of new structures in relation to old ones through "sequential nearest neighbor interactions" (Sonneborn, 1975).

It is worth pointing out that, even though the "nearest-neighbor" approach could perhaps be applied to *Stentor* and the gradient conceptualizations to *Paramecium* (e.g. Ng, 1976), it is no accident that these somewhat diverse viewpoints were first formulated as they were. *Stentor* is developmentally an extraordinarily flexible organism, with capacities for morphallactic restitution easily equal to those of *Hydra*. *Paramecium*, on the other hand, is the most "mosaic" of all ciliates that have been subjected to intensive investigation; as was pointed out by Tartar in the third Symposium of this Series [(1941), see also Schwartz (1963)], *Paramecium* is unusual among ciliates in being virtually incapable of morphallactic regulation. Hence it is not surprising that hypotheses of pattern regulation based on *Stentor* stress interaction whereas those that use *Paramecium* as a model stress structural continuity.

More recently still, several investigators have taken up the analysis of cortical pattern in the small and (for ciliates) relatively simple organisms of the *"Tetrahymena pyriformis"* species group. These organisms are almost as amenable to genetic analysis as are the *"Paramecium aurelia"* species, yet they approach *Stentor* in their capacity for morphallactic regulation and restitution of cell surface pattern. Investigations carried out on

Tetrahymena during the past 15 years have provided precise and detailed examples of both structural continuity and long-range interaction, and also have laid the groundwork for a developmental genetics of intracellular patterning.

In what follows I will review the recent analyses of intracellular patterning in *Tetrahymena*, with only passing reference to other ciliates. The conceptual framework of this review will be the dualistic view that qualitatively different "nearest-neighbor" and "gradient-field" mechanisms operate simultaneously, and jointly determine intracellular patterns. I will first introduce the system, then provide examples demonstrating the operation of both mechanisms, then supply evidence for the distinctness of the two, and finally consider the question of what sorts of measuring devices the cell might use and from where they measure.

II. DESCRIPTION OF THE SYSTEM

Tetrahymena thermophila [formerly *T. pyriformis*, syngen 1; cf. Nanney and McCoy (1976)] is a relatively small ciliate, averaging about 50 μm in length. It possesses a single macronucleus, one micronucleus, and a posteriorly situated pulsatile organelle involved in water and ion balance, the contractile vacuole (CV) (Fig. 1A). The structures that we will be considering are all located on the cell surface, which is schematically illustrated in a ventral view (Fig. 1B), a polar projection (Fig. 1C), and a "Mercator projection" (Fig. 1D). The cell typically possesses 17 to 21 longitudinal rows of ciliary basal bodies, here called ciliary meridians, all but two of which extend from the anterior to the posterior pole. The oral apparatus is situated near the anterior end of the cell and includes three sets of closely-apposed ciliary rows called membranelles, which are normally oriented from the anterior-right to posterior-left.* There is also a double row of basal bodies, only one row ciliated, named the undulating membrane, which is located to the right of the membranelles. The oral apparatus is so situated that generally two ciliary meridans fail to extend to the anterior pole but instead terminate on the posterior margin of the oral apparatus. The ciliary meridians are all arbitrarily numbered so that the right postoral meridian is no. 1, and numbering proceeds clockwise around the cell as viewed from the anterior pole; the highest numbered ciliary meridian ("n") is thus the left postoral meridian. These relationships are most clearly seen in a polar projection (Fig. 1C).

*Throughout this report, *right* and *left* refer to the observer's right and left, assuming that he stands inside the animal so that his anterior-posterior axis coincides with that of the animal, and he keeps turning around his own longitudinal axis to face the surface of the animal.

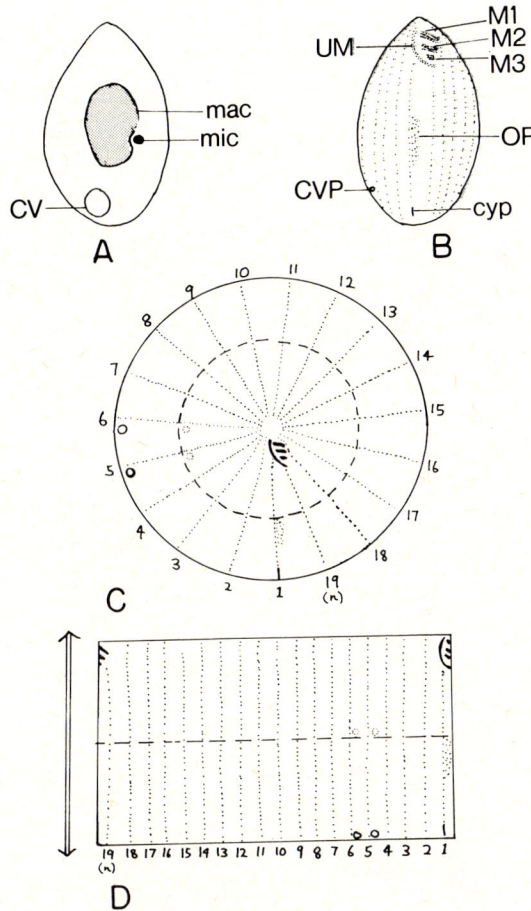

Fig. 1. Diagrammatic representations of the surface anatomy of *T. thermophila*. A. Longitudinal section, showing macronucleus (mac), micronucleus (mic) and contractile vacuole (CV). B. Surface view on oral side, showing features made visible by silver staining. Each small black dot indicates a ciliary basal body. Ciliary features include the oral apparatus, consisting of 3 membranelles (M1, M2, M3) and an undulating membrane (UM), longitudinal ciliary meridians (10 shown), and an oral primordium (OP) here illustrated in an early stage of development. Non-ciliary differentiations include a cytoproct (cyp) and contractile vacuole pores (CVP). C. A "polar projection" of the surface of this cell, with the anterior pole in the center. The 19 ciliary meridians are numbered according to the standard system. The dashed circle indicates the site of the fission zone, with the positions of the new CVPs indicated by dotted circles. Other features are drawn as in Fig. 1B. D. A "Mercator projection" of the surface of this same cell. The cell surface has been longitudinally "slit" between meridians 1 and 19 (n), and then "rolled out" with the same ensuing distortion as in a standard world map. All pictorial conventions as in Fig. 1C. The vertical arrows at the left indicate the direction of surface growth.

There are two special posterior structures. A slitlike cytoproct, or egestion pore, is located at the posterior pole of meridian no. 1. The cell also possesses one to three (typically two) contractile vacuole pores (CVPs), through which the CV empties its contents. These structures are situated just to the cell's left of the posterior ends of one or two ciliary meridians; when there are two "CVP meridians" they are almost invariably adjacent to one another. In cells with 19 ciliary meridians these CVP meridians are typically nos. 5 and 6 (Figs. 1C and 1D).

Two further sets of structures, not shown in Fig. 1, require mention. First, a longitudinal microtubule band (LM) is located directly under the innermost cell membrane to the cell's right of each ciliary meridian. Second, a large proportion of the cell's mitochondria are situated in more or less regular longitudinal rows just underneath the cell surface; there are one or two files of mitochondria between each pair of ciliary meridians, disposed in a definite asymmetrical arrangement (Fig. 4C) (Jerka-Dziadosz, personal communication; Aufderheide, 1978).

The ciliary rows elongate through addition of new ciliary units along the axis of the row itself. New basal bodies are generated anterior to old ones (Williams, 1964; Allen, 1969; Perlman, 1973); the cilia and accessory structures (rootlets and microtubule bundles) are formed afterwards (Allen, 1969). Although in exponentially growing cells new basal bodies are probably being added throughout the cell cycle, the most recent observations suggest that the majority of this increase is occurring in the second half of the cycle (Nanney, 1975).

The development of new OAs, cytoprocts, and CVPs typically differs from the propagation of the ciliary meridians in that new structures develop at a substantial distance from corresponding old ones. The new oral apparatus develops through the elaboration of a field of unciliated basal bodies immediately to the left of the equatorial region of the right postoral ciliary meridian (no. 1) (Fig. 1B). The field grows by addition of new basal bodies, and progressively becomes organized to produce the definitive arrangement of membranelles and undulating membrane (Fig. 2A). The details of this process are reviewed elsewhere (Frankel and Williams, 1973; see also McCoy, 1974). During late stages of oral development the fission zone appears as an equatorial band of gaps in the ciliary meridians, situated at a latitude just anterior to the developing oral apparatus (dashed line in Figs. 1C and 1D). The cell than cleaves along this fission zone while the final stages in the elaboration of the new (and remodelling of the old) oral structures are taking place (Fig. 2C). The substantial spatial separation of old and new oral structures found in dividing *Tetrahymena* cells is also characteristic of the majority of ciliates

CELL - SURFACE PATTERNING IN *TETRAHYMENA*

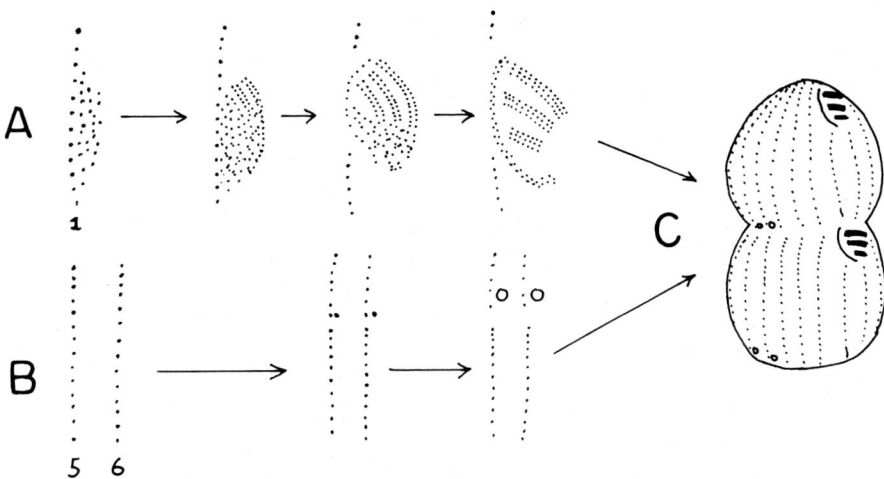

Fig. 2. Selected aspects of cell-surface development associated with cell division. *A.* Development of new oral apparatus adjacent to meridian no. 1. *B.* Appearance of new contractile vacuole pores along CVP meridians. Vertically aligned sketches in *A* and *B* represent concurrent events. C. Diagram of right oral side of cell during early division constriction.

(including *Stentor*) and differs from the situation in *Paramecium*, in which the new oral system develops within a specialized region of the old one (Kaneda and Hanson, 1974; Jones, 1976). However, in *Tetrahymena* there also exists an alternative process of oral development not associated with cell division, called oral replacement, in which new oral structures are formed in part within and in part adjacent to old ones (Frankel, 1969; Kaczanowski, 1976).

The new posterior structures, the cytoproct as well as the contractile vacuole and its pores, develop at the posterior end of the anterior division product just prior to the onset of division furrowing. The new CVPs become visible at the same time that the transverse gaps in the ciliary meridians are formed, always appearing just anterior to these gaps (Fig. 2B). The new CV becomes visible and functional at the same time. The formation of a new cytoproct and contractile vacuole system separate from the old ones is the rule in all ciliates, including even *Paramecium* (King, 1954; Sonneborn, 1963).

III. SHORT-RANGE POSITIONING

The combination of longitudinal perpetuation of ciliary meridians and transverse division of the cell creates the potential for immortality of each ciliary meridian (see Fig. 1D, in which the meridians continuously grow "north-south" and are bisected "east-west"). Within certain limits, such potential immortality might actually be realized. The limits are set by a cellular control of the differential stabilities of different numbers of ciliary meridians. Differences in initial ciliary meridian numbers can be vegetatively propagated with a stability that is inversely proportional to distance from a "stability center" (Nanney, 1966b,c; 1968b) that is probably under nuclear genic control (cf. Heckmann and Frankel, 1968). The fidelity of inertial maintenance of the pre-existing number of ciliary meridians is not affected by genetic exchange (conjugation) between cells of similar genotypes but of different pre-existing ciliary meridian numbers. Nanney's analysis of this maintenance was based on an experimental strategy of "cross-sectional" analysis that involved comparison of very many clones which were maintained for 20 fissions after their inception. We have followed this up with a "longitudinal" analysis of a smaller number of clones that were maintained and periodically sampled for a total of about 1000 fissions after their inception at conjugation. The time-course of changes in number of ciliary meridians in a typical pair of sister exconjugant clones derived from a mating between two parental clones manifesting 23-24 and 18-19

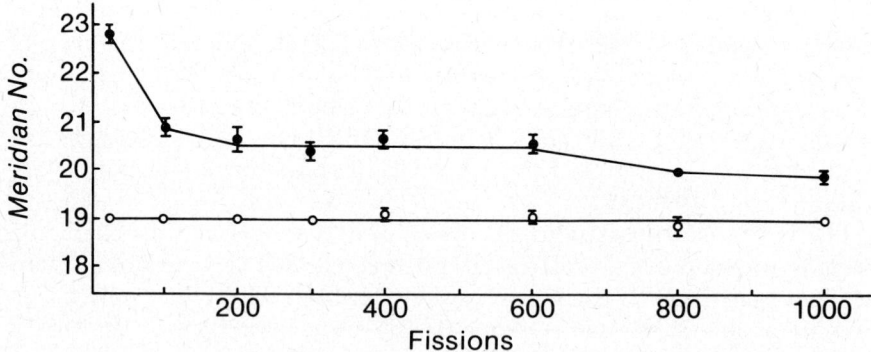

Fig. 3. Time-course of change in number of ciliary meridians in two sister exconjugant clones, represented by shaded and open circles respectively. The abscissa indicates the number of fissions after conjugation. The bars above and below the points indicate 95% confidence intervals; where bars are absent all of the cells tallied in the standard sample of 20 cells had the same number of ciliary meridians.

meridians respectively is shown in Fig. 3. The clone derived from the "low" parent continues to maintain predominantly 19 ciliary meridians, whereas the one derived from the "high" parent rapidly falls from 23 to 20-21 meridians within 100 fissions but changes little (or very slowly) thereafter. Our studies (Frankel and Nelsen, in preparation) suggest that 19 and 20 ciliary meridians are maximally stable, while 18 and 21 are nearly as stable. Within this range pre-existing differences can be propagated for very long times.

An altogether different demonstration of the substantial autonomy of ciliary meridians was provided in *Paramecium* by Beisson and Sonneborn (1965), who demonstrated that a clone could perpetuate a 180°-rotated configuration of one or more ciliary meridians. Stephen Ng succeeded in getting *Tetrahymena* to carry out the same feat after creating a temporary heteropolar configuration in a temperature-sensitive mutant conditionally blocked in cytokinesis (Ng and Frankel, 1977). A transmission electron microscopical analysis has demonstrated that all of the structural elements associated with the basal bodies as well as the adjacent microtubule bands are present in a 180°-rotated configuration (Ng and Williams, 1977; cf. Fig. 4A). Further, the direction of ciliary beating in the rotated ciliary meridians is inverted in *Tetrahymena* as it is in *Paramecium* (Tamm et al., 1975), and new basal bodies develop posterior (with reference to the cell) rather than anterior to old ones in the inverted rows (Ng and Frankel, 1977). The inversion is thus structural (Fig. 4A), functional, and morphogenetic.**

The results outlined above allow us to extend to *Tetrahymena* the conclusion arrived at by Beisson and Sonneborn (1965) and Sonneborn (1970) for *Paramecium*, that the local cortical environment created by the position and orientation of pre-existing ciliary units within a ciliary row is determinative of the position and orientation of new units within the same row; the ciliary meridian can thus be considered to be a system in which the old structures serve as scaffolds for the new. This "scaffolding" function does not, however, restrict itself to the sites of production and orientation of the ciliary units themselves. Ng (1977) has shown that "fine positioning" of CVPs is also determined by the orientation of ciliary meridians: the CVP always appears on the side of the ciliary meridian opposite to the LM. The CVP is therefore formed to the left side of normally oriented ciliary meridians, and to the right of inverted meridians (Fig. 4B), and is very precisely situated relative to a basal body within the meridian (Ng, 1979b). Furthermore, Aufderheide

---

**Ng has also recently demonstrated that propagation of the longitudinal microtubule band of *Tetrahymena* is largely autonomous (Ng, 1978a) and unidirectional (Ng, 1978b, 1979a).

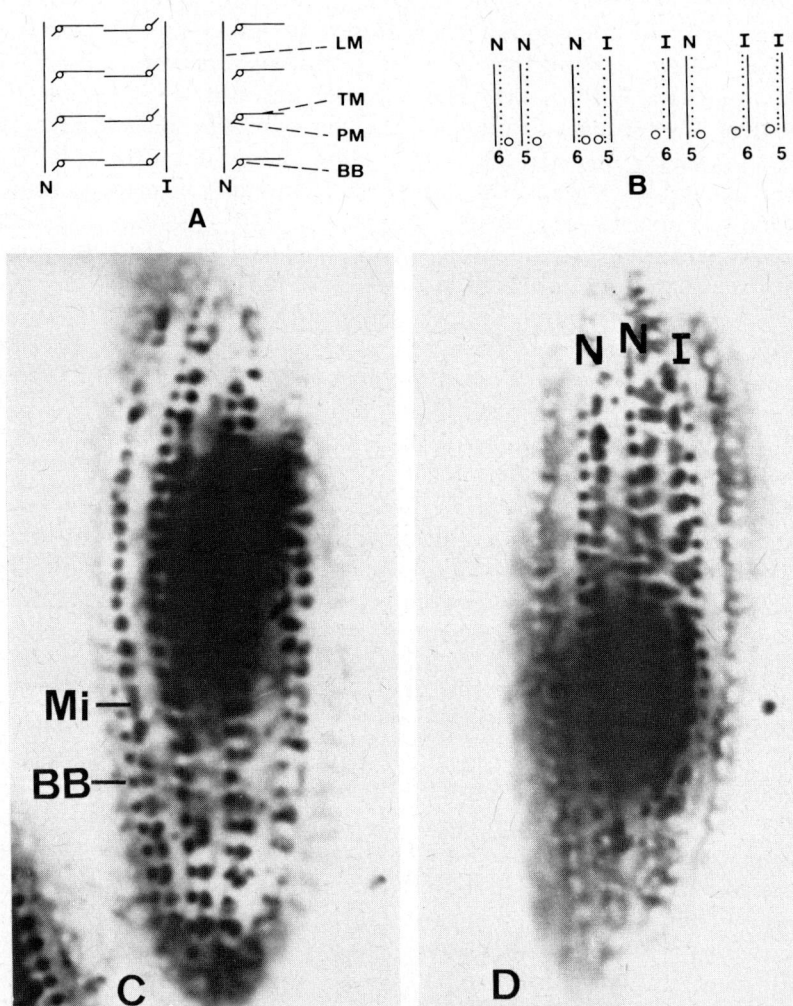

Fig. 4. Aspects of 180°-rotation of ciliary meridians in *T. thermophila*. A. A diagrammatic representation of part of the cell surface showing an inverted ciliary meridian (I) between two normal ciliary meridians (N). Basal bodies (BB) and three major cortical microtubule bands (longitudinal, LM; transverse, TM; postciliary, PM) are indicated. Redrawn from Fig. 5b of Ng and Frankel (1977). B. The fine-positioning of CVPs relative to normal and inverted CVP meridians. The four configurations illustrated correspond (left to right) to those photographically documented in Figs. 1, 10, 9, and 8, respectively, in Ng (1977). C. Geometrical relationship of cortically situated mitochondria to ciliary meridians. In this protargol-stained cell [a transformed "rapid swimmer"; cf. Nelsen and DeBault (1978)] a single row of mitochondria (Mi) is situated to the cell's left (viewer's right) of each ciliary meridian which is visible as a longitudinal row of basal bodies (BB). 2200X. D. A cell similar to C except for the presence of an inverted ciliary meridian (I). Note that mitochondria are located to the cell's *right* of this meridian, whereas they are situated to the cell's *left* of adjacent normally oriented ciliary meridians (N). 2200X.

(1978), following an earlier observation by Jerka-Dziadosz, demonstrated that the asymmetry of positioning of the cortically arrayed mitochondria was strongly correlated with, and presumably controlled by, the asymmetry of the neighboring ciliary meridians (Fig. 4C, D). Hence the ciliary meridian not only controls the geometry of its own perpetuation, but also has an "outreach" to adjacent cortically situated structures.

Before leaving the subject, it should be pointed out that the localized "scaffolding" system that we have been considering, though impressive in its short-term developmental effects, generally does not have the same long-term genetic stability as the molecular base-pairing template of DNA. This point has effectively been made by Nanney (1977), and we need here add only one further distinction. The cell appears to behave as if it has a genetically specified optimum of 19-20 normally oriented ciliary meridians. In clones bearing ciliary meridian *inversions*, these inversions must be maintained by constant selection of cells exhibiting certain characteristic abnormalities of swimming patterns that are diagnostic of the inversion; without (and sometimes even with) such selection clones lose the inversion. Exactly how such loss is accomplished is unknown, although both Ng (personal communication) and Jerka-Dziadosz (personal communication) have observed strong indications of resorption of existing basal bodies and/or of failure of formation of new basal bodies along inverted ciliary meridians, as well as hints of possible occasional re-rotations of entire rotated ciliary meridians. In analyses of the propagation of differences in *numbers* of normally oriented ciliary meridians, there is no known means of conscious selection by the experimenter. Hence the propagation over a long period of differences such as are shown in Fig. 3 suggests that the cell might truly be in-different as to whether it possesses 19 to 20 ciliary meridians. Within this narrow range, then, the "scaffolding" mechanism of longitudinal extension of ciliary meridians may indeed allow a non-genic perpetuation of a phenotypic difference that under suitable circumstances may approach the stability of differences based on alternative genic alleles.

IV. RELATIONAL SYSTEMS OF LONG-RANGE POSITIONING

We have already seen that some surface organelles, such as the primordium of the new OA during division, the new cytoproct, and the new CVP, develop at a considerable distance from the corresponding pre-existing structures. We will now consider how the positions of these structures are determined. Using the Mercator projection of the cell (Fig.

1D) as a provisional spatial framework, we first ask how structures are positioned along the vertical axis, i.e. how the cell controls the *latitudes* at which structures develop, and then ask how structures are positioned along the horizontal axis, i.e. how the cell determines the relevant *longitudes*. In this section we will review the phenomonology, and in the next section proceed to the difficult question of mechanism.

A. *Control of Latitudes*

As recognized very early by Morgan (1901) and emphasized recently by Wolpert (1969), the question of whether the size or position of a part is proportional to the dimensions of the whole is theoretically very important. A detailed analysis of the latitude at which the oral primordium develops during cell division of *T. thermophila* and the related *T. corlissi* has clearly demonstrated that positioning is relational (Lynn and Tucker, 1976; Lynn, 1977). As cell length increases, the distance between the old and new oral strucutures increases proportionately, and the number of basal bodies separating the anterior OA and the oral primordium along meridian no. 1 also increases. In his most recent study, Lynn (1977) compared the positioning of the oral primordium in wild type *T. thermophila* and in a "conical" form-mutant that at first glance appears to locate its oral primordium at a more posterior position than normal (Doerder *et al.*, 1975). Lynn found that *within* each genotype the position of the oral primordium shows a positive linear co-variation with cell length, but the regression lines are different for the two genotypes and do not extrapolate back to the origin of the graph. An *exact, common* proportionality, in which the points for both wild type and conical fit on the same regression line and extrapolate back to the origin, was obtained when the distance between old and new OAs was compared to an estimate of cell *surface area* rather than cell length (Lynn, 1977). These results demonstrate that the dividing cell does *not* determine the latitude of its new oral area by measuring a fixed distance backward from some anterior landmark, and strongly suggest that it is instead assessing some global property, possibly cell length but more likely some aspect of cell size.

A different type of evidence is available that also indicates that the cell is not positioning its oral primordium by counting a fixed number of basal bodies along the right postoral ciliary meridian. We have discovered a recessive "disorganized" mutation, *disA*, which when homozygous brings about a spatial disorganization of ciliary meridians that is particularly extreme at high temperatures (Frankel and Jenkins, unpublished). Despite the obviously very variable number and orientation of basal bodies in the region between the old and the new OA, the new oral

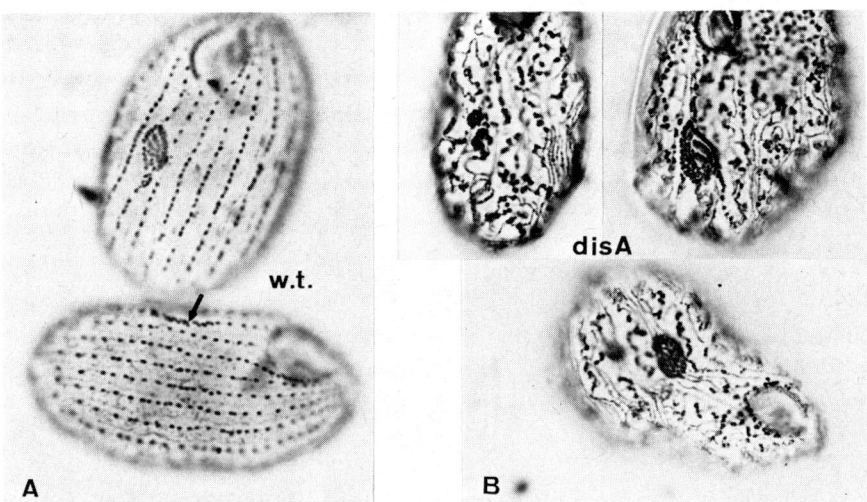

Fig. 5. Comparison of cortical configurations during oral development of wild type (WT) cells and of cells homozygous for the *disA* mutation. Cells were grown in a 2% proteose peptone — 0.5% yeast extract medium at 28° C and then shifted to 39.5° C and fixed after 3.5 hrs (w.t.) or 6 hours (*disA*) at the high temperature. *A.* Wild type cells, at 39.5° C, showing an oral primordium in the earliest stage (arrow) and during differentiation of membranelles (above). Note its approximately equatorial position adjacent to right postoral ciliary meridians. *B. disA* cells at 39.5° C in three stages of development of oral primordia. Note approximately normal location of primordium relative to anterior OA despite very abnormal pattern of ciliary meridians and somewhat abnormal cell shape. Chatton-Lwoff wet silver preparation, 850 X.

structures of mutant cells develop at approximately normal positions (Fig. 5).

The location of the fission zone is coordinated with that of the oral primordium, with the former appearing just anterior to the latter at an advanced stage of oral development. One of our temperature-sensitive "cell division arrest" mutations, *cdaE* (formerly *mo8*) seems to dissociate the two locations such that the fission zone develops at the latitude of the middle of the oral primordium rather than just anterior to it (Frankel *et al.,* 1977). However, a careful analysis by E.M. Nelsen has shown that the location at which the oral primordium begins to develop and the site of later appearance of the fission zone are both normal in *cdaE* cells grown at high temperature; the reason for the two systems coming to be latitudinally out of register is a localized failure of longitudinal surface extension along the oral meridian subsequent to the original appearance of the oral primordium (Frankel *et al.,* 1977). This apparent exception thus does not negate the rule of coordinate positioning of the new OA and the fission zone.

Both the cytoproct and the new CVP develop just anterior to the fission zone. For the CVP, there are two types of evidence that indicate that this spatial relationship is not coincidental. First, when CVP meridians are rotated 180°, the CVP appears to the cell's right rather than the usual left of a CVP meridian but it still forms at the fission zone rather than near the anterior end of the cell as would have been expected if the system determining the CVP were truly *rotated* through 180° along with the adjacent ciliary meridian (Ng, 1977). Second, temperature-sensitive cell division mutants at the *cdaA* locus (formerly *mo1*) are characterized by a failure of fission zone formation at nonpermissive temperatures and also manifest a coordinated failure in the elaboration of new CVPs (Frankel et al., 1977). When, in these mutants, some ciliary meridians become interrupted at the equator whereas others do not, the formation of a new CVP almost invariably succeeds or fails depending upon whether its adjacent CVP meridian becomes interrupted or not (Frankel, unpublished). However, following up on a prior observation that fission zones develop within inverted ciliary meridians later than within adjacent normal ciliary meridians (Ng and Frankel, 1977), Ng (1979c) has more recently observed that CVPs may develop adjacent to such inverted meridians before the fission zone gaps are evident in these meridians. These findings taken together suggest that the location of the fission zone and the CVPs do not influence each other directly but rather are separate outcomes of a preceding localized cellular activity that is common to both.

In exponentially growing tetrahymenas new basal bodies may appear within ciliary meridians at any latitude, but are more prevalent in the middle and posterior regions (Nanney, 1975; Kaczanowski, 1978). Kaczanowski (1978) detected a clear maximum in intensity of basal body proliferation in the equatorial region of cells that had been re-fed after prolonged starvation.

In general then, intense and apparently coordinated morphogenetic activity of various kinds is concentrated in the equatorial zone of cells preparing to divide. Lynn's analysis of oral primordium positioning taken together with the general association between the latitude of oral development and of the other manifestations of equatorial activity furthermore strongly suggest that the location of this active equatorial zone is determined by a cellular assessment of relative rather than absolute distance.

We might expect positioning mechanisms themselves to be subject to genic controls. There are currently two known perturbations of latitudes of development resulting from allelic substitutions. One, manifested under certain conditions in cells homozygous for *cdaA*, results in the appearance of supernumerary CVPs at an abnormal latitude, and will be described more fully in another context (section V,C). Another, more dramatic genically-controlled condition affects the latitude of oral development. Whereas wild type cells and the mutants discussed above (conical, disorganized) position their oral primordia at one of two discrete sites (mid-body, or adjacent to the old OA during oral replacement), cells that are homozygous for the temperature-sensitive recessive "pseudomacrostome" (*psmA*) mutation (Frankel et al., 1976) generate oral

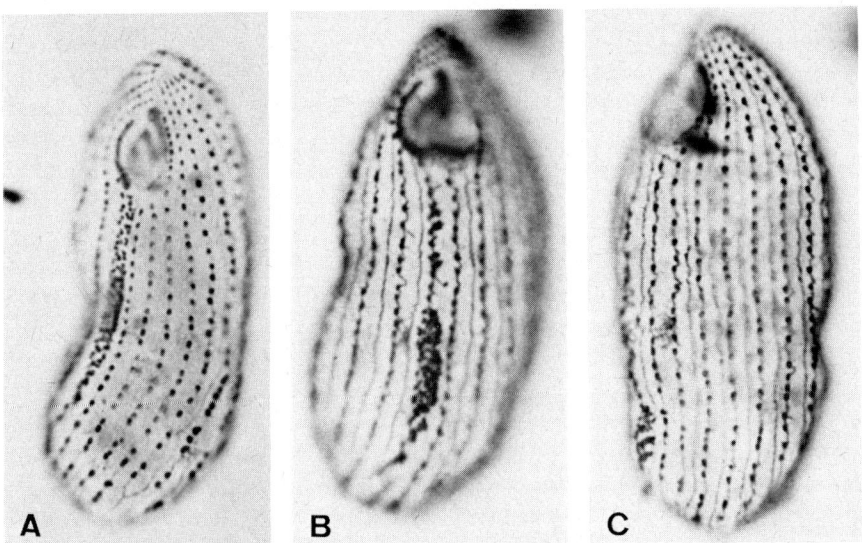

Fig. 6. Abnormal latitude of oral development in cells homozygous for the *psmA* mutation. Cells were grown in a 2% proteose peptone plus 0.5% yeast extract medium at 23° C (permissive) and then shifted to 39.5° C (nonpermissive). The cells illustrated were fixed 2 hours after the temperature shift, as they were forming the "first generation" of oral primordia at the nonpermissive temperature. The cell at the left (*A*) is developing an unusually long oral primordium along much of the length of the right postoral ciliary meridian, the center cell (*B*) is forming a major primordium at a more or less normal position with indications of a secondary oral primordium more anteriorly, while the third cell (*C*) is elaborating an oral primordium near the posterior end of the right postoral meridian. None of these configurations are observed in wild type cells (cf. Fig. 5A), and they are rare in *psmA* cells grown at 23° C or below. Chatton-Lwoff wet silver preparation, 1050 X.

primordia anywhere along meridian no. 1 at nonpermissive temperatures (Fig. 6; compare to Figs. 1B and 5A). The oral primordia usually develop into oversized feeding structures (hence the name) which, depending on their position, participate either in oral replacement or in abnormal cell division. We have recently uncovered a second and possibly a third locus generating a similar phenotype (Frankel and

Jenkins, unpublished). This phenotype partially mimics the macrostome phenotype of the polymorphic tetrahymenas, *T. patula* (Williams, 1960; Stone, 1963) and *T. vorax* (Buhse, 1966). The "pseudomacrostome" loci may thus be genic control switches analogous to the homoeotic loci in *Drosophila* (cf. Garcia-Bellido, 1975). Other genes affect the internal pattern of the OA with little or no effect on its size or position (Kaczanowski, 1975; 1976; Frankel *et al.*, 1977).

B. *Control of Longitudes*

We shift now to consideration of the determination of structures along the circular dimension of the cell, following the same order (OA — CVP — ciliary meridians) as we did above. First we ask, what determines the longitude at which the new oral system develops in dividing cells? At first sight the answer appears simple: the right postoral ciliary meridian is specialized as a "stomatogenic meridian" and thus serves as a spatial guide for the elaboration of the new OA. This simple answer is, however, almost certainly wrong. The oral primordium does not *always* develop along the right postoral ciliary meridian. Nanney (1967a) described two *Tetrahymena* clones in which formation of oral primordia along other meridians (usually but not always the *left* postoral meridian) is particularly common. This phenomenon, called by Nanney "cortical slippage", indicates that although the right postoral meridian normally accompanies the cell longitude along which the posterior OA is destined to develop, it does not determine the longitude. The same conclusion applies to the posteriorly situated cytoproct, which develops along the same longitude as the new oral area (Nanney, 1967a). Nanney (1967a) also pointed out that systematic cortical slippage in one direction, as encountered in one of the unusual clones, implies that any ciliary meridian can potentially serve as a "stomatogenic meridian".

As mentioned earlier, a *T. thermophila* cell with 19 ciliary meridians characteristically generates its CVPs adjacent to meridians 5 and 6. However, Nanney in 1966 demonstrated the dramatic fact that the position of the CVPs is proportional to the total number of ciliary meridians in the cell. For example, a cell with a total of 16 ciliary meridians will typically possess CVPs adjacent to meridians 4 and 5, while a cell with 23 ciliary meridians will usually have CVPs near meridians 6 and 7 (Nanney, 1966a, 1967b). Although the proportionality between the distance from the oral axis to the CVPs and the total cell circumference is not a simple linear one, both the general relationship and the specific irregularities in that relationship support a model in which the cell positions its CVPs within a zone centered at a specific proportion of the cell circumference to the right of the right postoral meridian (Nanney,

CELL - SURFACE PATTERNING IN *TETRAHYMENA* 231

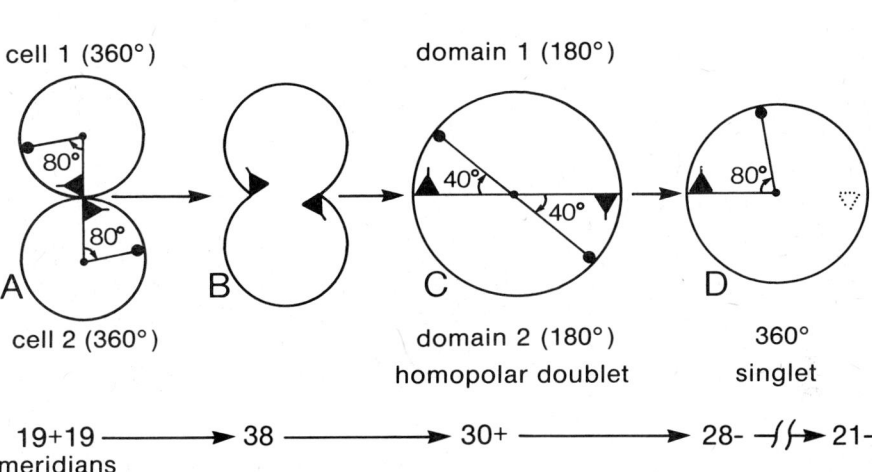

Fig. 7. Schematic illustration of formation of homopolar doublets from conjugating cells, and their subsequent reversion to singlets. Sketches are of highly schematized cross-sections. The triangles represent the OAs, with the sharp points indicating the direction of their asymmetry (undulating membrane side). The filled circles indicate the average position of the CVP *sets*. The angles are "central angles" [Nanney (1966a)], which indicate the locations of the CVP sets relative to the oral longitudes. A. Two conjugating cells. B. Permanent fusion of conjugating cells, after antiserum treatment. C. Homopolar doublet resulting after completion of fusion. D. Eventual singlet after loss of ciliary meridians from doublet. Based on results of Nanney *et al.*, (1975).

1966a). This proportion is most easily visualized as a "central angle" drawn with reference to a plane extending from the central longitudinal axis of the cell to the right postoral ciliary meridian (Fig. 7D; one can also easily visualize this angle in a polar projection of the *Tetrahymena* cell as is shown in Fig. 1C). In *T. thermophila* (then known as syngen 1 of *T. pyriformis*) this angle is near 80° (Nanney, 1967b). However, as Nanney emphasized from the start, the "central angle" is a largely fictional geometric construct. This is seen when parabiotic twins, known in the ciliate trade as "homopolar doublets", are generated as a consequence of division arrest or permanent fusion of conjugating pairs (Fig. 7). These initially contain a near double number of ciliary meridians and maintain and propagate two sets of oral structures; the CVPs now form 40° rather than 80° to the right of each oral longitude. Each component of the duplex is positioning its CVPs by measuring its own domain rather than the entire circumference. Nanney (1966a) concluded from this observation that "The cell in some way measures a fraction of the distance between one stomatogenic meridian and the next (or possibly between other cortical features) and regulates the field size in relation to this distance." This relationship is apparently independent of the shape of the surface being measured, as it is similarly manifested on the quasi-elliptical surface of a normal singlet cell, on the shorter and broader surface of cells homozygous for the "conical" gene (Doerder et al., 1975), and on the half-elliptical surface of a moiety of a homopolar doublet.

Turning finally to the ciliary meridians, there are three types of evidence indicating global coordination. First, Nanney (1971) demonstrated that as the number of ciliary meridians increases the number of ciliary units per meridian declines such that the total number of ciliary units in all of the ciliary meridians remains virtually constant. Second, when the number of ciliary meridians is initially above or below the stability center the number declines or increases at a rate proportional to the distance from that center (Nanney, 1966b,c, 1968b). The mechanism by which the number of ciliary meridians is reduced is not well understood, though it may be presumed to involve localized failures in production of new basal bodies, such as has been observed in *Euplotes* (Frankel, 1973b) and *Paramecium* (Chen-Shan, 1969, 1970; Suhama, 1975). E.M. Nelsen has recently discovered an unanticipated mechanism by which the number of ciliary meridians can increase. He observed that cells with 18 ciliary meridians that were completing oral replacement during the course of a morphogenetic transformation (Nelsen, 1978) to a "rapid swimmer" form (Nelsen and DeBault, 1978) generate a new, 19th ciliary meridian between the previous postoral meridians 1 and 18, with the new ciliary meridian appearing adjacent to the posterior portion of the newly organizing undulating membrane. Cells with 21 ciliary meridians that are carried through the same transformation sequence go through oral replacement but generate no new ciliary meridian. The cell thus calls this special mechanism of putting up a new ciliary meridian into play if and only if the previous number of meridians is on the low side of the stability center. Third, there is evidence for the existence of a circumferential gradient system that apportions basal body proliferation among different ciliary meridians. This was first implied by the results of Nanney and Chow's (1974) analysis of the distribution of ciliary units among meridians. In their analysis of the distribution of basal body numbers in seven meridians they found that each ciliary meridian is allotted a unique *proportion* of the total complement of ciliary units, irrespective of the size of the total complement. Assuming no resorption of basal bodies in the course of the cell cycle, this implies a corresponding difference among meridians in the number of units added in each cell cycle [this implication is treated more fully, for *Euplotes*, by Frankel (1975)]. A difference of this kind was demonstrated directly by Kaczanowski (1978) for *T. thermophila* cells undergoing intense basal body proliferation during refeeding after starvation. Kaczanowski observed a smooth gradation in number of *new* units per meridian, with a minimum mid-dorsally (around meridian 8) and a maximum mid-ventrally, near the oral longitude. Interestingly, the new oral apparatus develops at the intersection of the high points of the apical-basal gradient of proliferation within each meridian and the circumferential gradient of proliferation among meridians. Kaczanowski proposes that this relationship may be determinative rather than coincidental, and suggests that the initiation of the oral primordium results from a localized overcrowding of newly formed basal bodies (Kaczanowski, 1978).

V. WHAT ARE THE UNDERLYING MECHANISMS?

A. *The Duality of Positional Systems*

Can the global positional controls reviewed in section IV be conceived of as additive extensions of the "scaffolding" mechanisms presented in section III? I think not. Taking latitudes first, we have seen that dividing cells do not count basal bodies along the right postoral ciliary meridian when positioning their oral primordia, and that the polarity of the CVP meridian does not govern the latitude at which the CVP forms. As to control of longitudes, the phenotype of a unique mutant provides a clear demonstration that the global left-right "handedness" of the cell is not a mere additive expression of the left-right asymmetries of the individual ciliary meridians. The mutant, which we call *janus (jan)* (Frankel and Jenkins, 1979), when homozygous brings about a unique doublet condition that differs dramatically from the typical homopolar doublet state described earlier. A homopolar doublet consists essentially of two laterally fused cells with identical asymmetry; furthermore, this doublet condition is not stable, as doublet cells gradually lose ciliary meridians, and at a certain point during this loss the potential for oral development along one oral axis becomes suppressed and disappears (Fig. 7) (Fauré-Fremiet, 1948; Nanney, 1966b; Nanney *et al.*, 1975). The *janus* gene brings about the formation of atypical doublets that differ from the typical ones in two fundamental ways [Jerka-Dziadosz and Frankel (1979)]: (1) the oral structures produced along one oral longitude (the "primary axis") are completely normal, whereas those produced along the opposite longitude (the "secondary axis") are always abnormal and

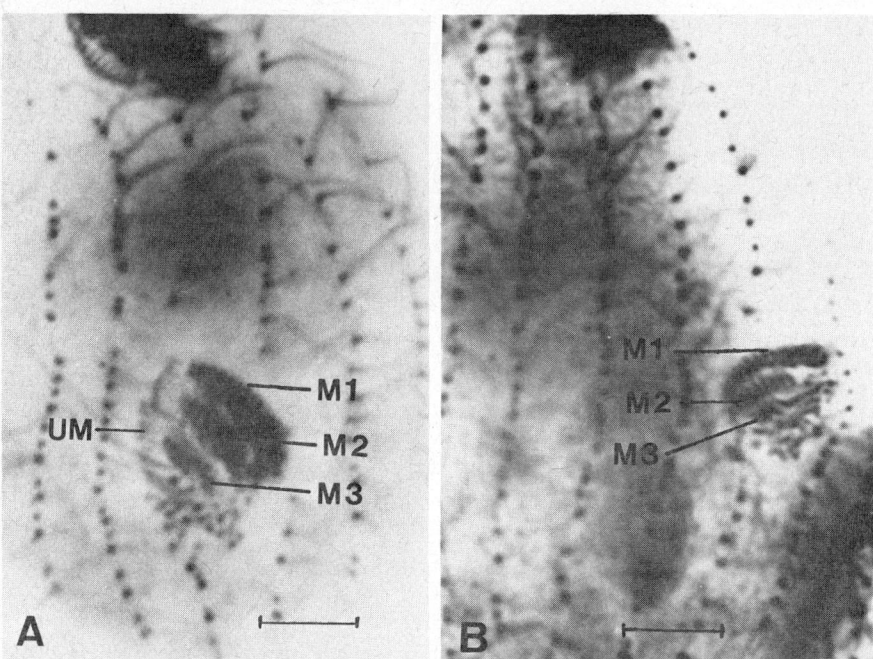

Fig. 8. A normal oral primordium (A) developing along the primary axis and an abnormal oral primordium (B) developing along the secondary axis of cells homozygous for the *jan* mutation. Note right-left reversal of orientation of membranelles (M1, M2, M3) of the abnormal oral primordium and lack of a differentiating undulating membrane (UM). Photographs taken by M. Jerka-Dziadosz. Protargol, 2500X (reference line indicates 5μm.)

frequently display a mirror image reversal of the normal arrangement of membranelles (Fig. 8), and (2) janus cells frequently manifest *two sets* of CVPs, one to the right (as usual) of the primary oral axis, the other to the cell's *left* of the secondary oral axis (Figs. 9, 10). Most importantly in this context, the coordinated reversal of oral asymmetry and of direction of CVP determination characteristic of the secondary axis of janus cells is manifested within a structural context that is otherwise completely *normal*: ciliary meridians of janus cells are everywhere of normal polarity and asymmetry, and both sets of CVPs develop on the "correct" (left) side of ciliary meridians (Jerka-Dziadosz and Frankel, 1979). The mitochondrial pattern is also generally normal (Aufderheide, personal communication). Even the number of ciliary meridians often remains unchanged as the reversed secondary axis becomes expressed after wild type cells become homozygous *jan/jan* (Frankel and Jenkins, 1979). Hence a reversal of a global asymmetry can become superimposed on a

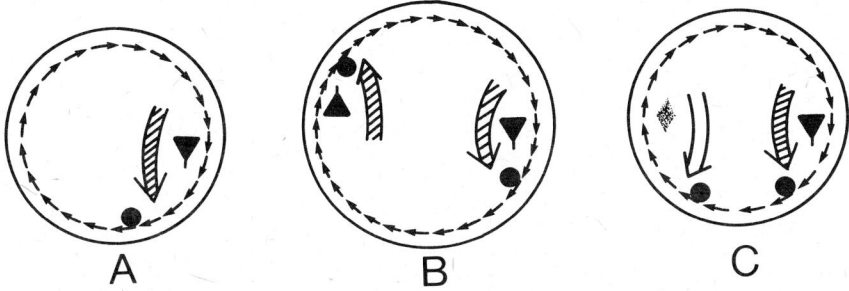

Fig. 9. Schematic cross-sections illustrating local and global asymmetries of *A*, normal singlets; *B*, wild type homopolar doublets, and *C*, janus cells. OAs and median positions of CVP sets are represented as in Fig. 7, with the stippled diamond-shape indicating the variable asymmetry and sporadic expression of secondary OAs in janus cells. The small arrows arrayed around the cell periphery indicate the asymmetry of the ciliary meridians, while the large arrows inside the cell indicate the global asymmetry as manifested in the direction of determination of the CVP longitude.

structurally normal local asymmetry (Fig. 9). The *janus* mutant thus reveals the distinctiveness of the long-range positional system that controls the general placement of the CVPs and also influences the asymmetry of the OAs. However, the details of the partial reversal of the secondary OAs suggest that the global and local asymmetries interact in such a way that complete and functional oral structures can only be produced when both are of the same handedness (Jerka-Dziadosz and Frankel, 1979). Ng (1977) has also noted some disturbance of the regional placement of CVPs in the neighborhood of inverted ciliary meridians.

B. *Do Pre-existing Ciliary Structures Serve as Reference Points?*

The issue of whether pre-existing ciliary structures themselves may serve as reference points for long-range positional systems has previously received considerable attention in the interpretation of combined microsurgical and cytological analyses of development in hypotrich ciliates, with arguments offered both pro (Grimes, 1976; Grimes and Adler, 1978) and con (Jerka-Dziadosz and Frankel, 1969; Frankel, 1973a). In my view, the information on *Tetrahymena* available to date suggests a tentative positive answer for control of latitudes, a negative answer for control of longitudes.

Considering first the latitude at which the oral primordium develops, Lynn (1977) observed exact and common proportionality in wild type and conical cells *only* for the regression of the distance between old and new oral structures on estimated cell size. The data clearly exclude the posterior end of the cell as a common reference point. As for the anterior end, the distance between the anterior end of the cell and the oral primordium is positively associated with cell size, but the regression lines are different for the two genotypes and do not extrapolate back to the origin (Lynn, 1977). We have also made qualitative observations, especially in pseudomacrostome cells returned to a permissive temperature after an interval at high temperature, that extreme variation in preoral distance resulting from earlier abnormal development at a restrictive temperature does not affect the placement of the oral primordium of dividing cells relative to the old oral area.

An observation that seems to challenge the referencing role of the OA is that doublets, both of the normal homopolar and janus varieties, may form an oral primordium along one oral axis despite the absence of an oral apparatus at the anterior end of that same axis (Nanney, 1966b; Jerka-Dziadosz and Frankel, 1979). However, such incomplete doublets *always* have a normal oral apparatus at the anterior end of the *other* oral axis. One oral apparatus may be sufficient, even in a duplex cell, to act as a reference point for a positional system that establishes a *latitude* for oral development, which can encircle the cell just as the Equator encircles the Earth. This deduction can be tested by finding out whether conditionally astomatous cells are able to divide while maintained under nutritionally adequate conditions that are non-permissive for oral development and then are able to develop oral structures at the right place after return to permissive conditions. Selective systems and media are now available which make the setting up of this sort of test potentially feasible (Orias and Pollock, 1975; Orias and Rasmussen, 1976).

Turning to the *longitude* at which new oral structures develop, it is clear that the maintenance of an oral axis at a definite position is independent of the presence of an oral apparatus on *that* axis. Nanney (1966b) made the first observations pointing in this direction. He noted that as homopolar doublets lose ciliary meridians they tend to revert to singlets (cf. Fig. 7). However, in the intermediate range of 22 to 27 ciliary meridians there is a substantial number of cells that possess two OAs but only one oral primordium or, more important here, one OA and two oral primordia. Cells possessing *one* OA but also *two* cytoprocts and CVP sets could be recognized as "monostome doublets", and the proportion of these that develop two equatorially situated oral primordia is similar to that of distome cells with the same ciliary meridian number. Nanney (1966b, p. 308) concluded ". . . that a pre-existing oral apparatus on a particular surface is not a sufficient, a necessary, nor perhaps even a relevant condition for the development of a new oral apparatus during fission".

Whereas in normal homopolar doublets this intermediate condition in the expression of distomy is a transient stage in the reversion of doublets to singlets, in janus cells it is a permanent condition as long as the genotype remains *jan/jan*. While oral areas develop along the primary oral axis with total reliability, development of new OAs along the secondary axis is sporadic, and does not depend on the presence of an anterior OA along that axis (Jerka-Dziadosz and Frankel, 1979). Nonetheless, the longitude along which the secondary oral primordium develops is remarkably constant, always close to 170° to the cell's right of the primary oral axis. This special longitude thus appears to be maintained in a manner that is independent of the expression of oral structures along the longitude. The possibility, however, remains that oral or other structures along the primary axis may somehow be involved in fixing the secondary axis at a near-antipodal position.

The evidence for dispensibility of pre-existing ciliary structures as positional reference markers is most unambiguous for the longitude of CVP development. Nanney's (1966b) observation of incomplete doublets with one OA and two CVP sets implies that one of these two CVP sets is positioned with reference to an oral axis lacking oral structures; however, no quantitative data on CVP positions in such cells were reported. We have carried out an analysis of numbers and positions of CVP sets in janus cells, and found that all possible combinations of one and two OAs and one and two CVP sets (Fig. 10) are found in frequencies that indicate random associations (Jerka-Dziadosz and Frankel, 1979). Furthermore, and crucially, the *position* of the CVP sets is independent of whether or not oral structures are present along the secondary axis (Fig. 10). The apparent influence of some property of this secondary axis on the placement of CVP sets thus does not depend on whether this axis is concurrently "occupied" by OAs. This in turn suggests that the *janus* gene is bringing to existence (or to expression) a morphogenetic field of reversed handedness that (a) is precisely centered in the cell relative to the other (normal) field, (b) is longitudinally propagated, and (c) controls the location of CVP sets in a manner that is independent of its (oral) expression. Taking the recent observations on janus cells together with Nanney's earlier work on wild type homopolar doublets, it is clear that the positional reference function of the longitudinal strip of cell surface that marks the oral-CVP "field center" is not dependent on the continuous presence of any structure along that longitudinal strip which is visible in the light microscope. These observations suggest that the reference axis, though it has visible expressions, is itself invisible.

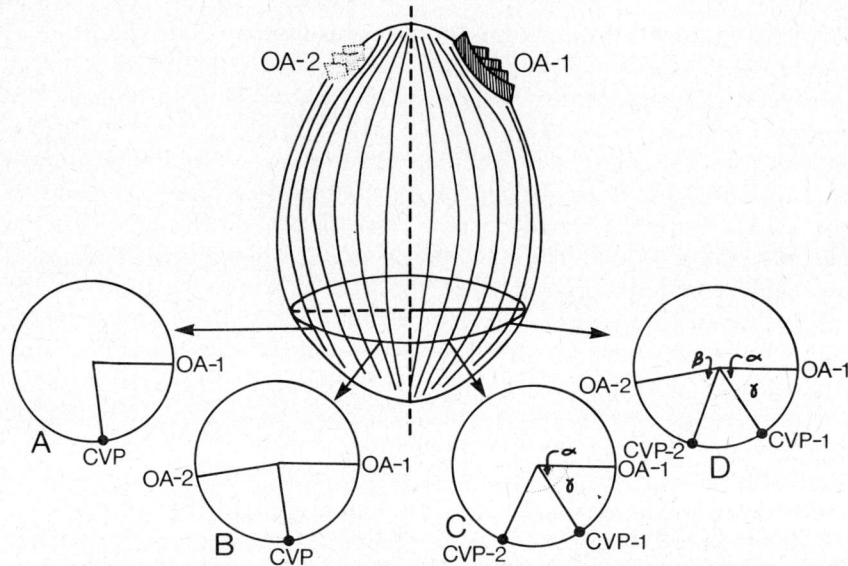

Fig. 10. The four configurations of OAs and CVP sets in janus cells. *A.* One OA, one CVP set. *B.* Two OAs, one CVP set. *C.* One OA, two CVP sets; *D.* Two OAs, two CVP sets. The average positions of the CVP sets observed for each of the four configurations is shown. Note especially that in cells with two OAs and two CVP sets (*D*), the average relative distance from the primary OA to CVP set 1 (given by angle α) is nearly equal to the average relative distance from the secondary OA to CVP set 2 (given by angle β). Angle β cannot be measured in cells that have two CVP sets but lack a secondary OA (*C*); however, note that the average relative distance from the primary OA to the secondary CVP set (angle γ) is the same in *C* and *D*. Hence the second CVP set manifests the same average position relative to the "non-expressed" secondary oral axis in *C* as it does relative to the "expressed" secondary oral axis in *D*. Modified from Jerka-Dziadosz and Frankel, 1979.

C. What Measuring System(s) Does the Cell Use?

A simple model for the long-range positional system of *Tetrahymena*, very similar to Uhlig's (1959) model for *Stentor*, can be synthesized from the work and ideas of Nanney (1968a, 1972), and Lynn (1977). This model postulates two orthogonal gradients, one parallel to each axis of the "Mercator projection" of the cell (Fig. 1D). The first gradient would thus be an anteroposterior gradient probably employing the OA or something closely associated with it as the primary reference marker; the second gradient would be a horizontal linear gradient wrapped around the cell, with an invisible "cliff" at the oral axis.† The coordinates for development

---

†An alternative would be to assume a spatially continuous gradient system with a ridge-crest along the oral axis and a mid-dorsal trough, analogous to circumferential-gradient formulations applied to the *Drosophila* egg by Kauffman, *et al.* (1978) and by Nüsslein-Volhard (1979). A model of this kind might account for the way in which proliferation of basal bodies is distributed around the cell (Kaczanowski, 1978) and it does away with the problematical "cliff". It fails, however, to account for the asymmetry of positioning of CVPs.

of specific structures such as the new OA and CVP would then be "read off" by the cell at appropriate levels of these two gradients, as postulated in Wolpert's (1969) "positional information" formulation. The horizontal gradient would additionally provide the cell with information about its asymmetry or "handedness" that is independent of, but could interact with, the information inherent in the ciliary meridians themselves (cf. Jerka-Dziadosz and Frankel, 1979).

While this conceptualization provides a reasonable abstract encapsulation of the by now rather extensive observations and analyses of pattern formation in *Tetrahymena*, considerable difficulty is encountered in visualizing what sorts of measuring devices the cell might use in making measurements along these gradients. The simplest and most popular type of model is that of chemical measurement involving at least one readily diffusible component, as in the models constructed for *Hydra* by Wolpert *et al.*, (1974) and by Gierer and Meinhardt (1972; cf. Gierer, 1974). A general difficulty in application of such models to ciliates is that the highly structured cell cortex is continuous with a cell interior characterized by rapid cyclosis (*Stentor*: DeTerra, 1971; *Paramecium*: Kuznicki *et al.*, 1972). Cortically located chemical gradients might thus be unstable. Lynn (1977) attempted to overcome this objection by arguing that according to Crick's (1971) estimate of intracellular diffusion rates it would require less than a minute to establish an effective chemical gradient along the length of a *Tetrahymena* cell, and that suspension of cyclotic movements for this short duration (such as has been observed in *Paramecium* by Kuznicki *et al.*, 1972) might, if it took place at the right stage in development, allow a chemical gradient to become established and be effective. Lynn further suggests that the anterior oral apparatus might be the source for such a diffusing morphogen. While it is difficult to refute this hypothesis, Lynn's analysis showed that the distance between old and new OA gave the best regression on a measure of cell surface area rather than on cell length, a circumstance that would tend to complicate a simple diffusion-based model; this observation is perhaps more readily accommodated within an idea of control based on mobile components within the cell membrane, as suggested by Kaczanowska (1974), which might distribute themselves over the cell surface.

The problems in interpreting the anteroposterior gradient in diffusional terms are minor compared to the difficulty in similarly conceiving of the horizontal gradient. How can one maintain a juxtapostion of the highest and lowest gradient values (the "cliff") at a longitude which lacks a partition? The difficulties are compounded when one considers interactions in homopolar doublet cells. Although at first the two components of a doublet seemed to behave as two independent morphogenetic systems (Nanney, 1966b), a more searching examination revealed that the site at which the CVP is formed in one component is influenced by the width of the *other* component (Nanney et al., 1975). In janus cells the interaction is even more profound: since the direction of CVP determination is opposite for the two components, two CVP-determining systems jointly occupy the same side of the cell (Fig. 10). If CVP determination is based on circular gradients, then the gradients associated with the two components must mutually interpenetrate each other. More detailed considerations (Jerka-Dziadosz and Frankel, 1979) indicate that the steepness of the putative gradients, and thus the degree of mutual interpenetration, is variable. It would clearly be difficult for a chemical-diffusional interpretation to explain these peculiar phenomena.

One possible response to these complexities would be to discard the conceptual framework that generates them. Can the observations perhaps be better accounted for within a non-Cartesian coordinate system? Following up on a suggestion by Sonneborn (1974, p. 346), we could adopt a polar coordinate system (Fig. 1C). The cell would then establish a "preferred radius", normally coinciding with meridian no. 1. The site of the oral primordium would be determined by measuring a certain distance "outward" along this preferred radius (i.e. posteriorward on the cell surface). This distance would also determine the fission zone, represented as a dashed circle in Fig. 1C. The crucial feature determining the site of CVP formation would then be a *cortical angle* that normally is measured in a clockwise direction outward from some point along the preferred radius. This formulation does away with the mysteries of a gradient horizontally wrapped around the cell, but has some difficulties of its own: (1) The evidence for the proportional positioning of the oral primordium (section IV, A) requires that the length of the preferred radius be a variable responsive to some global parameter such as the area of the whole circle. (2) The evidence from incomplete doublets (section V,B) indicates that the pre-existing anterior oral apparatus is not indispensible as a reference point for measurement of the putative cortical angle. (3) It is not obvious how one can construct a "cortical angle" that would bring about the positioning of CVPs at a near-quarter of the circumference in singlet cells of different shapes and sizes yet position CVPs at a near-eighth of the circumference of homopolar doublets. (4) In the polar construction, there is no unique CVP meridian. Instead, the postulated positioning of CVPs at an intersection between a diagonal arc and a circumferential latitude (the fission zone) implies that if CVPs were to form at different latitudes in the *same* cell they must

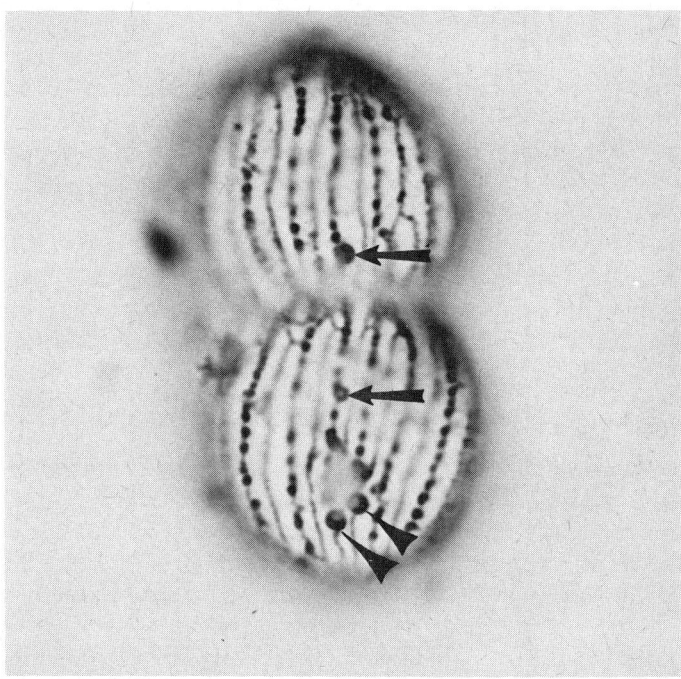

Fig. 11. A *cdaA2* cell manifesting an atypically positioned CVP. Cells were grown in 2% proteose peptone — 0.5% yeast extract medium at 28° C (permissive) and then shifted to 39.5° C (nonpermissive). They were fixed at 3 hours after the shift, when many cells were entering their first division at the high temperature. The *cdaA2* allele is one of the more weakly expressed alleles at this locus (cf. Frankel *et al.*, 1977, where it is indicated as *mo1*^{b}), and thus this cell has formed a division furrow. Note, however, that in addition to the normal new CVP anterior to the fission zone (upper arrow) there is a supernumerary CVP along the same ciliary meridian (lower arrow). The old CVPs are near the posterior end of the cell (arrowheads). Wet silver, 1400 X.

necessarily also develop at different longitudes: the further posterior the CVPs, the more distant they should be from the oral axis. It so happens that if cells homozygous for the *cdaA* (formerly *mo1*) mutations are maintained at temperatures allowing only partial expression of the defect in fission zone formation, many cells produce extra CVPs that are located halfway between the equator and the posterior end of the cell — along the *same* ciliary meridians as the typical CVPs that are formed anterior to the fission zone (Fig. 11) (Frankel, unpublished observations). These difficulties taken together warrant tentative rejection of this polar model.

Despite the difficulties of constructing a detailed model of cell surface patterning within an orthogonal framework, the capacity that does exist to bring about asymmetry reversals resulting in mirror-image patterns, both by quasi-operative means (for CVP fine-positioning associated with 180°-rotated ciliary meridians) and through mutant gene action (in janus cells), is most readily accounted for within a Cartesian reference system in which one axis can be reversed independently of the other. The generation of mirror-image configurations through selective reversal of a single axis has also been reported in folded-over longitudinal cell fragments in two very different larger ciliates (*Blepharisma*: Suzuki, 1957; *Stylonychia*: Grimes and L'Hernault, 1979). The axial independence encountered in ciliates is reminiscent of the separate and sequential fixation of orthogonal axes in developing amphibian limbs (Harrison, 1921; Swett, 1937) and eyes (Jacobson, 1968; Hunt and Frank, 1975).

VI. CONCLUSIONS

What advances have been made in the understanding of pattern formation in complex unicellular systems since Tartar's two pioneering presentations delivered at much earlier symposia of this series? In my view, there have been three such really important advances. The first is the rigorous verification of Tartar's early intuition of "cytoplasmic inheritance" of cortical pattern, particularly as is exemplified in the perpetuation of 180°-rotated ciliary meridians and also in the more or less stable perpetuation of differences in number of ciliary meridians in cells of identical genotype. Recent studies have shown that the "scaffolding" system associated with the ciliary meridians influences the positioning not only of new ciliary elements within these meridians but also of adjacent structures such as the CVPs and cortically situated mitochondria. The second advance is the extension of Tartar's concept of the "intimate relatedness of pattern components" to the system of positional values governing the establishment of new pattern elements as well as the mutually inhibitory interactions of already established elements. As exemplified by the passage from which Tartar's key phrase was extracted[†], the evidence amassed by Tartar and Uhlig for the reality of developmentally significant interactions over long intracellular distances relates largely (though by no means exclusively) to mutual inhibition among oral primordium sites. More recent analyses of positioning of ciliary primordia in hypotrich ciliates (e.g. Jerka-Dziadosz, 1974, 1977) as well as the analyses of oral and CVP positioning reviewed here demonstrate that long-range, relational systems operate to

[†]"Hence there seems to become established in any graft complex an intimate relatedness of the pattern components such that suppression of oral differentiation of loci of less sharp stripe-width contrast may occur in certain arrangements because of the presence of regions of greater contrast, though subsequent release and even exchange of dominance may later take place." (Tartar, 1956, p. 84).

determine the locations of new structural systems in the cell surface, not merely to suppress established systems. Ciliates thus, in the language of Wolpert et al., (1974) possess gradients of "positional value" as well as of "positional signals" (inhibitors). The results of morphometric analyses of positioning of contractile vacuole pores and of oral structures indicate that the application of the concept of "positional value" to ciliates is nontrivial, in the sense expressed by Sidney Brenner ("The terms positional information, positional value, and so on, should be reserved for cases where specification is a function of spatial relations, and should not be used when structures have trivial positional values but are generated by a completely different mechanism, such as a clock or a lineage" — p. 123 in discussion of paper by Wolpert et al., 1975). The third important advance is the demonstration that nuclear genes play a significant role in determining cell-surface patterns, both by setting limits to the structural divergence that can be achieved through the operation of extragenic "scaffolding" mechanisms and also by influencing and probably specifying the systems of positional value that underlie large-scale structural patterns. This third advance thus relates to both of the previous two. It also opens up the possibility, as yet still remote, of understanding intracellular structural patterns at a molecular level.

A final remark should be made about the relationship of the analysis of pattern in ciliates to that in multicellular organisms, for example *Hydra*. Almost all thinking about polarity in hydroids since Morgan's early investigations has concentrated on the idea of control involving gradients of chemical substances that are to varying degrees diffusible (e.g. Morgan, 1905; Gierer, 1974; Wolpert et al., 1974). Yet we have seen that the application of the concept of diffusion of a chemical "morphogen" as a basis of morphogenetic field systems in ciliates is at best problematical. One might therefore conclude that systems of positional values have very different bases in ciliates and multicellular organisms. However, not a single convincing example of a chemical morphogen that can generate positional values has been uncovered in any multicellular organism; the one plausible example in *Hydra*, the morphogenetic "activator" produced by nerve cells (Schaller and Gierer, 1973), has been rendered largely superfluous by the discovery of essentially normal pattern-regulation of nerve-free hydras (Marcum and Campbell, 1978; Campbell, 1979). It is thus still quite possible that all organisms, including ciliates, use some common means other than diffusion-gradients for assessing and regulating positional values.

ACKNOWLEDGEMENTS

I should like to thank the following persons who read a preliminary draft of this manuscript and provided valuable suggestions for its improvement: Drs. K. Aufderheide, A.W.K. Frankel, D. Lynn, D.L. Nanney, E.M. Nelsen, S.F. Ng, and T.M. Sonneborn. Research conducted in the author's laboratory and reported here has been supported by a grant by the National Science Foundation (GB-32408) and more recently by a grant from the National Institutes of Health (HD-08485).

REFERENCES

Allen, R.D. (1969). *J. Cell Biol.* **40**, 716-733.
Aufderheide, K.J. *J. Protozool.* **25**, 7A.
Beisson, J. and Sonneborn, T.M. (1965). *Proc. Nat. Acad. Sci. U.S.* **53**, 275-282.
Buhse, H.E. Jr. (1966). *Trans. Amer. Microsc. Soc.* **85**, 305-313.
Campbell, R. (1979. This volume.
Chen-Shan, L. (1969). *J. Exp. Zool.* **170**, 205-228.
Chen-Shan, L. (1970). *J. Exp. Zool.* **174**, 463-478.
Crick, F.H.C. (1971). *In* "Control Mechanisms in Growth and Differentiation" (25th Symposium of the Society for Experimental Biology), pp. 429-438. Cambridge University Press, Cambridge.
De Terra, N. (1971). *J. Cell Physiol.* **78**, 377-385.
Doerder, F.P., Frankel, J., Jenkins, L.M., and DeBault, L.E. (1975). *J. Exp. Zool.* **192**, 237-258.
Fauré-Fremiet, E. (1948). *Arch. Anat. Microsc. Morphol. Exp.* **37**, 183-203.
Frankel J. (1969). *J. Protozool.* **16**, 26-35.
Frankel, J. (1973a). *J. Protozool.* **20**, 8-18.
Frankel, J. (1973b). *Develop. Biol.* **30**, 336-365.
Frankel, J. (1974). *J. Theoret. Biol.* **47**, 439-481.
Frankel, J. (1975). *J. Embryol. Exp. Morphol.* **33**, 553-580.
Frankel, J. and Jenkins, L.M. (1979). *J. Embryol. Exp. Morphol.*, in press.
Frankel, J., Jenkins, L.M. and DeBault, L.E. (1976). *J. Cell Biol.* **71**, 242-260.
Frankel, J., Nelsen, E.M., and Jenkins, L.M. (1977). *Develop. Biol.* **58**, 255-275.
Frankel, J. and Williams, N.E. (1973). *In* "The Biology of Tetrahymena" (A.M. Elliott, ed.), pp. 375-409. Dowden, Hutchinson, and Ross, Stroudsburg, Pa.
Garcia-Bellido, A. (1975). *In* "Cell Patterning" (Ciba Foundation Symposium 29, new series), pp. 161-182. Associated Scientific Publishers, Amsterdam.
Gierer, A. (1974). *Sci. Amer.* **231**(6), 44-54.
Gierer, A. and Meinhardt, H. (1972). *Kybernetik* **12**, 30-39.
Grimes, G.W. (1976). *Genet. Res. Cambr.* **27**, 213-226.
Grimes, G.W. and Adler, J.A. (1978). *J. Exp. Zool.*, **204**, 57-80.
Grimes, G.W. and L'Hernault, S.W. (1979). *Develop. Biol.*, submitted.
Hanson, E.D. (1962). *J. Exp. Zool.* **150**, 45-68.
Hanson, E.D. and Ungerlieder, R.M. (1973). *J. Exp. Zool.* **185**, 175-187.
Harrison, R.G. (1921). *J. Exp. Zool.* **32**, 1-136.
Heckmann, K. and Frankel, J. (1968). *J. Exp. Zool.* **168**, 11-38.
Hunt, R.K. and Frank, E. (1975). *Science,* **189**, 563-565.

Huxley, J.S. and DeBeer, G.R. (1934). "The Elements of Experimental Embryology". Cambridge University Press, Cambridge.
Jacobson, M. (1968). *Develop. Biol.* **17**, 202-218.
Jerka-Dziadosz, M. (1974). *Acta Protozool.* **12**, 239-274.
Jerka-Dziadosz, M. (1977). *J. Exp. Zool.* **200**, 23-32.
Jerka-Dziadosz, M. and Frankel, J. (1969). *J. Protozool.* **16**, 612-637.
Jerka-Dziadosz, M. and Frankel, J. (1979). *J. Embryol. Exp. Morphol.*, in press.
Jones, W.R. (1976). *Genet. Res. Cambr.* **27**, 187-204.
Kaczanowska, J. (1974). *J. Exp. Zool.* **187**, 47-62.
Kaczanowski, A. (1975). *Genetics* **81**, 631-639.
Kaczanowski, A. (1976). *J. Exp. Zool.* **196**, 215-230.
Kaczanowski, A. (1978). *J. Exp. Zool.* **204**, 417-430.
Kaneda, M. and Hanson, E.D. (1974). *In* "Paramecium-A Current Survey", (W.J. Van Wagtendonck, ed). pp. 219-262. Elsevier, Amsterdam.
Kauffman, S.A., Shymko, R.M., and Trabert, K. (1978). *Science* **199**, 259-270.
King, R.L. (1954). *J. Protozool.* **1**, 121-130.
Kühn, A. (1955). "Vorlesungen über Entwicklungsphysiologie". Springer Verlag, Berlin.
Kuznicki, L., Sikora, J. and Fabczak, S. (1972). *Acta Protozool.* **11**, 237-242.
Lynn, D.H. (1977). *J. Embryol. Exp. Morphol.* **42**, 261-274.
Lynn, D.H. and Tucker, J.B. (1976). *J. Cell Sci.* **21**, 35-46.
McCoy, J.W. (1974). *Acta Protozool.* **8**, 155-159.
Marcum, B.A. and Campbell, R.D. (1978). *J. Cell Sci.* **29**, 17-33.
Morgan, T.H. (1901). *Biol. Bull.* **2**, 311-328.
Morgan, T.H. (1905). *J. Exp. Zool.* **2**, 495-506.
Nanney, D.L. (1966a). *J. Exp. Zool.* **161**, 307-317.
Nanney, D.L. (1966b). *Amer. Nat.* **100**, 303-318.
Nanney, D.L. (1966c). *Genetics* **54**, 955-968.
Nanney, D.L. (1967a). *J. Exp. Zool.* **166**, 163-169.
Nanney, D.L. (1967b). *J. Protozool.* **14**, 690-697.
Nanney, D.L. (1968a). *Science* **160**, 496-502.
Nanney, D.L. (1968b). *J. Protozool.* **15**, 109-113.
Nanney, D.L. (1971). *J. Exp. Zool.* **178**, 177-181.
Nanney, D.L. (1972). *Ann. N.Y. Acad. Sci.* **193**, 14-28.
Nanney, D.L. (1975). *J.Cell Biol.* **65**, 503-512.
Nanney, D.L. (1977). *J. Protozool.* **24**, 27-35.
Nanney, D.L. and Chow, M. (1974). *Amer. Nat.* **108**, 125-139.
Nanney, D.L., Chow, M. and Wozencraft, B. (1975). *J. Exp. Zool.* **193**, 1-14.
Nanney, D.L. and McCoy, J.W. (1976). *Trans. Amer. Microsc. Soc.* **95**, 664-682.
Nelsen, E.M. (1978). *Develop. Biol.* **66**, 17-31.
Nelsen, E.M. and DeBault, L.E. (1978). *J. Protozool.* **25**, 113-119.
Ng, S.F. (1976). *J. Exp. Zool.* **196**, 167-182.
Ng, S.F. (1977). *J. Cell Sci.* **25**, 233-246.
Ng, S.F. (1978a). *Protistologia.*, in press.
Ng, S.F. (1978b). *J. Cell Sci.* **33**, 227-234.
Ng, S.F. (1979a). *J. Cell Sci.*, submitted.
Ng, S.F. (1979b). *Acta Protozoologica.*, submitted.
Ng, S.F. (1979c). *Develop. Biol.*, to be submitted.
Ng, S.F. and Frankel, J. (1977). *Proc. Nat. Acad. Sci. U.S.* **74**, 1115-1119.
Ng, S.F. and Williams, R.J. (1977). *J. Protozool.* **24**, 257-263.

Nicolson, G.L., Poste, L., and Li, T.H. (1977). In "Dynamic Aspects of Cell Surface Organization" (G. Poste and G.L. Nicolson, eds.), pp. 1-73. North Holland, Amsterdam.
Nüsslein-Volhard, C. (1979). This volume
Orias, E. and Pollock, N.A. (1975). *Exp. Cell Res.* **90,** 345-356.
Orias, E. and Rasmussen, L. (1976). *Exp. Cell Res.* **102,** 127-137.
Perlman, B.S. (1973). *J. Exp. Zool.* **184,** 365-368.
Schaller, H. and Gierer, A. (1973). *J. Embryol. Exp. Morphol.* **29,** 39-52.
Schwartz, V. (1963). *Naturwissenschaften* **20,** 631-640.
Sonneborn, T.M. (1963). In "The Nature of Biological Diversity" (J.M. Allen, ed.), pp. 165-221. McGraw-Hill, New York.
Sonneborn, T.M. (1964). *Proc. Nat. Acad. Sci. U.S.* **51,** 915-929.
Sonneborn, T.M. (1970). *Proc. Roy. Soc. Lond. B.* **176,** 347-366.
Sonneborn, T.M. (1974). In "Actualités Protozoologiques" (P. dePuytorac and J. Grain, eds.) Vol. I, pp. 327-355. Universite de Clermont, Clermont-Ferrand.
Sonneborn, T.M. (1975). *Ann. Biol.* **14,** 565-583.
Stone, G.E. (1963). *J. Protozool.* **10,** 74-80.
Suhama, M. (1975). *J. Sci. Hiroshima Univ., Ser. B, Div. 1,* **26,** 37-51.
Suzuki, S. (1957). *Bull. Yamagata Univ., Nat. Sci.* **4,** 85-191.
Swett, F.H. (1937). *Quart. Rev. Biol.* **12,** 322-339.
Tamm, S.L., Sonneborn, T.M. and Dippell, R.V. (1975). *J. Cell Biol.* **64,** 98-112.
Tartar, V. (1941). *Growth* **5** (suppl.), 21-40.
Tartar, V. (1954). *J. Protozool.* **1,** 11-17.
Tarter, V. (1956). In "Cellular Mechanisms of Differentiation and Growth" (D. Rudnick, ed.), pp. 73-100. Princeton University Press, Princeton, N.J.
Tartar, V. (1961). "The Biology of Stentor". Pergamon Press, London.
Tartar, V. (1962). *Adv. Morphogen.* **2,** 1-26.
Uhlig, G. (1959). *Z. Naturforsch.* **14b,** 353-354.
Uhlig, G. (1960). *Arch. Protistenk.* **105,** 1-109.
Williams, N.E. (1960). *J. Protozool.* **7,** 10-17.
Williams, N.E. (1964). In "Synchrony in Cell Division and Growth" (E. Zeuthen, ed.), pp. 159-175. Interscience Press, New York.
Wolpert, L. (1969), *J. Theor. Biol.* **25,** 1-47.
Wolpert, L., Hornbruch, A. and Clark, M.R.B. (1974). *Amer. Zool.* **14,** 647-663.
Wolpert, L., Lewis, J. and Summerbell, D. (1975). In "Cell Patterning" (Ciba Foundation Symposium 29, new series), pp. 95-130. Associated Scientific Publishers, Amsterdam.

Intercellular Interactions and Pattern Formation in Filamentous Cyanobacteria

C. Peter Wolk

MSU-DOE Plant Research Laboratory
Michigan State University
East Lansing, Michigan 48824

I. Introduction 247
II. Physiology of Pattern Formation 249
III. Biochemical Studies of Differentiation and Intercellular Interactions 252
    A. The First Isolated Heterocysts 252
    B. Isolated Heterocysts which Retained Enzymes 255
    C. Structurally Intact, Metabolically Active Isolated Heterocysts 258
    D. Interactions between Vegetative Cells and Mature Heterocysts, and Control of Heterocyst Formation 260
    E. Control of Heterocyst Formation by Interactions of Vegetative Cells with Immature Heterocysts 262
IV. Concluding Remarks 264
    References 265

I. INTRODUCTION

Certain filamentous cyanobacteria (blue-green algae) provide an opportunity to elucidate, in morphologically simple organisms, the detailed biochemical mechanisms which govern the formation of multicellular patterns (Wolk, 1975). In these bacteria, vegetative cells can either divide to produce two vegetative cells; or differentiate to form heterocysts, cells that are specialized for aerobic fixation of dinitrogen; or differentiate to form spores (akinetes), which function in perennation. The heterocysts are normally formed either at semi-regular intervals along the filaments, as in *Anabaena* (Fig. 1a), or only from the terminal cells of filaments, as in *Cylindrospermum* (Fig. 1c). A revertible mutant of *Anabaena* has been isolated in which, as in a *Cylindrospermum*, heterocysts form only at terminal positions (Wilcox et al., 1975a). Single spores or strings of two or more spores normally form adjacent to heterocysts [as in *Anabaena cylindrica* (Fig. 1b) or *Cylindrospermum licheniforme* (Fig. 1d)], or midway between preexisting heterocysts (as in *Anabaena variabilis:* Geitler, 1932), depending upon the species. Under certain conditions, they may also form a definite small number of cells away from a heterocyst (observed by Canabaeus, 1929, in *A. Hassallii* and *A. flos-aquae*) or—as in *Anabaena doliolum* and *A. variabilis*—from all vegetative cells between two heterocysts (Wolk and Wojciuch, 1973). As will be

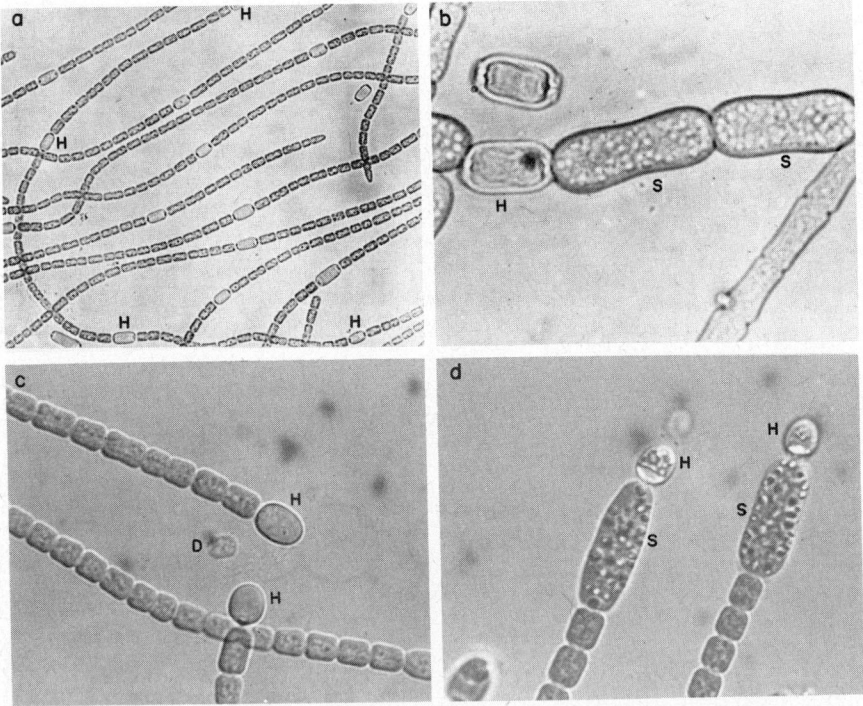

Fig. 1. Differentiated cells in cyanobacteria: (a) *Anabaena cylindrica*, with heterocysts (H) at semi-regular intervals along the filaments, and (b) with mature spores (S) shown adjacent to one heterocyst. (c) *Cylindrospermum licheniforme*, with terminal heterocysts (H) which have formed to either side of a vegetative cell which has died (D) approximately midway between two pre-existing heterocysts, and (d) subterminal spores (S) which are differentiating adjacent to heterocysts.

documented below, the mechanisms underlying pattern formation in cyanobacteria are analogous to embryogenetic fields: the heterocyst pattern corresponds to a field of inhibition, and the juxtaposition of spores and heterocysts corresponds to an embryogenetic induction.

II. PHYSIOLOGY OF PATTERN FORMATION

If filaments of *Anabaena cylindrica* are extensively fragmented, nearly every sequence of one or more vegetative cells—detached from heterocysts by the fragmentation procedure—forms one and only one heterocyst during an ensuing period of incubation (Wolk, 1967; see also Singh *et al.*, 1972, and Mitchison *et al.*, 1976). A net increase in the number of heterocysts can result, relative to an unfragmented control culture. Many of the supernumerary heterocysts form at terminal positions on the fragments, just as they often do in germlings arising from spores (see Fig. 2 of Fay *et al.*, 1968).* I argued that if the increased formation of heterocysts was the consequence of any effect of fragmentation other than detachment of vegetative cells from heterocysts, many of the fragments which would have been expected to produce one heterocyst during the incubation period should have produced a second heterocyst by reason of the same effect acting on them. Inasmuch as the frequency with which second heterocysts were found was similar to the frequency of equally closely spaced heterocysts in intact filaments, it was concluded that "no result of fragmentation other than separation of vegetative cells from heterocysts is responsible for stimulated heterocyst formation" and that, therefore, "heterocysts mediate the pattern of vegetative growth" (Wolk, 1967).

Growth of heterocyst-forming species with ammonium normally leads to a diminution of the relative frequency of heterocysts (Fogg, 1949; see Table I). When ammonium-grown *Cylindrospermum licheniforme* is subjected to nitrogen step-down, a pattern of heterocysts arises which includes intercalary heterocysts and thereby differs qualitatively from the pattern (only terminal heterocysts) which is normal in the absence of NH_4^+ (Wolk and Quine, 1975). This observation is not consistent with the concept, based on cultures grown for a relatively short time with NH_4^+,

---

*The argument by Mitchison *et al.* (1976) that "terminal heterocysts do not occur under normal growth conditions" in *Anabaena cylindrica* therefore applies only to long filaments, in which heterocysts are present, but at a distance (see below). On the other hand, the finding that killing certain cells leads to regression of nearby immature heterocysts (Wilcox *et al.*, 1973 a,b) may be attributable to substances released by the wound, and the observation that vegetative cells near killed heterocysts do not differentiate (Wilcox *et al.*, 1973a) may be attributable to the residual effect of the immature heterocyst and to such substances.

TABLE I

Effect of glutamine on heterocyst spacing in cyanobacteria grown for 7 days with N_2 plus different nitrogen sources in flasks containing an 8-fold dilution of the medium of Allen and Arnon (1955).

	Heterocyst/vegetative cell ratio, %		
Nitrogen source — N_2 plus	—	1 mM $(NH_4)_2HPO_4$	20 mM glutamine
Strain			
Anabaena cylindrica	8.1	0.6	0.7
Cylindrospermum licheniforme	2.5	0.8	0.6

that the normal pattern is merely latent and unexpressed during growth with NH_4^+ (Wilcox, 1970). Moreover, when ammonium-grown *Anabaena* strain 6411 is subjected to nitrogen step-down, the heterocysts which

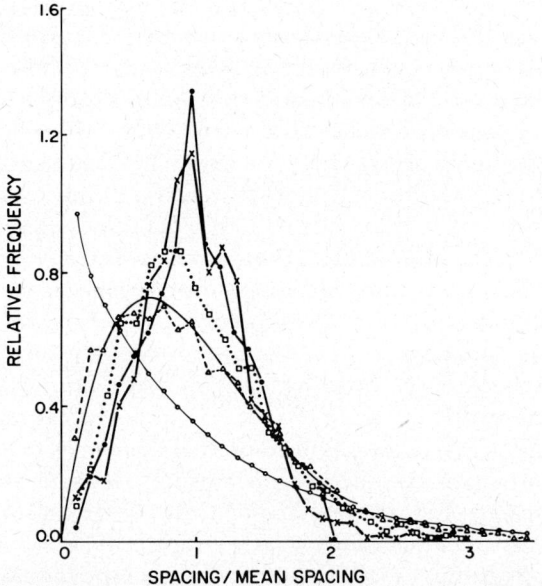

Fig. 2. Normalized frequency distributions of spacings arising in one-dimensional arrays. x: Spacings between heterocysts arising from heterocyst-free filaments during nitrogen-deprivation of nitrate-grown *Anabaena* strain 6411. o: The distribution expected if heterocysts arise randomly. Δ, □, •: Computer simulations of the distributions expected if heterocysts arise randomly and if they inhibit further heterocyst formation by (Δ) a process the effect of which spreads at constant velocity along the filaments from the heterocysts which have formed ("domino effect"; the solid line with no symbols is a theoretical curve for the continuous, rather than discrete, domino effect); (□) a diffusion process or (•) a modified diffusion process in which the diffusing substance is destroyed as it diffuses. Modified from Wolk and Quine (1975), with permission.

differentiate in what is therefore a *de-novo* pattern are not randomly spaced (Fig. 2, curve o-o); instead, the pattern corroborates the interpretation that immature heterocysts inhibit nearby cells from differentiating. The detailed distribution of spacings is inconsistent with the idea that inhibition spreads with constant velocity, as if it were a wave phenomenon, and accords with the idea that inhibition is propagated by a diffusion phenomenon (Wolk and Quine, 1975).

When a cell of a filament of *Cylindrospermum licheniforme* is relatively close to a heterocyst it will form a heterocyst if, through the experimental destruction of a neighboring cell, it becomes the terminal cell. Cells even farther from the nearest heterocyst would not differentiate in an intact filament. It therefore cannot be that differentiation is controlled primarily by the accretion, within vegetative cells, of a differentiation-stimulatory substance produced by those cells and destroyed by heterocysts. Had that interpretation been correct, intercalary heterocysts should normally have formed, but did not (Wolk and Quine, 1975). Rather, it appears that heterocysts produce, and transmit to the vegetative cells, a substance or substances which inhibit heterocyst differentiation, and the loss of which through the cut ends of the filaments can lead to differentiation (cf. also Wilcox *et al.*, 1973 a,b). Nonetheless, the general tendency against differentiation in the shortest filaments (Wolk, 1967; Wolk and Quine, 1975) may imply that other substances which can be lost from the ends of these filaments are essential for differentiation (Wolk and Quine, 1975).

Having thus far considered the role of intercellular interactions in the process of heterocyst formation, let us now very briefly consider the role of such interactions in sporulation. I showed some years ago, using *Anabaena cylindrica,* that vegetative cells gently detached from heterocysts — although appearing uninjured — do not differentiate into spores, whereas those vegetative cells remaining attached to heterocysts would sporulate (Wolk, 1966). As in species of *Anabaena* (Peterson and Wolk, 1978a), uptake hydrogenase activity is localized apparently exclusively in heterocysts, in *Cylindrospermum licheniforme;* and hydrogen can stimulate sporulation adjacent to heterocysts in the latter organism (Hirosawa and Wolk, unpublished observations). These results provided support for the idea that heterocysts play a role in sporulation. *C. licheniforme* secretes substances which greatly stimulate the sporulation of that same organism (Fisher and Wolk, 1976). One of these substances has been purified. Its molecular ion has the formula C_7SH_5NO, and its two major mass spectral fragments are formed by subtraction of CO and subsequent subtraction of CHN, so that it has a structure related to (Hirosawa and Wolk, unpublished observations).

III. BIOCHEMICAL STUDIES OF DIFFERENTIATION AND INTERCELLULAR INTERACTIONS

As discussed above, there is extensive documentation of the idea that the development of cyanobacteria depends upon interactions between heterocysts and vegetative cells. What are the development-controlling interactions? What are the major biochemical differences between the different types of cells? How do the substances which mediate the critical interactions control, within a vegetative cell, whether and to what endpoint that cell will differentiate?

The task of answering these questions has been paced by the availability of isolated heterocysts which retain the properties which they possessed *in vivo*. To a very gross approximation, there have been several "generations" of isolated heterocysts, each permitting analyses that were not possible with its predecessors, and at least one additional generation seems still to be in prospect.

A. *The First Isolated Heterocysts*

Heterocysts of the first generation had been extensively disrupted structurally by the isolation procedures employed (Fay and Lang, 1971). These heterocysts had no (Fay and Walsby, 1966; Fay et al., 1968) or extremely low (Stewart et al., 1969) nitrogenase activity, even in the presence of ATP and dithionite, and exhibited a light-dependent uptake of O_2 which was not diminished by boiling (Bradley and Carr, 1971). They were, however, adequate for studies of their structural components. For example, certain glycolipids could be shown to be localized in the heterocysts (Wolk and Simon, 1969; Walsby and Nichols, 1969), and to constitute the inner, laminated layer of their envelopes (Winkenbach et al., 1972); the chemical structures of these glycolipids were identified (Bryce et al., 1972; Lambein and Wolk, 1973: see Fig. 3). Although the principal "heterocyst glycolipid" from *A. cylindrica* [1-(0-α-D-glucopyranosyl)-3,25-hexacosanediol] is found in other species, it is often accompanied by larger quantities of other glycolipids not found in

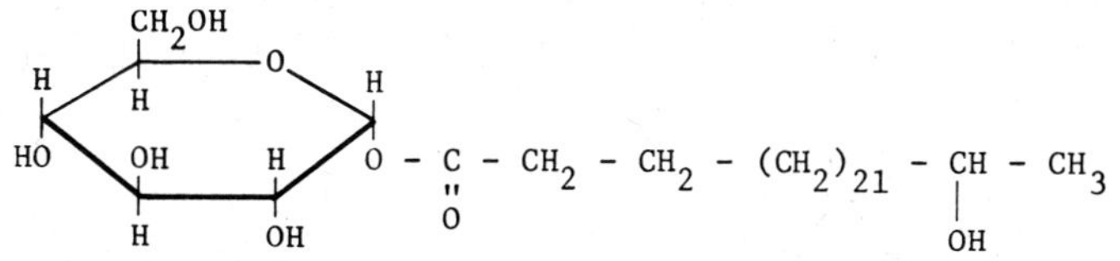

25-hydroxyhexacosanoic acid (1-α-D-glucopyranose) ester

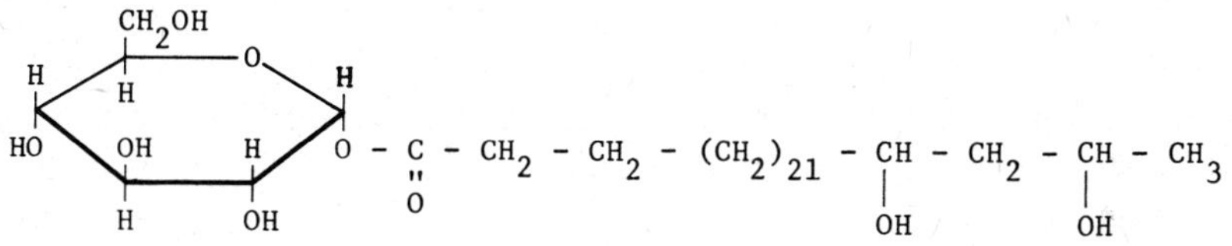

25,27-dihydroxyoctacosanoic acid (1-α-D-glucopyranose) ester

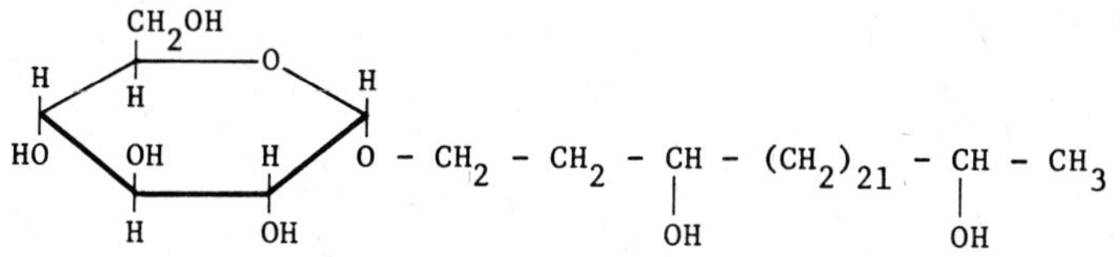

1 - (O-α-D-glucopyranosyl) - 3,25-hexacosanediol

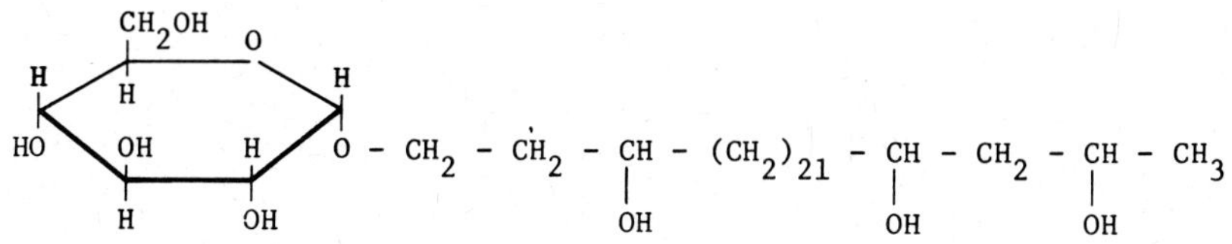

1 - (O-α-D-glucopyranosyl) - 3,25,27-octacosanetriol

Fig. 3. Envelope lipids of heterocysts of *Anabaena cylindrica* (Lambein and Wolk, 1973). The carbohydrate moiety is glucose in about 90% of the molecules and galactose in the remainder. From Wolk (1975), with permission.

A. cylindrica (Lorch and Wolk, 1974). Mutants deficient in envelope glycolipids cannot reduce acetylene under aerobic conditions (Haury and Wolk, 1978), an observation consistent with — but not proving — the idea that the envelope layers of heterocysts play a role in the protection of nitrogenase from oxygen (Stewart, 1973). The laminated layer of lipids which is absent only at the junctions to vegetative cells, presumably prevents the movement of lipid-insoluble substances between the protoplast of heterocysts and the medium directly, and thus probably constrains heterocysts to interact with their neighboring cells. The photosynthetic membranes of heterocysts retain the lipids characteristic of the membranes of vegetative cells (Winkenbach *et al.*, 1972), as well as certain but apparently not all of the proteins of those membranes (Fleming and Haselkorn, 1973; Sallal and Codd, 1977), and retain photosystem I activity (Wolk and Simon, 1969; Donze *et al.*, 1972; Tel-Or and Stewart, 1977).

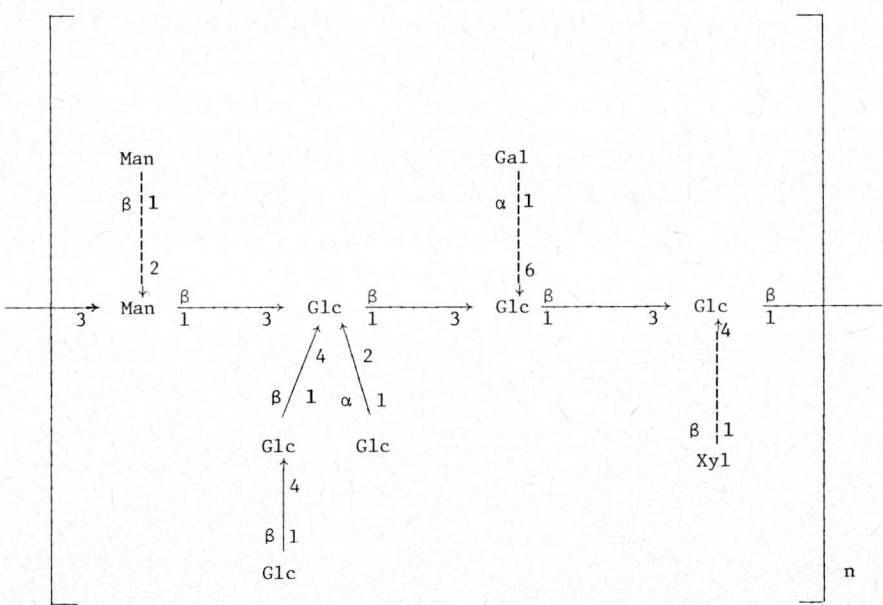

Fig. 4. Approximate subunit structure of the polysaccharides from the envelopes of heterocysts and spores of *Anabaena cylindrica*. Linkages to xylose, galactose and mannose indicated by dashed lines are present attached to only about half of the residual units shown. The terminal sugar of the disaccharide branch may in fact be linked to C-4 of the glucose linked to C-2 of the same glucosyl residue of the backbone; and galactose may possibly be attached to some of those glucosyl residues of the backbone, at the reducing end of the repeating unit shown, to which xylosyl residues are not attached (Cardemil and Wolk, unpublished observations).

The thick, outer, homogeneous layer of the envelope of heterocysts is diagnostic of their morphological differentiation. This layer consists of a complex polysaccharide, the structure of which (in *A. cylindrica*) has been elucidated (Cardemil and Wolk, 1976, and unpublished observations: Fig. 4). An identical polysaccharide constitutes much of the envelope of spores. Therefore, the same pathway of polysaccharide biosynthesis is activated during the two differentiation processes. The detailed structure of the heterocyst envelope polysaccharide appears to vary from species to species (Cardemil and Wolk, unpublished observations).

The occurrence of metabolic intercellular interactions in cyanobacteria was demonstrated by autoradiography of intact filaments. It could be shown that $^{14}CO_2$ is assimilated essentially only by vegetative cells, and that part of the fixed carbon moves through the filaments into the heterocysts. At least part of the fixed ^{14}C is then incorporated into the structural constituents of the heterocysts (Wolk, 1968). Certain of the interactions have subsequently been identified (see below).

B. *Isolated Heterocysts which Retained Enzymes*

The cells of a "second generation" of isolated heterocysts, compared with those of the first, were probably more intact, and retained water-soluble enzymes. It could be shown that the later heterocysts retained ca. 10 to 35 percent of the nitrogenase activity of the intact filaments (Wolk and Wojciuch, 1971a; Peterson and Burris, 1976; Thomas *et al.*, 1977). The fraction of nitrogenase retained in the heterocysts was even higher than 35 percent as assayed *in vitro* (Wolk and Wojciuch, 1971b) or—with the uncertainties attendant upon interpreting banding following electrophoresis in the presence of sodium dodecyl sulfate — as assayed in terms of protein bands on gels (Fleming and Haselkorn, 1973). Glutamine synthetase was shown to be present in both vegetative cells and heterocysts, although at slightly higher specific activity in heterocysts (Dharmawardene *et al.*, 1973; Thomas *et al.*, 1977), and glutamate synthase was found to be localized solely in vegetative cells (Thomas *et al.*, 1977).

Using intact filaments and nitrogen gas labeled with the radioisotope ^{13}N ($t_{1/2}$ = 10 min, maximum β^+ energy 1.25 MeV), we demonstrated unequivocally that N_2-derived ammonium is assimilated by the glutamine synthetase-glutamate synthase pathway (Wolk *et al.*, 1976; Meeks *et al.*, 1978; cf. also Stewart and Rowell, 1975, and Lawrie *et al.*, 1976). Isolated heterocysts metabolized [^{13}N]N_2 and $^{13}NH_4$ to glutamine but not to glutamate (Fig. 5); the formation of glutamine was greatly

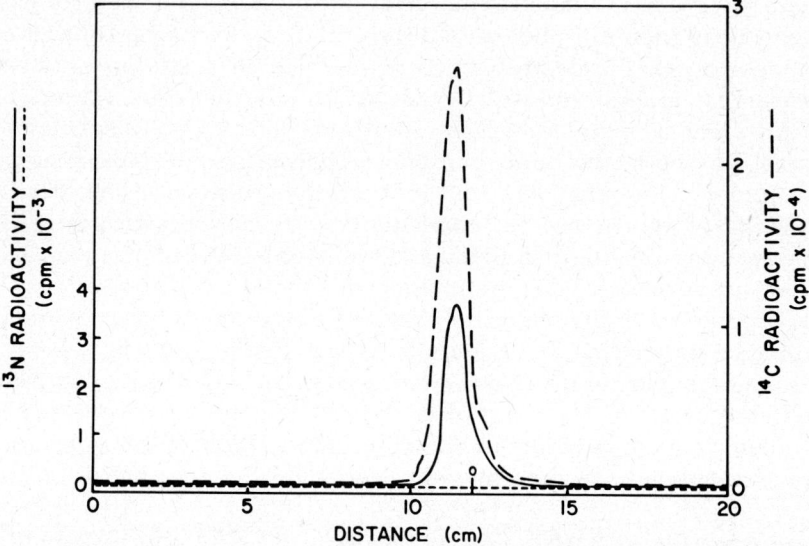

Fig. 5. Distribution of radioactivity from ^{13}N in an electrophoretogram of compounds extracted from a suspension of heterocysts isolated from *Anabaena cylindrica* and allowed to fix [^{13}N]N$_2$ for 15 min in the light in the presence of 1mM glutamate, 4mM MgCl$_2$, 4 mM ATP, an ATP-generating system, 10mM Na$_2$S$_2$O$_4$, and 24 mM TES buffer, pH 7.2. An 80% methanolic extract of the contents of the reaction vial, supplemented with unlabeled amino acids and [^{14}C]glutamine as markers, was applied at position O of a cellulose thin-layer plate, and was subjected to electrophoresis at 3000 v for 12 min in 70 mM sodium borate buffer, pH 9.2. The plate was scanned (———); rescanned after decay of the ^{13}N (----); and then, after removal of a thin layer of aluminum covering the detector, scanned for radioactivity from ^{14}C (—— ——). The plate was then dried, and sprayed with a solution of ninhydrin to localize the stable marker amino acids. Glutamate was present at the 6.7-cm position after electrophoresis. From Thomas *et al.* (1977), with permission.

TABLE II

Effect of added glutamate and methionine sulfoximine on the incorporation of ^{13}NH$_3$ into glutamine by heterocysts isolated from *Anabaena cylindrica* (modified from Thomas et al. (1977), with permission).

Additions to reaction vial[a]	$\dfrac{\text{Dpm of [}^{13}\text{N] glutamine recovered}^{b}}{\text{Dpm of }^{13}\text{NH}_3\text{ added}^{b}}$, %
1. 5 mM ATP, 8 mM MgCl$_2$, 13 mM TES, 50 μM DCPIP, 5 mM sodium ascorbate, 5 mM α-ketoglutarate	0.032
2. As in 1, plus 1 mM glutamate	1.21
3. As in 2, plus 1 mM methionine sulfoximine	<0.023

a. Reaction vials contained 2.27 × 10^6 heterocysts, and supplements, in a total volume of 0.4 ml. Reactions were initiated by the addition of ^{13}NH$_3$, 6.32 × 10^8 dpm[b], and were incubated in the light for 10 min.

b. Corrected to the time of distillation of ^{13}NH$_3$.

stimulated by the addition of glutamate (Table II; Thomas et al., 1977) derived, in vivo, from vegetative cells (see below). The results of the experiments with ^{13}N, together with the determinations of the localization of glutamine synthetase and glutamate synthase, showed that ammonium formed by fixation of nitrogen in heterocysts is transferred from heterocysts to vegetative cells in the form of glutamine (Thomas et al., 1977). If nitrogen fixed by heterocysts were to move into vegetative cells as ammonium, the ratios of the amounts of the products of assimilation of $[^{13}N]N_2$ and $^{13}NH_4^+$ would be expected to be similar. However, these ratios differed extensively, in just such a way as to corroborate the interpretation that nitrogen moves from heterocysts to vegetative cells in the form of glutamine (Meeks et al., 1977, 1978).

The "second generation" of heterocysts also provided evidence that photosystem II is absent from heterocysts (Bradley and Carr, 1971; Donze et al., 1972; Tel-Or and Stewart, 1977), an observation consistent with the reported absence of biliproteins from heterocysts (Fay, 1969; Wolk and Simon, 1969; Thomas, 1970; Fleming and Haselkorn, 1974) and with the idea that N_2 fixation would be favored in a cell which did not produce O_2. However, because the heterocysts would be unable to generate their own reductant for N_2 from water, they would have to receive a net flux of electron donors from vegetative cells. As noted above, just such a movement of carbon compounds had been shown by autoradiography (Wolk, 1968). A hint as to the nature of the electron donors moving from vegetative cells to heterocysts was obtained by the demonstration that although the oxidative pentose phosphate cycle is the principal oxidative pathway of vegetative cells, heterocysts of *A. cylindrica* have 60- to 70-fold higher activity of key enzymes — glucose 6-phosphate dehydrogenase and 6-phosphogluconate dehydrogenase — of this pathway than have vegetative cells, per cell. However, the heterocysts — in comparison with the photosynthesizing vegetative cells — have very low activity of ribulose bisphosphate carboxylase, an essential enzyme of the reductive pentose phosphate pathway, and of glyceraldehyde phosphate dehydrogenase, a key enzyme of the gluconeogenic pathway (Winkenbach and Wolk, 1973; cf. also Lex and Carr, 1974; Stewart and Codd, 1975; and Rowell and Stewart, 1976). Moreover, glucose-6-phosphate — in the presence of NADP — could be shown to be an efficient electron donor to nitrogenase in heterocysts (Peterson and Burris, 1978; Lockau et al., 1978) as well as an efficient electron donor to oxygen in those cells (Peterson and Burris, 1976). It was suggested, on the basis of this information, that heterocysts may receive from vegetative cells some precursor of the oxidative pentose phosphate cycle. Although the precursor was not identified, it was supposed not to be glucose *per se*, because hexokinase is present at relatively low activity in heterocysts (Winkenbach and Wolk, 1973).

Hydrogen is produced by nitrogenase as a side product of reduction of dinitrogen. The uptake hydrogenase which is restricted apparently completely to heterocysts (Peterson and Wolk, 1978a) is efficiently coupled to reaction with O_2 (Peterson and Burris, 1978). Reduction of oxygen by glucose 6-phosphate and H_2 might function to reduce the partial pressure of O_2 within heterocysts, thereby favoring N_2 fixation.

Wolk and Wojciuch (1971a) deduced from kinetic experiments that in the light, heterocysts can generate their own ATP and reductant for nitrogen fixation, and that the reduction can be stimulated by hydrogen gas. These capacities could be recovered, to a very limited extent, in (second-generation) purified heterocysts (Peterson and Burris, 1976; Lockau *et al.*, 1978). Moreover, the purified heterocysts could respond to exogenously supplied ATP and dithionite, and so were apparently permeable to those substances (Stewart *et al.*, 1969; Peterson and Burris, 1976; Thomas *et al.*, 1977). It will be recalled that second-generation heterocysts metabolized exogenously supplied glucose 6-phosphate, NADP and glutamate. They also metabolized exogenously supplied glutamine (as donor in an aminotransferase reaction), and 6-phosphogluconate, DL-isocitrate, dichlorophenolindophenol, and tetramethylphenylenediamine (as electron donors: Lockau *et al.*, 1978), and responded to aminooxyacetate (an inhibitor of aminotransferase reactions). The implication was that these heterocysts were "porous", permitting substances of low molecular weight to move relatively freely both inwards and outwards.

C. *Structurally Intact, Metabolically Active Isolated Heterocysts*

A "third generation" of heterocysts is one in which the heterocysts are "tight" in the sense of retaining substances of low molecular weight. In an abstract, Jüttner and Carr (1976) reported that if *Anabaena cylindrica* which had been pre-treated with lysozyme was pulse-labeled with $^{14}CO_2$ and heterocysts then isolated rapidly in the cold by rupturing vegetative cells with a French press at low pressure, a ^{14}C-labeled disaccharide accounted for the majority of ^{14}C recovered in the heterocysts. Vegetative cells, in contrast, had much more label in intermediates of the Calvin cycle. The disaccharide was not formed from $^{14}CO_2$ by isolated heterocysts, which — as noted above — lack ribulose bisphosphate carboxylase activity. The nitrogenase activity of heterocysts isolated in this manner was not reported, but heterocysts prepared somewhat similarly, by a combination of treatment with lysozyme and a Yeda press (Tel-Or and Stewart, 1976), have less than one percent of the nitrogenase activity of heterocysts prepared by other means (Thomas et al., 1977).

We have recently combined treatments of *Anabaena variabilis* by lysozyme and cavitation, followed by differential centrifugation, to produce purified heterocysts having the following properties: In the light, they usually account (on a per-heterocyst basis) for over 50% of the *in vivo* nitrogenase activity of intact filaments. This activity is extensively dependent upon H_2, but does not depend upon addition of another reductant, such as sodium dithionite, or of organic additives, such as ATP, and is constant for from 1 to 3 h. In the dark, the heterocysts express very slight, if any, nitrogenase activity in the presence of dithionite and ATP, and low, O_2-dependent nitrogenase activity (Peterson and Wolk, 1978b). It may be noted that a number of the properties found had been predicted on the basis of the kinetic experiments of Wolk and Wojciuch (1971a). Moreover, the metabolic intermediates in heterocysts isolated by the lysozyme-cavitation procedure from filaments which have assimilated $^{14}CO_3^=$ are extensively labeled. We interpret our observations as indicating that the isolated heterocysts are metabolically active and are not freely permeable to organic metabolites. The protein constituents of nitrogenase could be separated by anaerobic, non-denaturing gel electrophoresis and quantified using ^{55}Fe; the nitrogenase proteins in the isolated heterocysts accounted for 80 to 90% of the nitrogenase proteins in whole filaments (Peterson and Wolk, 1978b).

We have found two pools to be labeled extensively and consistently, in filaments which have assimilated $^{14}CO_3^=$ for 1 h, viz., sucrose and glutamate. Isolated heterocysts assimilate these substances from extracts of filaments, showing directly that vegetative cells supply heterocysts with sucrose and glutamate. Concordantly, sucrose phosphorylase, which metabolizes sucrose to glucose-1-phosphate and fructose, has been assayed in extracts of heterocysts. We propose that two quantitatively important interactions between heterocysts and vegetative cells are the interchange of glutamate and glutamine, already discussed, and an assimilation of sucrose by heterocysts (Schilling and Wolk, unpublished observations: see Fig. 6). These studies are only the beginning of investigation of intercellular interactions using isolated heterocysts.

Fleming and Haselkorn (1974) showed that heterocysts isolated from

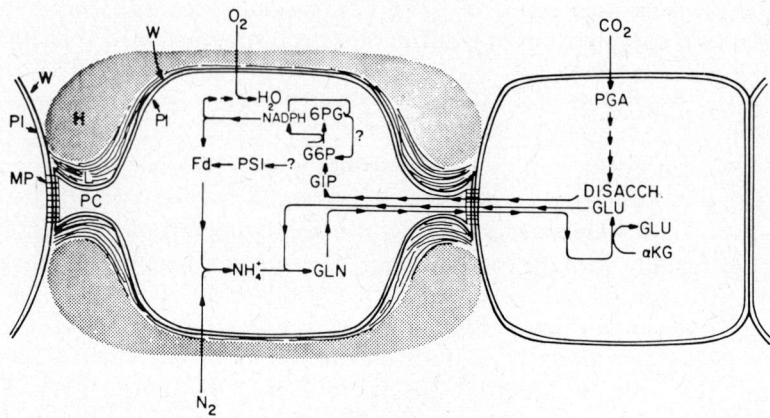

Fig. 6. Diagram showing the principal known interactions between a heterocyst (at left) and a vegetative cell (at right). Outside of the wall (W) of the heterocyst is an envelope consisting principally of a laminated, glycolipid layer (L) and a homogeneous, polysaccharide layer (H). Microplasmodesmata (MP) join the plasma membranes (Pl) of the heterocyst and vegetative cell at the end of the pore channel (PC) of the heterocyst. A disaccharide (in *A. variabilis,* apparently sucrose) formed by photosynthesis in the vegetative cells moves into heterocysts, and is thought then to be cleaved (in part, phosphorolyzed), metabolized to glucose 6-phosphate, and oxidized by the oxidative pentose phosphate pathway. Pyridine nucleotide (NADPH) reduced by this pathway can react with O_2 to maintain reducing conditions within the heterocysts, and can reduce ferredoxin (Fd). Ferredoxin can also be reduced by photosystem I. Reduced ferredoxin can donate electrons to nitrogenase, which reduces N_2 to NH_4^+ with concomitant production of hydrogen. Glutamate produced by vegetative cells reacts in heterocysts with NH_4^+ to form glutamine. The glutamine moves into the vegetative cells, where it reacts with α-ketoglutarate (αKG) to form two molecules of glutamate.

filaments pre-labeled with $^{35}SO_4^=$ contain newly synthesized proteins. Once it is known what set of substances heterocysts normally receive from vegetative cells, it should be possible — by supplying those substances during and after the isolation procedure — to maintain the metabolic activity of the isolated heterocysts for many hours, and perhaps for indefinitely long periods of time. Heterocysts isolated and maintained in this fashion would belong to a "fourth generation", a generation which might be put to use as catalysts of N_2 fixation in practical applications.

D. *Interactions between Vegetative Cells and Mature Heterocysts, and Control of Heterocyst Formation*

Elucidation of the chemical interchanges between heterocysts and vegetative cells is clearly of great importance relative to an understanding not only of the physiology of heterocysts, but also of the developmentally important interactions which control heterocyst formation and sporulation. As discussed above, heterocysts inhibit nearby vegetative cells from differentiating into heterocysts, and induce adjacent vegetative cells to differentiate into spores. We have seen that during aerobic N_2 fixation, heterocysts are net producers of glutamine and net consumers of glutamate, giving rise to gradients of nitrogen which have been directly demonstrated (Gorkom and Donze, 1971; Wolk et al., 1974). Although exogenously supplied ammonium normally inhibits heterocyst formation (Fogg, 1949), it is prevented from doing so by the presence of methionine sulfoximine, an inhibitor of glutamine synthetase (Stewart and Rowell, 1975; Wolk et al., 1976). Glutamine, added exogenously in the presence (Ownby, 1977) or absence (Table I) of methionine sulfoximine, itself inhibits heterocyst formation. These results are consistent with the idea that, at least to some extent, it is glutamine which mediates the intercellular inhibition of heterocyst formation in N_2-fixing cultures. However, because it is unclear whether the effects of exogenously supplied glutamine are due in the first instance to a competition with methionine sulfoximine rather than to a reversal of its action, and in the second instance to possible partial breakdown of glutamine to ammonium, the idea that glutamine in part mediates the inhibition cannot be taken as proven.

We hope to test further the idea that glutamine is important in the control of pattern formation by making use of mutants, which are relatively easy to isolate from *A. variabilis*. This organism can be broken into viable fragments as close as desired to one cell in length by the simple expedient of cavitating a flask culture in a sonic cleaning bath (Wolk and Wojciuch, 1973), and can thereupon be treated as a unicellular bacterium. For example, it has been possible to isolate, by means of penicillin enrichment and replica plating, mutants altered in development and nitrogen fixation, auxotrophic mutants (Currier et al., 1977), and mutants defective in photoautotrophy (Shaffer et al., 1978). Two types of available mutants are potentially useful for analysis of the role of glutamine in pattern formation. One, isolated from an ammonium-grown fermentor culture, produces heterocysts constitutively, in a normal pattern, in the presence of 2mM ammonium, although its nitrogenase activity is repressed under these conditions. The other grows in the presence of methionine sulfoximine, provided that glutamate is also present. Study of these mutants has only been initiated.

E. *Control of Heterocyst Formation by Interactions of Vegetative Cells with Immature Heterocysts*

As noted above, when ammonium-grown filaments lacking heterocysts are transferred to medium lacking fixed nitrogen, a pattern of spaced proheterocysts arises — and therefore, proheterocysts start to inhibit nearby cells from differentiating — before they are capable of fixing nitrogen, i.e., before they can serve as a net source of N_2-derived nitrogen (Kulasooriya *et al.*, 1972). In fact, the pattern of spaced proheterocysts is developed in wild-type filaments transferred from ammonium to $Ar-CO_2$ (95:5, v/v) (Bradley and Carr, 1976) and in filaments of mutants unable to fix N_2 upon their removal from medium containing fixed nitrogen (Currier *et al.*, 1977).

Micrographs of Wilcox *et al.* (1973a, 1975b) show that during normal growth of *Anabaena*, two or more contiguous vegetative cells may start to differentiate into heterocysts, and then all but one revert to being vegetative cells. More generally, it has been observed that when ammonium-grown cultures are deprived of fixed nitrogen, and heterocysts start to develop, the phycocyanin content and the rate of CO_2 fixation—both of which are very low in mature heterocysts—decline markedly in the entire cultures (Bradley and Carr, 1976), and subsequently increase in the cells that do not go on to become heterocysts. The synthesis of a variety of proteins characteristic of heterocysts is initiated in all cells, but continues only in cells that are to complete the differentiation process (Fleming and Haselkorn, 1974). These results have been interpreted as indicating that all of the cells start to differentiate, but that only a minority completes the process and the remainder revert to being vegetative cells. What can be the mechanism of inhibition by the proheterocysts?

As indicated above, heterocysts are very deficient in certain major proteins of vegetative cells — viz., phycocyanin and ribulose bisphosphate carboxylase — and in addition, glutamate synthase and glyceraldehyde phosphate dehydrogenase. In fact, the most conspicuous proteins of vegetative cells have apparently already been lost from very immature heterocysts nine hours after transfer to nitrogen-free medium (Fleming and Haselkorn, 1974). Two proteases, one of which is active against phycobiliproteins, including phycocyanin, and the other of which degrades many of the other proteins which are turned over during the differentiation of heterocysts, have been detected in *Anabaena* 7120; activity of the latter protease increases within one h after nitrogen stepdown, with the increase in activity inhibitable by chloramphenicol (Wood and Haselkorn, 1978). Fleming and Haselkorn (1974) therefore suggested that it may be products of proteolysis released from proheterocysts which inhibit nearby cells from differentiating.

This is an engaging hypothesis, which has the virtue that morphological differentiation would amplify or stabilize the pattern-determining, inhibitory, intercellular interactions, but not be a necessary prerequisite for those interactions (compare Wolk, 1973). However, it poses the problem: if that is how a proheterocyst "turns off" the differentiation of its neighbors, why does it not first "turn off" its own differentiation? One speculative answer is that nitrogen deprivation leads to derepression not only of protease but of nitrogen-assimilatory systems as well, including uptake systems. It could then be that of a group of cells which is deprived of nitrogen, the *last* cell to be so affected has as neighbors cells which are actively removing nitrogenous substances from it, so that while that cell helps to repress protease in its neighbors, it is unable to re-repress its own protease. It would then be this cell (of the group) which would proceed to differentiate as a heterocyst. Such an account provides an explanation of why *single* vegetative cells, which would be able to re-repress themselves, do not ordinarily differentiate. Moreover, provided that "parasitism" by only one adjacent cell were normally insufficient to prevent the products of autodigestion of a cell from re-repressing that cell it would explain why *terminal* vegetative cells of an *Anabaena* normally fail to differentiate. Only in short filaments which, because of a lack of any heterocysts, are exceptionally nitrogen stressed, would terminal vegetative cells often differentiate. Terminal cells of *Cylindrospermum* would be expected to differentiate if, as previously suggested, they are particularly leaky*.

How would nitrogen deprivation lead to derepression of protease and other enzymes, and why would the cell that "loses the race" end up synthesizing the enzymes required for morphological differentiation? The former question appears to be concerned with physiological responses not directly related to differentiation. The simplest answer to the latter question would appear to be that the enzymes related to cellular morphology may be regulated coordinately with nitrogenase (with the regulation modulated by O_2), in order that the support systems required for nitrogenase to function under aerobic conditions would be present (compare Singh et al., 1977).

Instances are known in which spores arise at a place where, in the normal pattern of vegetative growth, a heterocyst would have differentiated. Such a positioning of spores is the normal event in species in which spores form approximately midway between preexisting heterocysts. Instances are also known in which heterocysts arise at a

---

* Haselkorn (1978), on whose data the above idea is based, has independently suggested that amino acid "pumps" are important in the control of differentiation.

place where, in the normal pattern of sporulation, a spore would have differentiated. Geitler (1925) has suggested that these instances are indicative of a relationship between the metabolic processes leading to the two types of differentiated cells. Cyanobacteria may therefore also be highly useful as organisms with which to ask: what are the relationships between alternative differentiation processes within a single organism? Because the "only known biosynthetic process associated with *sporulation* which appears to differ qualitatively from processes of growth of vegetative cells is deposition of the spore envelope" (Wolk, 1975; italics added), the control of sporulation may turn out to be particularly simple.

IV. CONCLUDING REMARKS

In certain filamentous cyanobacteria, heterocysts inhibit nearby vegetative cells from differentiating into heterocysts and induce nearby vegetative cells to differentiate into spores. Some of the principal structural (especially, envelope layers) and physiological (especially in relation to dinitrogen fixation and its supporting systems) differences between vegetative cells and differentiated cells have been characterized in biochemical terms. These characterizations, plus the availability of isolated heterocysts which are increasingly undamaged, have permitted the identification of what may be the principal interactions between heterocysts and vegetative cells. Glutamine, produced by heterocysts from N_2 which they fix and from glutamate synthesized in vegetative cells, may be the inhibitory-field morphogen produced by mature heterocysts. It appears that sucrose assimilated from vegetative cells provides reductant to reduce substrates of nitrogenase and to reduce oxygen that enters the heterocysts. A hitherto unknown sulfur-containing compound is somehow involved in the sporulation of cells adjacent to heterocysts, in *Cylindrospermum*.

It has now become possible to *attempt* (i) to study the developmental genetics of these cyanobacteria; (ii) to maintain, for possible practical applications, the metabolic activity of isolated heterocysts; and (iii) to formulate plausible, testable hypotheses about the means by which certain intercellular interactions control the biochemical processes which result in differentiation.

ACKNOWLEDGEMENTS

I thank Mr. Paul W. Shaffer for excellent technical assistance. This work was supported by the U.S. Department of Energy under Contract EY-76-C-02-1338.

REFERENCES

Allen M.B. and Arnon, D.I. (1955). *Plant Physiol.* **30,** 366-372.
Bradley, S. and Carr, N.G. (1971). *J. Gen. Microbiol.* **68,** xiii-xiv.
Bradley, S. and Carr, N.G. (1976). *J. Gen. Microbiol.* **96,** 175-184.
Bryce, T.A., Welti, D., Walsby, A.E. and Nichols, B.W. (1972). *Phytochem.* **11,** 295-302.
Canabaeus, L. (1929). *Pflanzenforschung* **13,** 1-48.
Cardemil, L. and Wolk, C.P. (1976). *J. Biol. Chem.* **251,** 2967-2975.
Currier, T.C., Haury, J.F. and Wolk, C.P. (1977). *J. Bacteriol.* **129,** 1556-1562.
Dharmawardene, M.W.N., Haystead, A. and Stewart, W.D.P. (1973). *Arch. Mikrobiol.* **90,** 281-295.
Donze, M., Haveman, J. and Schiereck, P. (1972). *Biochim. Biophys. Acta* **256,** 157-161.
Fay, P. (1969). *Arch. Mikrobiol.* **67,** 62-70.
Fay, P. and Lang, N.J. (1971) *Proc. Roy. Soc. B.* **178,** 185-192.
Fay, P. and Walsby, A.E. (1966). *Nature* **209,** 94-95.
Fay, P., Stewart, W.D.P., Walsby, A.E. and Fogg, G.E. (1968). *Nature* **220,** 810-812
Fisher, R.W. and Wolk, C.P. (1976) *Nature* **254,** 394-395.
Fleming, H. and Haselkorn, R. (1973). *Proc. Nat. Acad. Sci. U.S.* **70,** 2727-2731.
Fleming, H. and Haselkorn, R. (1974). *Cell* **3,** 159-170.
Fogg, G.E. (1949). *Ann. Bot. n.s.* **13,** 241-259.
Geitler, L. (1925). *Beih. Botan. Centralbl.* **41,** 163-294.
Geitler, L. (1932). Cyanophyceae, Aufl. 2, Akademische Verlagsgesellschaft, Leipzig.
Gorkom, H.J. van and Donze, M. (1971). *Nature* **234,** 231-232.
Haselkorn, R. (1978). *Ann. Rev. Plant Physiol.* **29,** 319-344.
Haury, J.F. and Wolk, C.P. (1978). *J. Bacteriol.* **136,**
Jüttner, F. and Carr, N.G. (1976). *Proc. 2nd Int. Symp. Photosyn. Prok.* 121-123.
Kulasooriya, S.A., Lang, N.J. and Fay, P. (1972). *Proc. Roy. Soc. B* **181,** 199-209.
Lambein, F. and Wolk, C.P. (1973). *Biochem. (A.C.S.)* **12,** 791-798.
Lawrie, A.C., Codd, G.A. and Stewart, W.D.P. (1976). *Arch. Microbiol.* **107,** 15-24.
Lex, M. and Carr, N.G. (1974). *Arch. Microbiol.* **101,** 161-167.
Lockau, W., Peterson, R.B., Wolk, C.P. and Burris, R.H. (1978). *Biochim. Biophys. Acta.* **502,** 298-308.
Lorch, S.K. and Wolk, C.P. (1974). *J. Phycol.* **10,** 352-355.
Meeks, J.C., Wolk, C.P., Thomas, J., Lockau, W., Shaffer, P.W., Austin, S.M., Chien, W.S. and Galonsky, A. (1977). *J. Biol. Chem.* **252,** 7894-7900.
Meeks, J.C., Wolk, C.P., Lockau, W., Schilling, N., Shaffer, P.W. and Chien, W.-S. (1978). *J. Bacteriol.* **134,** 125-130.
Mitchison, G.J., Wilcox, M. and Smith, R.J. (1976). *Science* **191,** 866-868.
Ownby, J.D. (1977). *Planta* **136,** 277-279.
Peterson, R.B. and Burris, R.H. (1976). *Arch. Microbiol.* **108,** 35-40.
Peterson, R.B. and Burris, R.H. (1978). *Arch. Microbiol.* **116,** 125-132.
Peterson, R.B. and Wolk, C.P. (1978a). *Plant Physiol.* **61,** 688-691.
Peterson, R.B. and Wolk, C.P. (1978b). *Abst. Steenbock-Kettering Internat. Symp. Nit. Fix.,* p. 18.
Rowell, P. and Stewart, W.D.P. (1976). *Arch. Microbiol.* **107,** 115-124.
Sallal, A.-K.J. and Codd, G.A. (1977). *Br. Phycol. J.* **12,** 163-169.
Shaffer, P.W., Lockau, W. and Wolk, C.P. (1978). *Arch. Microbiol.* **117,** 215-219.
Singh, R.N., Tiwari, D.N. and Singh, V.P. (1972). *In*"Taxonomy and Biology of Blue-Green Algae" (T.V. Desikachary, ed.), 27-37. Univ. Madras Centre for Advanced Study in Botany, Madras.

Singh, H.N., Ladha, J.K. and Kumar, H.D. (1977). *Arch. Microbiol.* **114,** 155-159.
Stewart, W.D.P. (1973). *Ann. Rev. Microbiol.* **27,** 283-316.
Stewart, W.D.P. and Codd, G.A. (1975). *Br. Phycol. J.* **10,** 273-278.
Stewart, W.D.P. and Rowell, P. (1975). *Biochem. Biophys. Res. Comm.* **65,** 846-856.
Stewart, W.D.P., Haystead, A. and Pearson, H.W. (1969). *Nature* **224,** 226-228.
Tel-Or, E. and Stewart, W.D.P. (1976). *Biochim. Biophys. Acta* **423,** 189-195.
Tel-Or, E. and Stewart, W.D.P. (1977). *Proc. Roy. Soc. B* **198,** 61-86.
Thomas, J. (1970). *Nature* **228,** 181-183.
Thomas, J., Meeks, J.C., Wolk, C.P., Shaffer, P.W. and Austin, S.M. (1977). *J. Bacteriol.* **129,** 1545-1555.
Walsby, A.E. and Nichols, B.W. (1969). *Nature* **221,** 673-674.
Wilcox, M. (1970). *Nature* **228,** 686-687.
Wilcox, M., Mitchison, G.J. and Smith, R.J. (1973a). *J. Cell Sci.* **12,** 707-723.
Wilcox, M., Mitchison, G.J. and Smith, R.J. (1973b). *J. Cell. Sci.* **13,** 637-649.
Wilcox, M., Mitchison, G.J. and Smith, R.J. (1975a). *Arch. Microbiol.* **103,** 219-223.
Wilcox, M., Mitchison, G.J. and Smith, R.J. (1975b). *In* "Microbiology-1975" (D. Schlessinger, ed.), 453-463. Am. Soc. Microbiol., Washington, D.C.
Winkenbach, F. and Wolk, C.P. (1973). *Plant Physiol.* **52,** 480-483.
Winkenbach, F., Wolk, C.P. and Jost, M. (1972). *Planta* **107,** 69-80.
Wolk, C.P. (1966). *Am. J. Bot.* **53,** 260-262.
Wolk, C.P. (1967). *Proc. Nat. Acad. Sci. U.S.* **57,** 1246-1251.
Wolk, C.P. (1968). *J. Bacteriol.* **96,** 2138-2143.
Wolk, C.P. (1973). *Bacteriol. Rev.* **37,** 32-101.
Wolk, C.P. (1975). *In* "Spores VI"(P. Gerhardt *et al.,* eds.) 85-96. Am. Soc. Microbiol., Washington, D.C.
Wolk, C.P. and Quine, M.P. (1975). *Develop. Biol.* **46,** 370-382.
Wolk, C.P. and Simon, R.D. (1969). *Planta* **86,** 92-97
Wolk, C.P., Austin, S., Bortins, J. and Galonsky, A. (1974). *J. Cell Biol.* **61,** 440-453.
Wolk, C.P., Thomas, J., Shaffer, P.W., Austin, S.M. and Galonsky, A. (1976). *J. Biol. Chem.* **251,** 5027-5034.
Wolk, C.P. and Wojciuch, E. (1971a). *Planta* **97,** 126-134.
Wolk, C.P. and Wojciuch, E. (1971b). *J. Phycol.* **7,** 339-344.
Wolk, C.P. and Wojciuch, E. (1973). *Arch. Mikrobiol.* **91,** 91-95.
Wood, N.B. and Haselkorn, R. (1978). *In* "Limited Proteolysis in Microorganisms: biological function, use in protein structural and functional studies" (H. Holzer and G.N. Cohen, eds.), Fogerty Internat. Center Publns., Bethesda, Md. In press.

Development of Hydra Lacking Interstitial and Nerve Cells ("Epithelial Hydra")

Richard D. Campbell
Department of Developmental and Cell Biology
University of California
Irvine, California 92717

I.	Introduction	267
II.	Background of Hydra Development	268
	A. Tissue Structure and Growth	268
	B. The Interstitial Cell Theory	270
	C. Role of Nerve Cells in Development	271
III.	Selective Elimination of Interstitial Cells	272
	A. Colchicine	272
	B. Other Methods	274
	C. Structure of Epithelial Hydra	275
	D. Development of Epithelial Hydra	277
IV.	Hydra Chimeras	282
V.	Other Aspects of Epithelial Hydra	285
	A. Cell and Tissue Movements	287
	B. Cell Differentiation	288
	C. Histological Structure	288
	D. Electrophysiology and Behavior	289
VI.	Control of Pattern in Hydra: Conclusions	289
	References	291

I. INTRODUCTION

Many developing tissues are composed of several cell types, and it may be essential to identify which cells regulate patterning before one can understand the actual mechanisms of pattern formation. This is particularly true in hydra where the two prime candidates for the governors of pattern — nerve cells and epithelial cells — have rather different implications for patterning mechanisms. Nerve cells are scattered as a loose nerve net in hydra, and they would presumably regulate patterning through secretion of diffusible morphogens. Epithelial cells provide the mechanical foundation of hydra, and they could presumably control patterning in quite different ways, such as by causing and organizing physical forces in the tissue.

In studying hydra developmental mechanisms one has a rich literature in methods, observation and experimentation to turn to (see Baker, 1952; Kanaev, 1952; Lenhoff and Loomis, 1961; Lentz, 1966; Burnett, 1973). However, many investigations of hydra have been led into pitfalls, as described below, by presuming the existence of a particular developmental mechanism and then gathering evidence for it. Since it is usually possible to obtain data consistent with almost any proposal, there is a danger to working solely upon a particular hypothesis.

Our approach to the analysis of hydra developmental mechanisms involves assigning patterning roles to the various cell types by methods which make no presumption of a mechanism. The methods involve selectively eliminating cell types and looking for resultant developmental deficiencies, and by forming chimeric hydra whose various cell types have different genetic constitutions to see which cell types dominate in controlling development. Because our work involves considerable manipulation of cell types, and has resulted in the first complete denervation of a viable organism, I will first give some background concerning the cellular dynamics of hydra and concerning methods for selective cell removal.

II. BACKGROUND OF HYDRA DEVELOPMENT

A. *Tissue Structure and Growth*

Hydra is a column, consisting of two concentric epithelial layers, ectoderm and endoderm, separated by the mesolamella (Fig. 1). The tissue grows continuously with a doubling time of about 4 days, and at the same rate tissue is lost through the processes of budding and atrophy. The balance between cell production and loss results in a complete and systematic renewal of tissue (Campbell, 1974a). Thus there are no permanent cells in hydra.

Hydra tissue is composed of about 12 cell types. Only three of these are mitotically self-perpetuating: ectodermal epithelial cells, endodermal epithelial cells, and the undifferentiated interstitial cells (I cells). The remaining cells, including the nerve cells, the nematocytes (stinging cells) and gametes are derived, by differentiation, from the I cells*. David (1975) and others (see David and Gierer, 1974; David and Murphy, 1977; Bode and David, 1978) have rigorously demonstrated the multipotency of the I cells and have worked out the kinetics with which I cells differentiate into nerves and nematocytes.

The terminology of these cell types is confusing and lacks uniformity. In this paper the term "I cell" refers to the multipotent, stem cell that gives rise to the lineage of cells that includes nerves, nematocytes, gametes and the abundant cells at intermediate stages of differentiation. Since all of these cells reside in the interstitial spaces, between epithelial cells, the term "interstitial cells" will be used to refer to this entire cell lineage. Interstitial cells are so abundant in well fed hydra that they obscure other aspects of tissue histology (see Fig. 1).

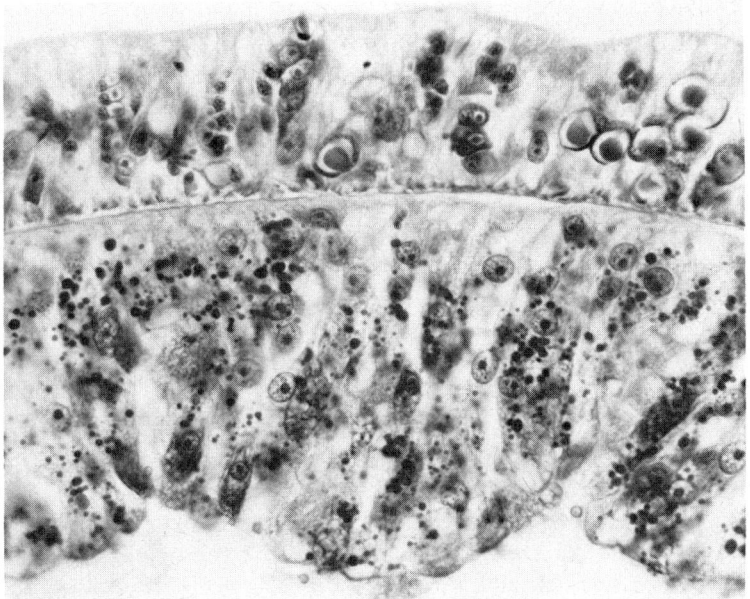

Fig. 1. Epithelial structure of hydra. Ectoderm (top) and endoderm (bottom) are separated by a thin mesolamella (center). The abundant, darkly staining ectodermal cells represent the interstitial cell lineage. (*Hydra littoralis*, hematoxylin staining, x 440)

* Endodermal mucous cells and zymogen gland cells comprise another population of cells whose origins are not completely understood at this time. While they are mitotically active (McConnell, 1933b; Campbell, 1967; Challoner, 1975), their proliferation rate may be insufficient to account for their renewal. Hence it is possible that this population of cells requires a slow replenishment from either the epithelial or, more likely, the interstitial cell lineages.

B. *The Interstitial Cell Theory*

Three serious misunderstandings of histological data followed Kleinenberg's (1872) discovery of interstitial cells in hydra. One misunderstanding concerned the abundant, small, undifferentiated cells typically occuring in clusters and frequently seen in division. These cells were usually referred to as "interstitial cells" in the older literature, and were assumed to represent the undifferentiated I cell line. We now know (see David and Gierer, 1974) that these "little interstitial cells" (David, 1973) are determined, differentiative intermediates undergoing several final rounds of synchronized mitosis before synthesizing nematocysts. Another misunderstanding concerned the less abundant, larger, often solitary undifferentiated cells. These were sometimes described as intermediates in the differentiation of (small) interstitial cells into epithelial cells. We now know that some of these "big interstitial cells" (David, 1973) are the actual multipotent stem cells (David, 1975; David and Murphy, 1977). A third misunderstanding concerned the mitotic activity of epithelial cells, which make up the mechanical foundation of hydra. These cells were thought to not divide mitotically, an idea whose origin is not surprising considering that mitotic little interstitial cells are so much more abundant and conspicuous. As a result of these three misunderstandings the "interstitial cell theory" was elaborated (see Kanajew, 1930; Tardent, 1954; Brien and Reniers Decoen, 1955; Diehl, 1973), holding that the interstitial cell would be the only cell type in hydra with any developmental capacity. Other cells were considered incapable of proliferation, for example. Budding and regeneration were understood to result from local accumulation of migratory interstitial cells, which then underwent differentiation to form the new tissue. Interstitial cells were thought to be graded in abundance along the column, most concentrated near the hypostome and tentacle whorl, and thus provided a ready explanation for the axial gradients of hydra's developmental capacity. The only two parts of a hydra lacking all regenerative capacity are the basal disc and the tentacles, and these are the only regions which lack interstitial cells.

The interstitial cell theory, always hotly contested, was seriously eroded through the discovery that regeneration does not involve blastema formation but rather involves morphallactic remodeling of adjacent epithelia (Kanajew, 1930), and by the growing realization that epithelial cells were capable of mitotic proliferation (McConnell, 1933a; Campbell, 1967; David and Campbell, 1972). With Tripp's (1928) and Brien and Reniers-Decoen's (1949) demonstrations that tissue renewal in hydra is continuous and complete and that individual polyps are as immortal as a colony of bacteria, the interstitial cell was considered as a constant but low level source of fresh cells for growth and development (in addition to supplying nerves and nematocytes).

The crucial test of the interstitial cell theory appeared at hand with the discovery that X-irradiation destroys hydra interstitial cells (Strelin, 1929; Zawarzin, 1929). Could X-irradiated hydra still bud and regenerate? Unfortunately the resulting interstitial cell-free hydra neither proved nor disproved the theory: irradiated hydra had *almost* no developmental capacity. They could form buds, but only once or twice at most. They could regenerate, but the resulting hypostome and tentacles were severely hypomorphic. Irradiated hydra lived, but only for a few weeks.

The nearly complete loss of developmental capacities following X-irradiation was first interpreted (see Brien and Reniers-Decoen, 1955) as proving that interstitial cells are key elements in development. To later investigators, however, the ability to form even a single bud or to regenerate a single tentacle in the absence of interstitial cells showed that other cells can organize and effect development. Diehl and Burnett (who developed an excellent chemical treatment — Nitrogen mustard — to substitute for X-irradiation) convincingly supported this latter interpretation, and ascribed the low developmental capacities of X-irradiated hydra to previously unsuspected damage to epithelial cells (Diehl and Burnett, 1964, 1965 a,b, 1966; Diehl, 1973). We now know that X-irradiation does "kill" all cells in the sense of limiting their mitotic cycle. Interstitial cells respond to such a limit by swelling and then disappearing (Diehl and Burnett, 1964). Epithelial cells, on the other hand, remain in the form of a hydra and are capable of full hydra morphogenesis, as David (David, 1975; David and Murphy, 1977) has so elegantly shown by dissociating and reaggregating killed epithelial cells.

C. *Role of Nerve Cells in Development*

At the time when confidence in the I cells' developmental capacities was waning, interest in hydra's nerve cells was growing. Since I cells are the stem cells for hydra's nervous system, this recent interest represents a new interpretation of the interstitial cell theory. Chemical destruction of the nervous system is said to eliminate regenerative capacity in hydra (Bursztajn, 1974). Neuropharmacological agents have strong influences over regeneration of hydra tissue (Ham and Eakin, 1958; Lentz, 1966). Electron microscopy suggests that at least some hydra nerve cells are neurosecretory (Davis *et al.*, 1968) and that the contents of the neurosecretory vesicles are released at times of regeneration (Burnett *et al.*, 1964; Lentz, 1965a; Bursztajn, 1974). Attempts to extract materials from nerve cells have resulted in the isolation of a number of developmentally potent factors (Lentz, 1965a, Lesh and Burnett, 1966; Schaller, 1978). The most highly purified of these factors has, at extremely low concentrations, numerous effects on hydra's development ranging from altering the interstitial cell cycle (Schaller, 1976) to affecting tentacle number during regeneration (Schaller, 1973; see Schaller, 1978, for review).

The idea that nerves control developmental patterns through secretion of morphogens fits in with the distribution of nerves in a hydra: nerve cells are most abundant in those body regions (Bode *et al.*, 1973) that exhibit the most dominant developmental properties. The three tissue regions capable of induction (hypostome, bud tip and basal disc) are the sites of greatest nerve cell density. Furthermore, neurosecretory control could explain why some aspects of development occur on a time scale of hours (for example, release of regenerative inhibition following amputation of a hydranth) while other aspects occur over the course of days (for example, repolarization of a tissue column segment following reversal of its orientation in hydra). Neurosecretory release, diffusion and breakdown of a morphogen would presumably occur in hours, but setting up a new pattern of nerve cells, through differentiation of interstitial cells, would require days (Gierer and Meinhardt, 1972; Wolpert *et al.*, 1974). There are also numerous other lines of evidence linking nerve cells to development in hydra (See Lentz, 1966; Burnett, 1973; Bursztajn, 1974; Schaller, 1978; Bode and David, 1978).

III. SELECTIVE ELIMINATION OF INTERSTITIAL CELLS

Several new methods result in removal of interstitial cells without damaging epithelial cells. We propose to term the resulting hydra that are entirely free of all cells except viable epithelial cells as *epithelial hydra*. By studying epithelial hydra, we can assess unequivocally the developmental capacities of nerve (and interstitial) cell-free tissue.

A. *Colchicine*

Colchicine is one agent that can be used to eliminate the interstitial cell lineage, because it results in epithelial cells removing nonepithelial cells

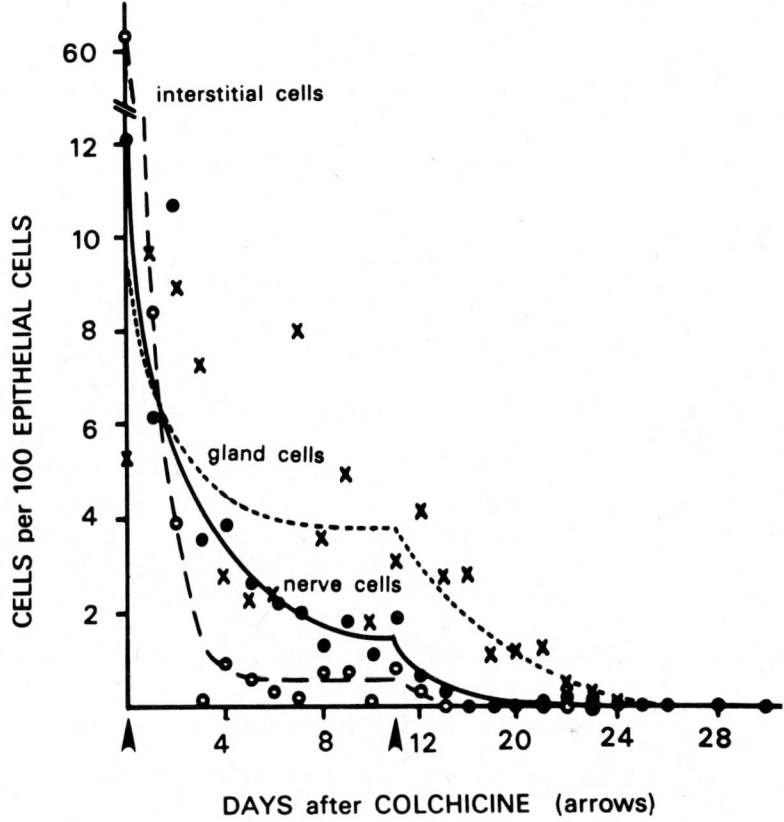

Fig. 2. Cell loss following colchicine treatment applied at times indicated by arrows. Density of interstitial, gland and nerve cells is expressed relative to epithelial cells. (From Marcum and Campbell, 1978a)

by phagocytosis (Campbell, 1974a, 1976; Marcum and Campbell, 1978a). Within a few days after soaking for 8 hours in 0.4% colchicine, hydra lose over 90% of interstitial and nerve cells (Fig. 2). A second treatment eliminates most of the remaining ones. We find that between one-third and one-half of doubly treated hydra are completely free of interstitial cells (Marcum and Campbell, 1978a). Yet the hydra remain fully viable. If force-fed (they have no feeding behavior) they grow and even bud at approximately normal rates. Thus large clones of hydra can be developed from individual hydra. By determining the cell compositions of hydra in each clone of treated hydra, it is possible to select those clones that are completely free of interstitial cells. Within a few weeks the last vestiges of the interstitial lineage — the nerves and nematocytes — die or are diluted out by growth. With continued growth, over the course of more than a year (about 50-100 tissue doubling times) the interstitial cells do not reappear in either the original polyps or in buds. Thus the nerve cell and interstitial cell-free condition is stable. In this paper we will term hydra entirely lacking the interstitial cell lineage as *epithelial* hydra.

B. *Other Methods*

The effects of colchicine treatment are highly reproducible with *Hydra attenuata*, but not with other species. Several other methods have been developed to get around this problem. Of these, γ-radiation (Fradkin *et al.*, 1978) is the most useful. Irradiation twice with 4,000 rad has a low animal lethality but great destructive action on interstitial cells. Treated animals lose most of their interstitial cells over the course of 2 weeks, and a fraction of these animals lose all interstitial cells. Again, cloning and selection allow one to obtain epithelial hydra. Gamma irradiation does not induce an immediate depression in the hydra, as does colchicine, and it does not immediately eliminate nerve cells (but these disappear within a few weeks because the stem cells are eliminated). Another means for obtaining interstitial cell-free hydra has been reported by Sugiyama and Fujisawa (1978a) who find many spontaneous interstitial cell-free polyps in a strain that has been highly inbred in the laboratory. We have also found spontaneously arising interstitial cell-free hydra in mass cultures (Campbell *et al.*, 1977). Also, Colcemid can be used to remove interstitial cells in the same way as colchicine (Campbell, 1976).

Another novel method for removing the interstitial cell lineage from a hydra involves construction of a chimeric hydra whose interstitial cells are more temperature sensitive than the epithelial cell. We have produced such a chimera by repopulating epithelial cells of *H. attenuata* (whose lethal temperature is 35°C) with interstitial cells of *Pelmatohydra oligactis* (whose lethal temperature is 30°C). A 30 minute heat shock of the chimera renders it free of interstitial cells within a few weeks (Fradkin, Lee and Campbell, unpublished).

A number of other methods have been reported but do not yet have complete efficacy. X-irradiation and Nitrogen Mustard are, as described above, effective at eliminating interstitial cells but the remaining polyps are not viable (Brien and Reniers-Decoen, 1955; Burnett and Diehl, 1964; Campbell *et al.*, 1977). Brien (1961) observed that *Hydra pirardi* lose all interstitial cells during sexual reproduction; however these polyps die shortly after reproduction. An unusual report described how regeneration from basal disc yielded interstitial cell-free hydra although these polyps were not viable (Burnett and Lambruschi, 1973). Methylene blue and eserine are reported to specifically destroy nerve cells (Saffitz et al., 1972; Bursztajn, 1974) but the nervous system is quickly regenerated from I cells in treated polyps. Hydroxyurea (Bode et al., 1976; Sacks and Davis, 1977) reduces interstitial cell numbers but conditions have not yet been worked for complete elimination.

With this wealth of effective and prospective methods, the experiments described in this paper will almost certainly be extendable to other hydra species, each of which offers some special variety, cell type or unique developmental trait which could aid in this study.

C. *Structure of Epithelial Hydra*

Hydra deprived of the entire interstitial cell lineage have the normal morphology of hydra, with the following peculiarities. The gastric region and hypostome are distended and the tissue is transparent and thin (Fig. 3b,c). The tentacles are straight, motionless and rigid, and are

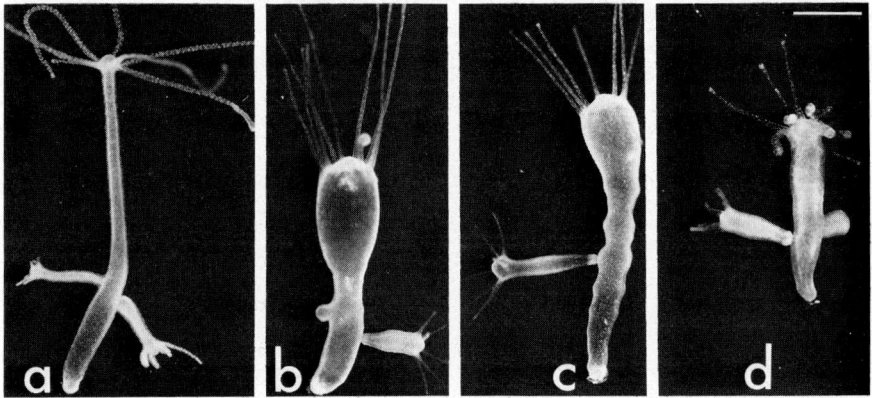

Fig. 3. Appearance of epithelial hydra. a, Normal hydra. b, c, typical appearance of epithelial hydra. d, epithelial hydra shortly after artificial deflation. (H. attenuata derived from colchicine treatment. Bar represents 1 mm b-d; 1.2 mm a).

more numerous and often not in a perfectly regular whorl. The peduncle is slenderer than normal, and the basal disc is often bulbous rather than flat. Buds progress through the normal stages although after completion they often remain resident on the parent for weeks. The gastric swelling probably results from hydra's osmoregulatory mechanism which involves accumulation of brine in the gastric cavity (Macklin et al., 1973); nerve-free hydra lack periodic spitting behavior and thus become swollen. If the mouth is opened by prodding it with a blunt needle, the nerve-free hydra collapse and look somewhat more normal before redistending in 30-90 minutes (Fig. 3d).

Thus all body parts are present, but they are not perfectly proportioned and the degree of abnormality is quite variable. Histologically, epithelial hydra consist only of two cell types: ectodermal and endodermal epithelial cells (Fig. 4). Extensive histological and maceration studies verify the complete absence of all other cell types, even during

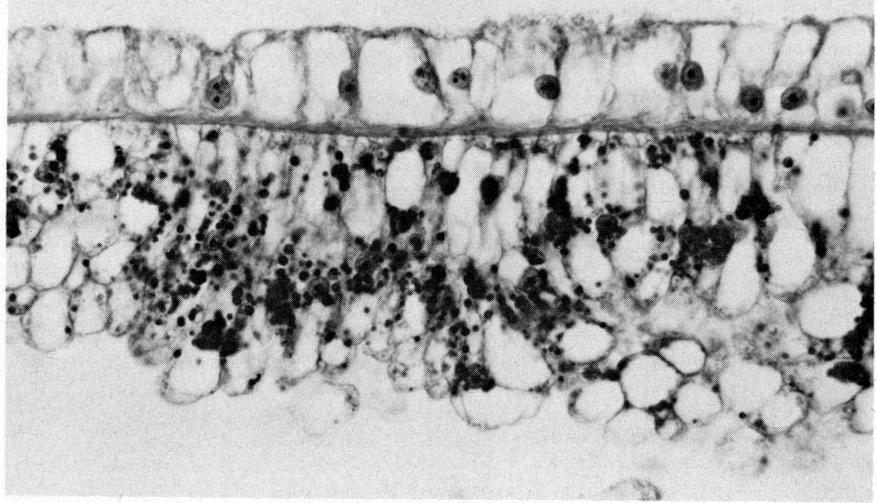

Fig. 4. Epithelial hydra tissue structure. Darkly staining structures are: in ectoderm, epithelial cell nuclei; in endoderm, epithelial cell nuclei and food granules (smaller). (x400).

more than a year of active growth and budding (Marcum and Campbell, 1978a; Wanek, Marcum and Campbell, 1978). The epithelial cells look approximately normal, and exhibit the specialized states in particular body regions as is typical in normal hydra. For example, ectodermal epithelial cells become glandular in the basal disc and low and squamous in the tentacles, and endodermal cells take up varying heights to produce the longitudinal ridges (taeniolae) of the hypostome. Thus, there are no obviously missing trophic influences on the epithelial cells themselves.

The possibility exists that neurotrophic stimuli may be eliminated with the nerve cells and that epithelial cells assume compensatory functions. Marcum (1978) has therefore examined epithelial hydra using electron microscopy, to determine if their epithelial cells contain structures resembling neurosecretory vesicles of normal hydra. Surprisingly, epithelial cells contain neurosecretory-like vesicles in both normal and nerve-free hydra (Table I). These vesicles are not nearly as densely abundant as are vesicles in nerve cells, but represent a conceivable ultrastructural basis for epithelial cells taking over nerve cell functions. However, Marcum's data show no overall increase in abundance of these vesicles in epithelial hydra compared to normal hydra (Table I).

TABLE I

Abundance of Vesicles in Nerve and Epithelial Cells.

Hydra type	Abundance of Vesicles (number/μm^2)	
	In epithelial cells	In nerve cells
normal	0.06	0.75
epithelial	0.07	—

Vesicles with dense cores, resembling the vesicles present in nerve cells, were counted in measured cytoplasmic areas of electron micrographs of normal and epithelial *Hydra attenuata*. (Marcum, unpublished)

D. Development of Epithelial Hydra

Our most surprising discovery was that nerve-free hydra are nearly normal in all aspects of development. However they typically do exhibit quantitative deviations from normal development.

Epithelial cell cycle, tissue growth and atrophy are normally paced and patterned in epithelial hydra. The balance of these processes results in continuous tissue movements throughout the polyp, and these have a normal pattern in epithelial hydra (Marcum and Campbell, 1978a). Thus the epithelial cells are capable of supporting morphogenesis, because as they pass from one body region to another (for example, from body column to tentacles) the tissue changes its shape.

Epithelial hydra regenerate in a nearly normal fashion. When a hydranth is removed, a new one regenerates. A basal disc regenerates following removal of the original base. The rate of tentacle regeneration is shown in Fig. 5. Although tentacle regeneration is at first slightly slower in epithelial hydra than in normal animals, regeneration is completed in the normal time. Epithelial hydra restore the same percentage of original tentacles as do normal hydra, but since they have more tentacles than normal hydra they actually regenerate more tentacles than do normal hydra.

Hypostomal tissue has inductive capacities. A small fragment of hypostome implanted into the gastric region of another hydra induces

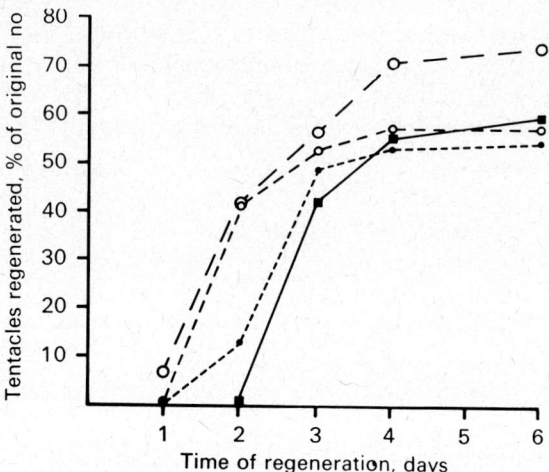

Fig. 5. Tentacle regeneration in epithelial hydra (■ - ■) and in control hydra starved for 1 (0--0), 5 (o--o) and 10 days (●---●). Original tentacle numbers were 9.6, 5.9, 5.6, and 6.1, respectively. (From Marcum and Campbell, 1978a).

TABLE II

Induction of Secondary Axes by Tissue Implants

Donor	Host	Number of Grafts	Type of Induction	
			positive	negative
A. Hypostomal Tissue Implants				
Nerve-free	Nerve-free	8	8	0
Nerve-free	Normal	18	18	0
Normal	Nerve-free	11	9	2
Normal	Normal	8	7	1
B. Gastric Region Tissue Implants				
Nerve-free	Nerve-free	9	0	9
Nerve-free	Normal	9	0	9
Normal	Nerve-free	11	1	10
Normal	Normal	9	0	9

Donor tissue, taken from hypostome (A) or gastric region (B) was implanted into the gastric column of a host hydra. (Results partly from Marcum and Campbell, 1978a).

the surrounding host tissue to form a secondary axis. A control tissue fragment taken from elsewhere in the hydra generally will not provoke secondary axis formation. Table II shows that nerve-free hypostomal tissue, implanted into the gastric region of a nerve-free hydra, induces a new axis. Nerve-free gastric tissue does not have this capacity.

Thus nerve-free tissue shows inductive capacity. Even more interesting is the finding that nerve-free hypostomal tissue implanted into the gastric region of a *normal* hydra induces a secondary axis (Fig. 6). In fact, considering all possible combinations of epithelial and normal hosts and donors, induction is independent of whether the tissue contains interstitial and nerve cells; in our experiments hypostomal tissue almost always induced and gastric tissue almost never induced. This important result shows that epithelial hydra have normal levels of patterning information.

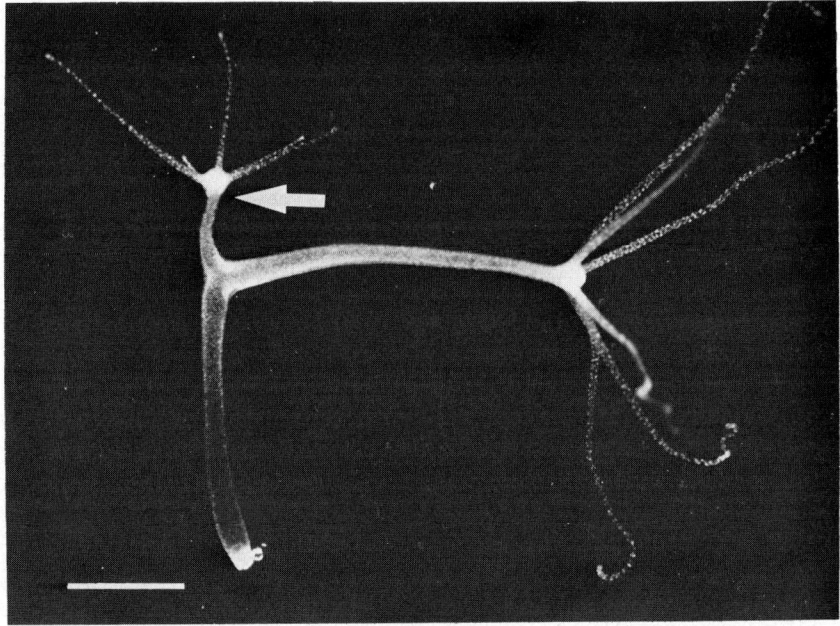

Fig. 6. Induction of a secondary axis (arrow) in a normal hydra by a small implant of hypostomal tissue taken from an epthelial hydra. (3 days after grafting, *H. attenuata*, bar at left indicates 1 mm)

These experiments oversimplify our understanding of inductive ability because there are inductive and susceptibility gradients running, respectively, down and up the body column. Our experiments involved extreme levels of induction and non-induction because of the body regions used. In a detailed quantitative study, Sacks and Davis (1977,

1978) found that developmental gradients are similar, but not quite identical, in treated and normal hydra.

Another developmental character we have examined is tissue determination. As mentioned above, a small implant of hypostomal tissue will induce secondary axis but a small implant of subhypostomal tissue (tissue below the tentacles) will not. However, if the hydranth is cut off of a hydra, the remaining subhypostomal tissue differentiates into hypostome. Webster (1966) discovered that the newly regenerating tissue becomes *determined* to become hypostome well in advance of actual differentiation. To demonstrate this they allowed subhypostomal tissue to regenerate for varying periods of time and then implanted them into the gastric regions of host hydra. Within 6 hours of amputation the tissue exhibited inductive properties. This is also true in epithelial hydra: subhypostomal tissue from epithelial hydra lacks inductive activity, but if tested 6 hours after amputation the subhypostomal tissue possesses inductive activity (Table III). Thus nerve-free tissue is capable of undergoing a hypostomal determination step.

TABLE III

Determination of Regenerating Subhypostomal Tissue.

Source of subhypostomal tissue	Time allowed to regenerate h.	No. of cases	% of cases showing determination
Epithelial hydra	0	11	0
Epithelial hydra	6	42	64
Normal hydra	0	7	0
Normal hydra	6	13	23

Subhypostomal tissue was isolated from hydra, allowed to regenerate for 0–6 hours, and tested for hypostome determination by implanting into the side of an intact hydra. (from Marcum and Campbell, 1978a)

The developmental patterns (regeneration and induction) studied above are usually interpreted as being controlled by diffusible morphogens. These controls represent labile developmental states and can change in a few hours. Tissue polarity, on the other hand, is a more fixed property of tissue and has often been ascribed to the graded distribution of a cell type along the body column (sometimes to nerve cells or interstitial cells). Polarity can be changed experimentally but it requires several days to do so, consistent with the supposition that cell distributions must be changed through differentiation and death.

Nerve-free column tissue shows normal developmental polarity. A ring of tissue cut out of the gastric region regenerates a new hydranth at the end closest to the original hydranth and a base regenerates at the originally basal end. Furthermore, polarity reversal imposed experimentally shows the same kinetics in nerve-free as in normal tissue. In one experiment (Marcum et al., 1977), gastric regions of epithelial and of normal control hydra were cut out and grafted back in place in an inverted position. Under these conditions the polarity of the inverted gastric tissue slowly reversed in accord with the new positions of the hydranth and base. One can monitor this reversal by simply cutting out the gastric columns at various times and seeing which end of each gastric column regenerates a hydranth. Table IV shows that, in control hydra, gastric regions reisolated after 12 hours in the inverted graft orientation regenerate according to their original polarity. Those left in the inverted grafts for 48 hours regenerate with reversed polarity. The time required for half of the gastric regions to reverse polarity is 36 hours, a time similar to that found by other workers (Wilby and Webster, 1970). Nerve-free gastric regions reverse their polarity with the same kinetics (Table IV): by 36 hours half show reversed polarity, by 48 hours all have reversed.

There is one difference between epithelial and control hydra tissue in these experiments. Most of the control regions reisolated 24 hours after

TABLE IV

Polarity of Regeneration in Normal and Nerve-free Hydra.

Time in reversed orientation (hours)	Number of Hydra	Polarity of regeneration		
		Original	Mixed	Reversed
A. Normal Hydra				
12	11	11	0	0
24	11	0	10	1
36	9	0	3	6
48	12	0	2	10
60	15	0	1	14
B. Nerve-free Hydra				
12	12	12	0	0
24	13	8	1	4
36	8	1	2	5
48	11	0	0	11
60	10	0	0	10

Gastric columns were grafted in inverted position between hypostome and peduncle for 12-60 hours. They were then tested for the polarity in which they regenerated. (From Marcum et al., 1977).

inversion showed "intermediate" polarities — hypostomes and tentacles arose at both ends or in the middle. Nerve-free tissue rarely showed this phenomenon (Table IV). Thus tissue polarity reversal in epithelial hydra is nearly but not entirely identical to polarity reversal in normal hydra.

Our conclusions from these analyses is that hydra deprived of all interstitial and nerve cells have essentially normal development. Thus, tissue composed only of epithelial cells is capable of controlling all of the major developmental phenomena in hydra. In addition, there are consistent deviations from normalcy in the quantitative details of each developmental trait. Epithelial hydra have the strategic framework of developmental controls, but seem to have lost some of the fine tuning.

IV. HYDRA CHIMERAS

After finding that epithelial cells are competent to pattern development in the absence of nerves, we sought to examine which developmental processes are being affected by nerves and interstitial cells when they are present. We constructed chimeric hydra by repopulating epithelial hydra of one strain with interstitial cells from a strain or species of hydra which was developmentally distinct. We could then assess directly which developmental phenomena were under control of nerve cells, by determining those aspects of the chimeras that resembled the interstitial cell donor. Chimeras are quite simple to produce: one grafts, to an epithelial hydra called the "epithelial cell host", a piece of normal tissue from an "interstitial cell donor". Within a few days interstitial cells will invade the epithelial cell host due to the migratory tendencies of these amoeboid cells (Brien and Reniers-Decoen, 1955; Tardent and Morgenthaler, 1966). After removal of the donor graft, the introduced interstitial cells rapidly repopulate the host to normal levels of both stem cells and differentiated product cells, due to homeostatic cell density control mechanisms (Bode et al., 1976; David and MacWilliams, 1978).

The choice of hydra strains to use as host and donor is a critical one: only certain species can be grafted together (Campbell and Bibb, 1970), yet closely related species frequently do not differ significantly in development. Thus, this analysis of chimeras has been greatly aided by Sugiyama and Fujisawa's (1977) extensive genetic program in which mutant strains of hydra have been isolated from inbreeding of wild populations.

In one chimera study (Marcum and Campbell, 1978b) we have analyzed the determinants of size. We used Sugiyama and Fujisawa's strains *mini-4* and *maxi-1* of *Hydra magnipapillata*. These mutant strains are normal in nearly every respect except in size: *maxi* polyps are about ten times the size of *mini* polyps. As a *normal* strain we used wild type *Hydra attenuata* which is closely related to (if not conspecific with) *H. magnipapillata* and which is intermediate in size between *mini* and *maxi* (Fig. 7). We prepared epithelial clones of *mini, normal* and *maxi*, and then

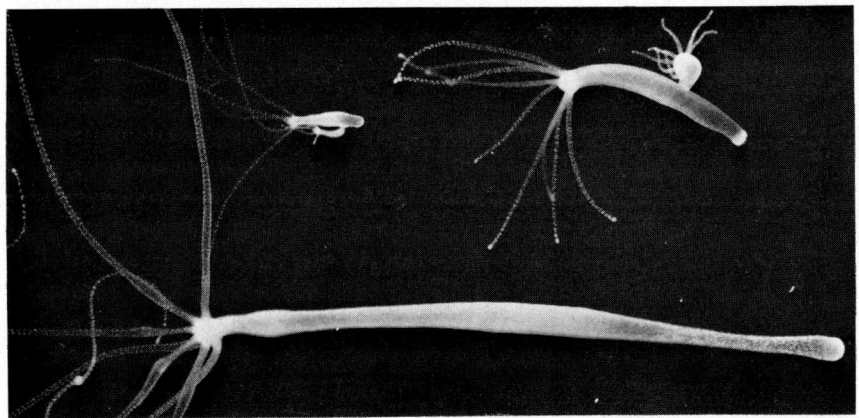

Fig. 7. Size variants of hydra used to produce chimeras. Small polyp (top center); *mini, H. magnipapillata;* middle-sized polyp (top right): *normal H. attenuata;* large polyp (bottom): *maxi, H. magnipapillata.* (x5)

repopulated each type with each of the three types of interstitial cells. The profile area ("size") of each of the nine resulting chimeras was compared to the sizes of the epithelial and interstitial (nerve) cell parents. Fig. 8 shows that chimera size is about the same as that of the epithelial cell parent in each case. For example, chimeras containing *mini* epithelial cells were small regardless of the type of interstitial (and hence nerve) cell that they contain (Fig. 9). Thus the differences in size between these three strains are attributable almost solely to epithelial cell differences. Sugiyama and Fujisawa (1978a) also concluded, from similar experiments, that dry weight in about 10 strains of hydra was determined by epithelial rather than interstitial cells.

In our chimeras between *mini, normal* and *maxi* hydra there were some size characters which seemed affected by the interstitial cell lineage. For example, *normal* epithelial cells repopulated by *maxi* interstitial cells were conspicuously slender so that although in overall size they resembled the epithelial cell donor (*normal*) in simple length they resembled the interstitial cell parent (*maxi*). Similarly, chimeras composed of *mini* interstitial cells in *normal* epithelial cells, although *normal* in size, were conspicuously short, thus resembling the interstitial cell parent.

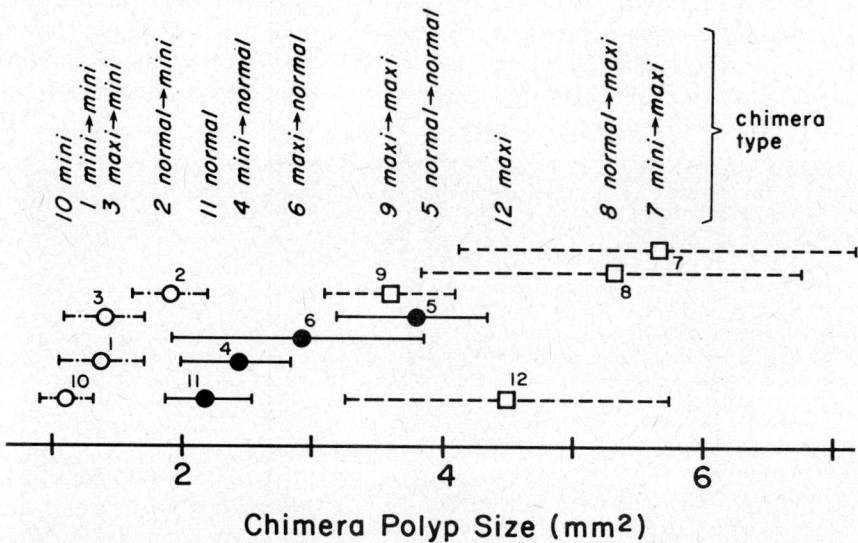

Fig. 8. Profile size of 9 chimeras and 3 control hydra strains (numbered and identified at top as [I cell donor → epithelial cell host]). Symbols indicate type of epithelial cell: O, *mini*; ●, *normal*; □, *maxi*. Dotted lines show standard deviations of measured sizes (length x width); vertical position has no significance. (From Marcum and Campbell, 1978b)

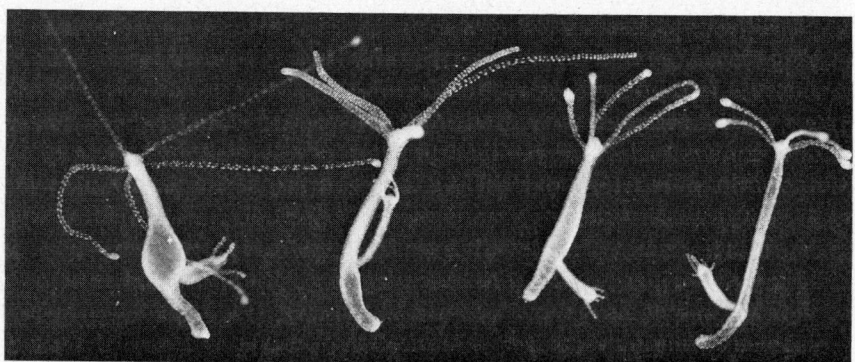

Fig. 9. Chimeras containing *mini* epithelial cells. Left: *mini* control. Other polyps, left to right, are chimeras containing *mini*, *normal* and *maxi* interstitial cells. (From Marcum and Campbell, 1978b) (x12)

In another experiment we produced a chimera of *Pelmatohydra oligactis* interstitial cells in *H. attenuata* epithelial cells. These two species differ in the pattern in which tentacles arise on a bud. Young *P. oligactis* buds have two lateral tentacles, which become quite long, which are followed by the appearance of two intercalated tentacles (Fig. 10a), while in the *H. attenuata* bud the tentacles are more uniform in time of origin and length (Fig. 10c). The chimera (Fig. 10b) has bud tentacles of the *H. attenuata* pattern, thus indicating that this pattern is determined by epithelial cells.

There are a number of other characters that have now been analyzed in chimeras. These are listed in Table V. Morphological characters tend to follow the epithelial cell parent. A number of cellular characters (including cell densities) resemble the interstitial cell parent showing that the method of chimera analysis is sensitive to influences of interstitial cell lineage. We thus conclude that patterning at the morphological level is controlled predominantly by epithelial cells.

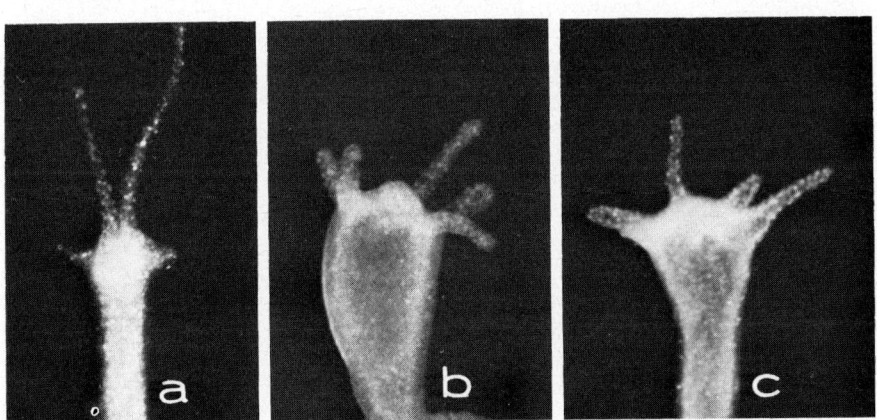

Fig. 10. Tentacles on young buds. *a*, *P. oligactis* with 2 long lateral tentacles and 2 short tentacles; *c*, *H. attenuata* with all tentacles approximately the same length; *b*, chimera (*P. oligactis* interstitial cells in *H. attenuata* epithelial cells) with all tentacles the same length.

V. OTHER ASPECTS OF EPITHELIAL HYDRA

Interstitial and nerve cell-free hydra offer new approaches to many old and difficult problems, largely because of their simple histological structure and their transparent, motionless stance. The following are four subjects of research where epithelial hydra are already being used to advantage.

TABLE V

Characters of Chimeric Hydra.
A summary of conclusions regarding which chimeric traits resemble the epithelial vs interstitial cell parents.

Character	Strains Represented in Chimera	Parent more closely resembled	Reference*
A. Morphological			
Overall size	H. magnipapillata; H. attenuata	Epithelial	(1)
Length	H. magnipapillata; H. attenuata	Epithelial and interstitial	(1)
Dry weight	H. magnipapillata (numerous strains)	Epithelial	(2)
Growth rate	H. magnipapillata (numerous strains)	Epithelial	(2)
Budding rate	H. magnipapillata (numerous strains)	Epithelial	(2)
	H. magnipapillata; H. attenuata	Epithelial	(1)
Pattern of bud tentacles	P. oligactis; H. attenuata	Epithelial	(4)
Irregular body column	"twisted column" (H. magnipapillata); H. attenuata	Epithelial	(6)
Tentacle number	H. magnipapillata (numerous strains)	Epithelial	(2)
Position of budding region	maxi (H. magnipapillata) and other strains.	Interstitial	(1)
B. Cytological			
Nematocyst morphology	P. oligactis; H. attenuata	Interstitial	(4)
	Strain nem1; wild type (H. magnipapillata)	Interstitial	(2)
Nerve and I cell densities	H. magnipapillata (numerous strains)	Interstitial	(2,3)
Epithelial cell color	strain 105; nf1 (H. magnipapillata)	Epithelial	(2)
I cell temperature sensitivity	P. oligactis; H. attenuata	Interstitial	(5)

*(1) Marcum and Campbell (1978b)
(2) Sugiyama and Fujisawa (1978b)
(3) Sugiyama and Fumisawa (personal communication)
(4) Lee and Campbell (in preparation)
(5) Fradkin, Lee and Campbell (in preparation)
(6) unpublished.

A. Cell and Tissue Movements

The frequent contractions and somersaulting of normal hydra preclude the observation of any finer movement, and the many studies made on cell migrations and tissue morphogenesis have relied on indirect methods. Nerve-free hydra are sufficiently immobile that many minor movements are visible. For example, individual nematocytes can be introduced and their migrations through the tissue followed under continuous, high magnification microscopy. Thus for the first time it is possible to directly examine the possibility that nematocyte migration is directed towards the tentacles, to measure migration velocities, and to watch the process of locomotion. We find that these cells are highly polarized and move by means of a single, large anterior pseudopod at rates as high as 30 μm/min (Fig. 11). They travel axially along the hydra but they move both distally and proximally without regard for tissue polarity (Campbell and Marcum, 1978), apparently guided by the muscle

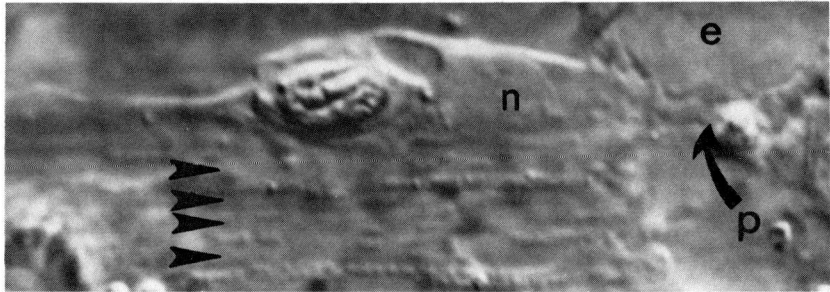

Fig. 11. Introduced nematocyte migrating along tentacle of an epithelial hydra. Cell is migrating toward the right. P, pseudopod tip; n, nucleus; capsule (not marked) trails; e, nucleus of epithelial cell. Arrowheads indicate muscle processes of epithelial cells. (Living preparation, Nomarski optics, x2,600).

processes of the epithelial cells (see Fig. 11). It is also possible to examine the budding process in its entirety, as a timelapse photograph of the process can be made with relative ease using eipthelial hydra. We have also been able to detect a previously unsuspected column movement. Fine annulations (conspicuous in Fig. 3c) arise in the peduncle and are propagated upwards along the body column with a periodicity of a few minutes. The waves also pass outwards along young buds, and at a particular bud developmental stage a new train of waves is initiated on the bud. It is possible that these waves represent periodic morphogenetic behavior of the epithelial cells.

B. Cell Differentiation

Epithelial hydra are providing data on the question of transformations between cell types. In our investigations, I cell-free hydra have not regained their interstitial cells during 23 months of culture. Thus, epithelial cells apparently do not constantly, or even occasionally, give rise to interstitial cells. These hydra also remain free of endodermal gland cells suggesting that these cells also do not arise from epithelial cells (as has been suggested, see Kanaev, 1952). By reintroducing interstitial cells into these hydra, we have learned two things about interstitial cell differentiation. First interstitial cells do not differentiate into, and thus gradually renew, the epithelial cells. In a chimera of *H. attenuata* epithelial cells and *P. oligactis* interstitial cells, the epithelial tissue retained its *attenuata* healing characters for at least 6 months. A chimera constructed of cells from two color varieties of hydra maintains its epithelial color, also demonstrating that interstitial cells do not renew the epithelial cell line (Sugiyama and Fujisawa, 1978a). Second, if fewer than 1 percent of the normal number of interstitial cells are introduced they do not differentiate into nerves and nematocytes but maintain a stable population level, by proliferating in the growing host, for many months (Marcum and Campbell, 1978a). These animals provide the only way we currently have of growing interstitial cells in the complete absence of their product cells, and thus should be an important tool in sorting out the developmental feedback signals that probably pass from the differentiated cells to the interstitial cells (Bode, 1973; Bode and David, 1978). Perhaps sparse interstitial cells do not differentiate because the hydra lack all such positive feedback signals.

C. Histological Structure

The ectoderm of normal hydra is so crowded with interstitial cells and nematoblasts that some of the epithelial cell organization is obscured, and thus interstitial cell-free hydra offer promise in the study of epithelial cell arrangement. Otto (1977) discovered that the muscle processes of the epithelial cells can be seen in live epithelial hydra using either Normarski differential interference or polarization optics. She was thus able to map out muscle processes in developing buds and during some experimental conditions. Since muscle processes probably play a key role in morphogenesis (Campbell, 1974b; Otto, 1977), the ability to study the processes in whole hydra represents an important advance in understanding the development of hydra.

Another interesting feature of epithelial cells has come to light in histological sections of I cell-free hydra. A small fraction of epithelial cells in every polyp are withdrawn from the surface of the hydra and rest on the mesoglea (Wanek et al., 1978). They are quite compact and the large central vacuole is seen in various stages of disappearance. Sometimes these cells are seen spanning the two tissue layers as though passing through the mesolamella.

D. *Electrophysiology and Behavior*

Epithelial hydra have little or no spontaneous contractile activity (Campbell et al., 1976) supporting the suggestion (Passano, 1963) that the nervous system initiates spontaneous activity. These nerve-free hydra are excitable, however, and can conduct impulses that lead to contraction of the whole animal (Campbell et al., 1976) consistent with the widespread occurrence of epithelial conduction in coelenterates. The study of behavior of chimeric hydra, whose nerve and epithelial cells (or ectoderm and endoderm) are from strains with different behavior, should show the roles of nerve and epithelial cells in the different forms of hydra's behavior (Saffitz et al., 1972; Marcum and Campbell, 1978b).

VI. CONTROL OF PATTERN IN HYDRA: CONCLUSIONS

Returning to the question of which cell type controls morphogenetic patterning in hydra, we can conclude:
1. Epithelial cells are capable of regulating all aspects of hydra morphogenetic pattern formation, and hydra consisting only of epithelial cells can grow and proliferate indefinitely.
2. Hydra composed only of epithelial cells show quantitative abnormalities in shape and in developmental controls, suggesting that pattern formation mechanisms have lost fine tuning.
3. In complete hydra, epithelial cells have predominant control over morphogenetic patterns. Thus the interstitial cell lineage, including nerve cells, exerts only a small but detectable patterning influence in chimeric hydra.

How are these conclusions reconciled with the evidence supporting a patterning role of the nervous system in hydra? It is possible that neurosecretory factors in hydra are primarily governors of interstitial cell determination. Certainly most of the purported effects of these factors deal with the interstitial cell lineage rather than morphogenesis (Lesh and Burnett 1966, Lesh, 1970; Berking, 1977; Schaller, 1978; Bode and David, 1978). and the morphogenetic effects present are of a more quantitative than qualitative nature. The effects on morphogenesis might even be secondary ones. Alternatively the neurosecretory factors might exert strong influences over morphogenesis, but these are overpowered by even stronger influences from epithelial cells. Certainly developmental mechanisms are highly multiple, as would be expected since evolution should result in increased complexity at the mechanistic level as it does at the morphological level. A third way to reconcile these divergent views of the nervous system is that epithelial cells efficiently compensate for the lack of a nervous system in interstitial cell-free hydra. This view, although not in line with the chimeric data presented above, has precedent in the case of amphibian regeneration (Thornton, 1970; Singer, 1974) and must be carefully examined experimentally.

Our studies on epithelial hydra demonstrate that epithelial cells can regulate pattern formation, and actually appear to be the prime regulators in intact hydra. What is so unique about these epithelial cells that evolution should have relegated such an important role to them? The most obvious peculiarities of epithelial cells are their mechanical attributes: they are strongly bound to one another; they rest on a substratum, the mesolamella; and they have muscle processes capable of exerting forces on the tissue and perhaps moving the cells as well. No other cell type of hydra has any of these three mechanical attributes. Therefore it would be enlightening to consider mechanical models of pattern formation, in contrast to the chemical models which are currently receiving attention. A few rudimentary formulations of hydra's tissue mechanics and patterning are available to catalyze this new approach to the problem (Hausman and Burnett, 1970; Hausman, 1973; Campbell, 1968, 1974b; Gierer, 1977). We must consider that the mesolamella might confer patterning on overlying cell movements, that epithelial cells in the ectoderm and endoderm may strike mechanical balances by virtue of their orthogonal muscular arrangements, that hydrostatic pressure and tissue tensions offer effective vehicles for signalling throughout the polyp, and that local changes in cell relationships may lead to changes in tissue morphology. The exploration of such ideas will be greatly aided by the simplicity of epithelial hydra.

ACKNOWLEDGEMENTS

I express warm appreciation to many colleagues who have contributed to this work, including M. Chow, M. Fradkin, H. Lee, B. Marcum, T.

Sugiyama, N. Wanek and D. West. Supported in part by research grants NS-12446 from the Public Health Service and PCM 77-00276 from the National Science Foundation.

REFERENCES

Baker, J.R. (1952). "Abraham Trembley of Geneva, Scientist and Philosopher, 1710-1784". Arnold, London.
Berking, S. (1977). *Arch. Entwicklungsmech.* **181**, 215-225.
Bode, H.R. (1973). *In* "Humoral Control of Growth and Differentiation, Vol. II Nonvertebrate Neuroendocrinology and Aging" (J. Lobue and A.S. Gordon, eds.), pp. 35-57. Academic Press, New York.
Bode, H.R. and David, C.N. (1978). *Progr. Biophys. Mol. Biol.* **33**, 189-206.
Bode, H., Berking, S., David, C.N., Gierer, A., Schaller, H., and Trenkner, E. (1973). *Arch. Entwicklungsmech.* **171**, 269-285.
Bode, H.R., Flick, K.M., and Smith, G.S. (1976). *J. Cell Sci.* **20**, 29-46.
Brien, P. (1961). *Bull. Biol. France Belgique* **95**, 301-364.
Brien, P. and Reniers-Decoen, M. (1949). *Bull. Biol. France et Belg.* **83**, 293-386.
Brien, P. and Reniers-Decoen, M. (1955). *Bull. Biol. France et Belg.* **89**, 258-325.
Burnett, A.L. (ed.) (1973). "Biology of Hydra". Academic Press, New York.
Burnett, A.L. and Lambruschi, P.G. (1973). *In* "Biology of Hydra" (A.L. Burnett, ed.), pp. 239-247. Academic Press, New York.
Burnett, A.L., Diehl, N.A. and Diehl, F. (1964). *J. Exp. Zool.* **157**, 227-235.
Bursztajn, S. (1974). Studies on the nervous system during regeneration and growth in hydra. Ph.D. Thesis, Syracuse University.
Campbell, R.D. (1967). *Develop. Biol.* **15**, 487-502.
Campbell, R.D. (1968). *In Vitro* **3**, 22-32.
Campbell, R.D. (1974a). *Amer. Zool.* **14**, 523-535.
Campbell, R.D. (1974b). *In* "Coelenterate Biology: Reviews and Perspectives." (L. Muscatine and H.M. Lenhoff, eds.), pp. 179-210. Academic Press, New York.
Campbell, R.D. (1976). *J. Cell Science* **21**: 1-13.
Campbell, R.D. and Marcum, B.A. (1978)., in preparation.
Campbell, R., Fradkin, M., Kakis, H., Marcum, B., and Wanek, N. (1977) *Amer. Zool.* **17**, 923 (abstract).
Campbell, R.D. and Bibb, C. (1970). *Transplant. Proc.* **2**, 202-211.
Campbell, R.D., Josephson, R.K., Schwab, W.E., and Rushforth, N.B. (1976). *Nature* **262**, 388-390.
Challoner, D. (1975). M. Sci. Thesis. University of Newcastle-Upon-Tyne.
David, C.N. (1973). *Arch. Entwicklungsmech.* **171**, 259-268.
David, C.N. (1975). *In* "Microbiology 1975", American Society for Microbiology, Washington, D.C. pp. 434-441.
David, C.N. and Campbell, R.D. (1972). *J. Cell Sci.* **11**, 557-568.
David, C.N. and Gierer, A. (1974). *J. Cell Sci.* **16**, 359-375.
David, C.N. and MacWilliams, H. (1978). *Proc. Nat. Acad. Sci. U.S.* **75**, 886-890.
David, C.N. and Murphy, S. (1977). *Develop. Biol.* **58**, 372-383.

Davis, L.E., Burnett, A.L. and Haynes, J.F. (1968). *J. Exp. Zool*, **167**, 295-331.
Diehl, F.A. and Burnett, A.L. (1964). *J. Exp. Zool*. **155**, 253-259.
Diehl, F.A. and Burnett, A.L. (1965a). *J. Exp. Zool*. **158**, 238-297.
Diehl, F.A. and Burnett, A.L. (1965b). *J. Exp. Zool*. **158**, 299-317.
Diehl, F.A. and Burnett, A.L. (1966). *J. Exp. Zool*, **163**, 125-139.
Diehl, F.A. (1973). *In* "Biology of Hydra" (A.L. Burnett, ed.), pp. 109-141. Academic Press, New York.
Fradkin, M., Kakis, H. and Campbell, R.D. (1978). *Radiat. Res.* **76**, 187-197.
Gierer, A. (1977). *Quart. Rev. Biophys.* **10**, 529-593.
Gierer, A. and Meinhardt, H. (1972). *Kybernetik* **12**, 30-39.
Ham, R.G. and Eakin, R.E. (1958). *J. Exp. Zool.* **139**, 33-53.
Hausman, R.E. (1973) *In* "Biology of Hydra" (A.L. Burnett, ed.), pp. 393-453. Academic Press, New York.
Hausman, R.E. and Burnett, A.L. (1970). *J. Exp. Zool.* **173**, 175-185.
Kanaev, I.I. (1952) "Hydra, Outlines of the Biology of fresh-water polyps" translated (1969) from Russian edition by E.T. Burrows and H.M. Lenhoff, edited and published by H.M. Lenhoff.
Kanajew, J. (1930) *Arch. Entwicklungsmech.* **122**, 736-759.
Kleinenberg, N. (1872). "Hydra. Eine anatomisch-entwicklungsgeschichtliche Untersuchung." Englemann, Leipzig.
Lenhoff, H.M. and Loomis, W.F. (eds.) (1961). "The Biology of Hydra and of Some Other Coelenterates: 1961", University of Miami Press, Coral Gables, Florida.
Lentz, T.L. (1965a). *Science* **150**, 633-635.
Lentz, T.L. (1965b). *J. Exp. Zool.* **159**, 181-193.
Lentz, T.L. (1966). "The Cell Biology of Hydra". North Holland Publ. Co., Amsterdam.
Lesh, G.E. and Burnett, A.L. (1966). *J. Exp. Zool.* **163**, 55-77.
Lesh, G.E. (1970). *J. Exp. Zool.* **173**, 371-382.
Macklin, M., Roma, T., and Drake, K. (1973). *Science*, **179**: 194-195.
Marcum, B.A. (1978)., in preparation.
Marcum, B.A. and Campbell, R.D. (1978a). *J. Cell Sci.* **29**, 17-33.
Marcum, B.A. and Campbell, R.D. (1978b). *J. Cell Sci.* **32**, 233-247.
Marcum, B.A. and Campbell, R.D. & Romero, J. (1977). *Science* **197**, 771-773.
McConnell, C.H. (1933a). *Biol. Bull.* **64**, 86-95.
McConnell, C.H. (1933b). *Biol. Bull.* **64**, 96-102.
Otto, J.J. (1977). *J. Exp. Zool.* **202**, 307-321.
Passano, L.M. (1963). *Proc. Nat. Acad. Sci. U.S.* **50**, 306-313.
Sacks, P.G., and Davis, L.E. (1977). *J. Cell Biol.* **75**, 404A (abstract)
Sacks, P.G., and Davis, L.E. (1978). 10th Miami Winter Symposium on Differentiation and Development p. 106 (abstract).
Saffitz, J.E., Burnett, A.L. and Lesh, G.E. (1972). *J. Exp. Zool.* **179**, 215-223.
Schaller, H.C. (1976). *Cell Diff.* **5**, 1-11.
Schaller, H.C. (1978). *Symp. Soc. Develop. Biol.* **35**, 231-241.
Singer, M. (1974). *Ann. N.Y. Acad. Sci.* **228**, 308-322.
Strelin, G.S. (1929). *Arch. Entwicklungsmech.* **115**, 27-51.
Sugiyama, T. and Fujisawa, T. (1977). *Develop. Growth & Diff.* **19**, 187-200.
Sugiyama, T. and Fujisawa, T. (1978a). *J. Cell Sci.* **29**, 35-52.
Sugiyama, T. and Fujisawa, T. (1978b). *J. Cell Sci.* **32**, 215-232.
Tardent, P. (1954). *Arch. Entwicklungsmech.* **146**, 593-649.
Tardent, P. and Morgenthaler, U. (1966). *Rev. Suisse Zool.* **73**, 468-480.
Thornton, C.S. (1970). *Amer. Zool.* **10**, 113-118.

Tripp, K. (1928). *Zeit. Wiss. Zool.* **132,** 476-525.
Wanek, N., Marcum, B.A., and Campbell, R.D. (1978). in preparation.
Webster, G. (1966). *J. Embryol. Exp. Morphol.* **16,** 91-104.
Wilby, O.K. and Webster, G. (1970). *J. Embryol. Exp. Morphol.* **24,** 595-613.
Wolpert, L., Hornbruch, A. and Clarke, M.R.B. (1974). *Amer. Zool.* **14,** 647-663.
Zawarzin, A.A. (1929). *Arch. Entwicklungsmech.* **115,** 1-26.

Pattern Formation, Growth Control and Cell Interactions in *Drosophila* Imaginal Discs

Peter J. Bryant
Center for Pathobiology
University of California, Irvine
Irvine, California 92717

I.	Introduction	295
II.	Imaginal Discs	296
III.	Pattern Formation and Regulation in Imaginal Discs	298
IV.	Wound Healing in Imaginal Discs	301
V.	The Polar Coordinate Model	303
VI.	Intercalation in Imaginal Discs	305
VII.	Bilaterally Symmetrical Fields	306
VIII.	Interpretation	308
IX.	Disc Interactions and the Universality of Positional Information	312
X.	Prospects	313
	References	314

I. INTRODUCTION

The complex patterns of hairs, bristles, and sense organs which are found on the exoskeleton of *Drosophila* are attracting increasing attention for the study of the way in which spatial organization arises during development. These patterns are produced suddenly during metamorphosis, but it is clear that the presumptive pattern is not suddenly imposed upon virgin tissue late in ontogeny, but develops gradually in the imaginal discs as they grow by cell multiplication in the larval stage (see Schubiger, 1974; Gateff and Schneiderman, 1975). Furthermore it appears that during pattern regulation (regeneration, duplication, etc.) in imaginal disc fragments, the new patterns that develop are also gradually built up during the growth of the cell populations that ultimately differentiate them (see Ursprung, 1959; Lüönd, 1961; Gehring, 1966; Schubiger, 1971). Because insect integumentary pattern elements such as bristles and hairs are constructed, in general, by individual cells or small well-defined cell clusters, it is essential that the growth control mechanism operating in imaginal discs must not only deploy the correct total number of cells for the pattern, but must also allocate the appropriate number of cells to each part of the developing pattern. These general considerations, as well as the experimental evidence to be discussed later, suggest that growth control and pattern formation must be intimately associated in these insect tissues. In fact, it seems likely that both of these controls on developing tissues are mediated by the same cell interactions and that growth control may, in fact, be merely one facet of the problem of pattern formation.

Cell interactions of two rather different kinds have been considered as possible mediators of pattern specification. In one type of model it is proposed that a special region or regions in the field is or are responsible for signalling positional information to the other cells. Examples are the gradient models of Stumpf (1966), Wolpert (1968) and Crick (1970) and the phase-shift model of Goodwin and Cohen (1969). On the other hand, it is conceivable that reproducible global patterns could arise as a result of cell activities or interactions in which all cells of the field participate equally. The gradient model of Meinhardt and Gierer (1974), the cell-contact model of McMahon (1973) and the polar coordinate model of French, *et al.*, (1976) are of this general type. In addition, models have been proposed which combine elements of both kinds of mechanism (Rose, 1952; Webster, 1971; Lawrence *et al.*, 1972). In the following, I shall argue that pattern formation and regulation in *Drosophila* imaginal discs is controlled by a mechanism of the second type mentioned above; that is, a system of strictly local cell interactions which generates global patterns without the need for special signalling regions.

II. IMAGINAL DISCS

The integument of the *Drosophila* adult is derived during metamorphosis from nests of cells in the larva known as imaginal discs, or in the case of the abdomen, from smaller groups of cells called histoblast nests (see Bryant, 1978). There is an imaginal disc for each wing, leg, haltere, eye, antenna, genital apparatus, and for other components of the adult. For our purposes, it is important to point out

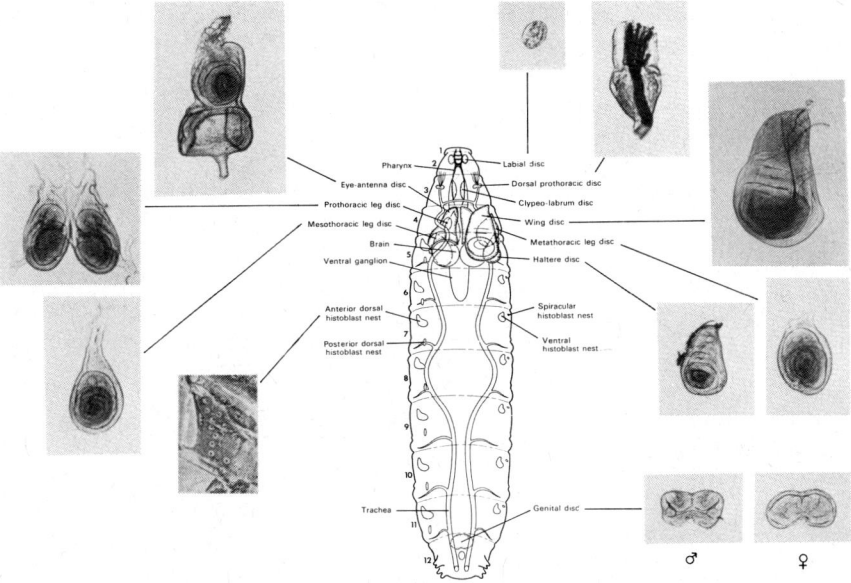

Fig. 1. The imaginal discs and abdominal histoblast nests of a third-instar larva of *Drosophila melanogaster*. The information on the histoblast nests is from Roseland (1976).

that most of these discs are asymmetrical and paired, but that the genital discs are symmetrical, unpaired structures located on the midline of the larva (Fig. 1).

Each imaginal disc consists fundamentally of a sac-like single layer of epithelial cells (Poodry and Schneiderman, 1970; Ursprung, 1972), which originates in the embryo from the surface layer of hypodermis (Auerbach, 1936; Madhavan and Schneiderman, 1977). In the case of the wing disc and most other discs that have been studied, one side of the disc (the "disc epithelium") consists of a thickened layer of columnar cells while the other side is a thin layer of extremely squamous cells known as the peripodial membrane (Fig. 2; Poodry and Schneiderman, 1970; Reinhardt, 1978). At the margin of the disc is a transition zone where the two layers meet. It is thought that most of the adult structures are derived from the disc epithelium, although contributions from the transition zone and perhaps the peripodial membrane have been demonstrated in the case of the labial disc (Kumar and Ouweneel, personal communication) and first leg disc (Fristrom and Fristrom, 1975) and the wing disc of a larger fly, *Calliphora* (Sprey and Oldenhave, 1974). In the genital discs, no part of the epithelium is sufficiently squamous to be considered as peripodial membrane (Littlefield and Bryant, 1979) and this

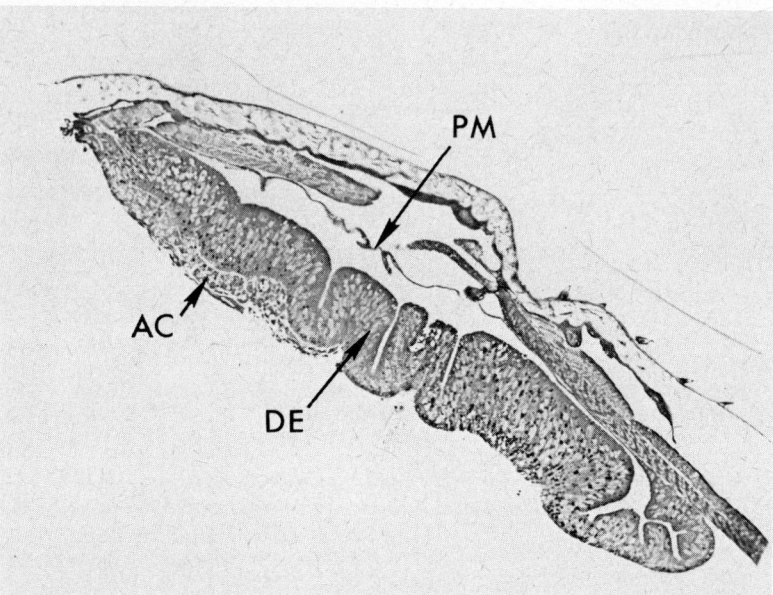

Fig. 2. Longitudinal section of a wing disc *in situ* in a third-instar larva, stained with toluidine blue. D.E., disc epithelium; P.M., peripodial membrane; A.C., adepithelial cells. From Reinhardt (1978).

might have important consequences for the developmental behavior of this disc, as will be discussed later. The surface of the epithelium which faces the lumen of the disc is the presumptive outer surface of the adult structure, which will synthesize the cuticular pattern elements. During normal development this surface comes to face the outside through a morphogenetic process termed evagination or eversion, but transplanted imaginal discs or disc fragments go through metamorphosis as inside-out vesicles in which the bristles, hairs, etc., are produced on the inside (Loosli, 1959). Imaginal discs also contain other cell types which give rise to muscles and nerves (see Poodry and Schneiderman, 1970), but these will not concern us further.

III. PATTERN FORMATION AND REGULATION IN IMAGINAL DISCS

Pattern formation and regulation have been investigated in imaginal discs by three experimental procedures. In the first procedure a number of different specific fragments of the disc are implanted into mature larvae, where they undergo metamorphosis with the host. The transplanted fragments produce distinct sections of the pattern, permitting the establishment of a detailed fate map of the disc (see, for example, Schubiger, 1968; Bryant, 1975a). The fact that such fate maps can be made shows that the presumptive pattern of differentiation has already been established in the imaginal disc, before differentiation occurs. Furthermore, when discs are taken from younger larvae and implanted into hosts which are ready to metamorphose, they produce partial patterns which suggest that the pattern is built up in a specific sequence during growth of the tissue (Schubiger, 1974; Gateff and Schneiderman, 1975; Ginter and Kuzin, 1970).

In the second experimental procedure specific disc fragments are given the opportunity for growth and pattern regulation before metamorphosis. This is done simply by culturing the fragments in adult female hosts, where they grow by cell multiplication but do not differentiate until transferred to larval hosts for metamorphosis (Hadorn et al., 1949). It has been shown now for several imaginal discs that some fragments can regenerate missing parts during adult culture, but that other fragments, in contrast, undergo pattern duplication (Vogt, 1946; Gehring, 1966; Schubiger, 1971; Van der Meer and Ouweneel, 1974; Bryant, 1971, 1975a). The results of a large series of adult-culture experiments on the wing disc (Bryant, 1975a; Haynie and Bryant, 1976) are summarized in Fig. 3. In most cases the disc was cut into three pieces by two parallel cuts and the resulting three fragments cultured separately. In many of these series, the central fragment was able to regenerate the missing parts of the disc in both directions, whereas the two edge-pieces underwent pattern duplication during culture. This confirmed the principle, recognized earlier (Bryant, 1971) that when one fragment of an imaginal disc shows regeneration during culture, then the complementary piece will usually show duplication.

These three-segment experiments were carried out with cuts at several levels in orthogonal and diagonal directions, with similar results. However, in two series where one of the edge pieces was large, a different result was obtained (Fig. 3e and 3h). In these cases, the large edge piece could regenerate the remainder of the disc, but the central piece underwent duplication and at the same time regenerated the small edge piece. We assume that regeneration and duplication occur by growth from opposite edges of these central pieces, thus preserving the rule of regeneration and duplication from complementary fragments.

The above set of experiments suggested a model in which regeneration always proceeded outward from a point in the approximate center of the disc (Bryant, 1975b). Fragments with cut edges facing away from this point would show regeneration, whereas fragments with cut edges

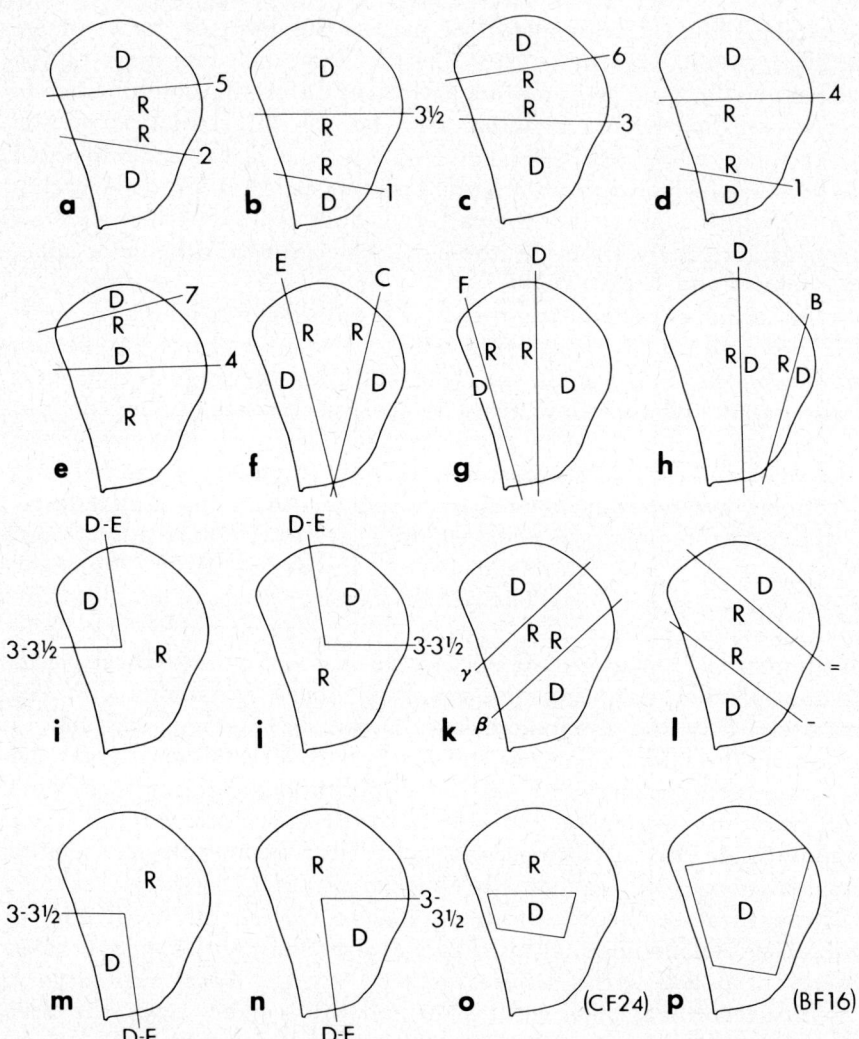

Fig. 3. Summary of the observed regulative behavior of various fragments of the imaginal wing disc. R., regeneration; D., duplication. The numbers and letters outside the disc outlines indicate cutting levels. From Bryant (1978).

facing the central point would duplicate. However, this simple model was eliminated by the finding that central squares of the wing disc fail to show the expected regeneration, but duplicate instead (Bryant, 1975a). In one series which was tested, the complementary annulus could regenerate the center. Furthermore, the model makes no prediction about the behavior of 3/4 and 1/4 sectors, which were shown to regenerate and duplicate, respectively (Haynie and Bryant, 1976).

The behavior of 3/4 and 1/4 sectors of the wing disc first suggested that pattern regulation in this system could not be explained by specific kinds of regulation being associated with particular cut edges, since identical cut edges are at the margins of regenerating 3/4 sectors and duplicating 1/4 sectors. Our new interpretation, suggested by the work of French and Bullière (1975a, b) on cockroach legs, is based on the principle that pattern regulation is controlled by interactions between cells originating in different parts of the presumptive pattern, when they are brought into contact during wound healing.

IV. WOUND HEALING IN IMAGINAL DISCS

A commonplace observation in imaginal disc work is that fragments of discs tend to round up to form spheroidal vesicles when they are put through metamorphosis (see, for example, Loosli, 1959). We have studied directly the behavior of such fragments in adult hosts and have shown that the rounding-up process accompanies wound healing and that these events occur quite rapidly within the first one or two days of culture in an adult host (Reinhardt et al., 1977; Reinhardt, 1978).

The events which occur during wound healing in the wing disc can best be illustrated by reference to a 3/4 sector. In producing such a fragment, orthogonal cuts are made through both the disc epithelium and the peripodial membrane. In theory, such a fragment could heal into a spheroid vesicle either by fusing the cut edges of the disc epithelium to those of the peripodial membrane, or by fusing the two cut edges of the disc epithelium to each other and the two cut edges of the peripodial membrane together. Our studies show that, although there is a transient contact between the edges of the two layers, wound healing itself in these 3/4 fragments eventually restores the continuity of each layer separately (Figs. 4 and 5).

In 1/4 sectors and in segments of the wing disc, the evidence indicates that the same final result (fusion of each layer separately) is achieved, although greater distortion of the epithelium is necessarily involved in the healing process with these fragments. In central squares of the wing disc, all four edges appear to contract until the sheet is converted into a hollow ball where the healed wound is at the opposite pole from the center of the original square (Reinhardt et al., 1977).

In some wing disc fragments wound healing is complete in one day, whereas others take two days. In either case, it is known from short-

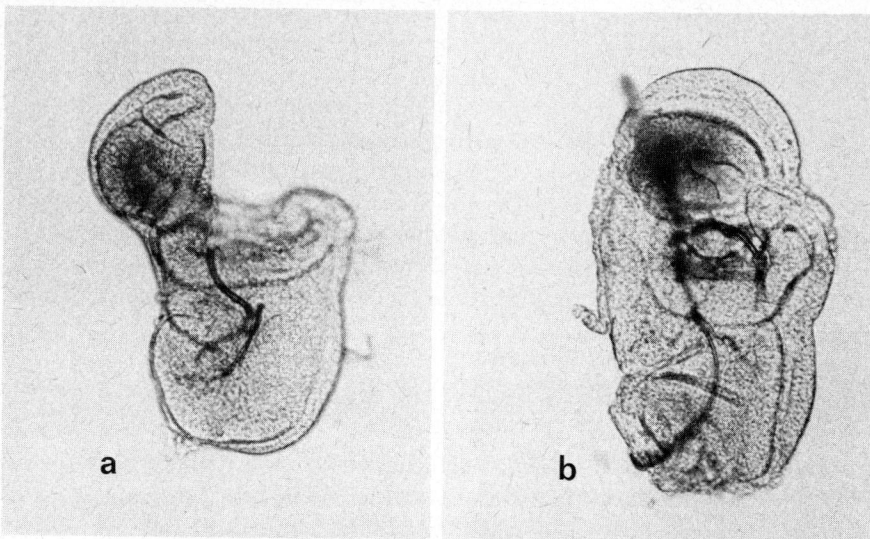

Fig. 4. Wound healing in a 3/4 sector of the wing disc (photographs of fresh specimens, unstained). (a) after one day of adult culture. (b) after two days of adult culture. From French et al., (1976).

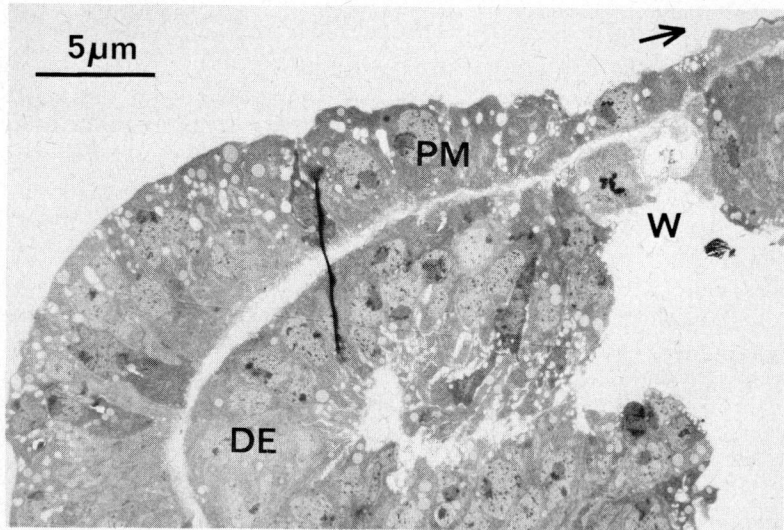

Fig. 5. Electron micrograph of a section through the wound region of a 3/4 sector of a wing disc, after one day of adult culture. This disc had reached a stage of healing between the stages shown in Figs. 4a and 4b; continuity of the disc epithelium (D.E.) and of the peripodial membrane (P.M.) has already been established across the wound region. The position of the wound (W) can be recognized by the dead cell in the disc epithelium. The wound junction in the peripodial membrane has migrated in the direction of the arrow and is not visible here. From Reinhardt (1978).

culture experiments on wing disc fragments (M. Bownes and A. James, personal communication) that little or no regeneration is detectable at two days, so it is clear that wound healing precedes regulative growth in these fragments.

V. THE POLAR COORDINATE MODEL

The regulative behavior of imaginal disc fragments, as well as that of several other developing systems such as regeneration blastemas in amphibians and the appendages of cockroach nymphs, can be understood in terms of a single formulation called the "polar coordinate model" (French et al., 1976). In this model the positional information of cells is considered to be specified by a two-dimensional array of polar coordinates, so that each cell would have information with respect to both radial and circumferential position in the field (Fig. 6). In the case of

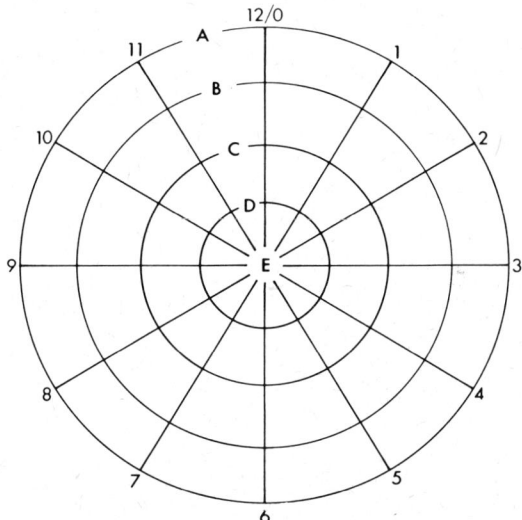

Fig. 6. The idea of polar coordinates in a positional information field. Cells are supposed to be specified with respect to circumferential (1-12/0) and radial (A-E) positions. From French et al., (1976).

the wing disc and other discs which produce appendages, the central point of the polar coordinate system would correspond to the presumptive distal tip of the appendage, while the field boundary would correspond to the margin of the presumptive pattern in the disc. Circumferential positional information is assumed to be specified without any discontinuities in spite of our labelling convention which is discontinuous at 12/0. The most important feature of this model is the proposition that these tissues have the general property of intercalation.

That is, when any two different positions are confronted with one another, as a result of either grafting or wound healing, the positional value discontinuity at the site of the confrontation stimulates local growth by cell multiplication, and during this growth those positional values are generated which normally lie, either in the radial or circumferential direction, between the confronted positions. If the confronted cells differ in circumferential position, then it is proposed that intercalation generates the set of positional values that separates the two points by the shorter, rather than the longer of the two possible circumferential routes (the "shortest intercalation rule"). This proposition is supported by direct evidence in the case of the cockroach leg (French and Bulliere, 1975a, 1975b), and by indirect evidence in imaginal discs and other systems. As discussed elsewhere (French et al., 1976), the shortest intercalation rule gives us a simple way of understanding the regulative behavior of sectors and segments of imaginal discs.

The apparently paradoxical behavior of central squares of the wing disc is correctly predicted by the second rule of the polar coordinate model—the "complete circle rule." In this rule, it is proposed that whenever a complete circumference is generated (by grafting, wound healing or intercalation) at a given proximal-distal (=radial) level, then growth occurs and during this growth all of the more distal presumptive parts are generated. This rule provides for normal, reversed and supernumerary regeneration in other systems, and accounts for duplication of the central wing-disc square and regeneration of the complementary annulus.

According to this interpretation of pattern regulation in imaginal discs, growth should stop when either duplication or regeneration is complete. This is, in general, true (Hadon et al., 1949; Ursprung, 1962; Wildermuth, 1968; Bryant, 1975a), and in fact some fragments often begin to shrink when cultured for a long period in adult hosts (Ursprung, 1962; Van der Meer and Ouweneel, 1974). A few cases have been reported in which duplication is followed by limited regeneration (Lüönd, 1961; Bryant and Hsei, 1977; Duranceau, 1977) but this does not seem to be a general property. The fact that cultured intact discs usually show little or no growth (Schläpfer, 1963; Ginter and Kusin, 1969; Tobler, 1966; Bryant, 1975a) further supports the idea that positional information discontinuities are responsible for stimulating cell proliferation in these tissues. There are some mutations *(1(2)gd,* Bryant and Schubiger, 1971; *1(3)c43^{hs1},* Martin et al., 1977) that release imaginal discs from these intrinsic growth controls, allowing them to develop into enormous structures in which either the whole presumptive pattern, or parts of it, are duplicated. We hope that such mutations will, in the future, provide an avenue of experimental approach to identifying the kinds of cell interaction required for growth control.

The polar coordinate model suggests a way in which normal development could be related to pattern regulation. When imaginal discs are first set aside in the embryo they consist of small groups of about 10-20 cells (Bryant and Schneiderman, 1969; Bryant, 1970; Nöthiger, 1972). If some of the peripheral positional values of the field were allocated to cells at this early stage, then positional value discontinuities would be established, and these could lead to intercalation of intermediate values as in pattern regulation. Growth would then occur until the pattern was complete, and then stop. This model is consistent with the finding (see, for example, Bryant and Schneiderman, 1969) that the cell lineage of imaginal discs is indeterminate, since the cells produced during intercalation could arise from cells on either side of the positional information discontinuity.

VI. INTERCALATION IN IMAGINAL DISCS

We recently performed (Haynie and Bryant, 1976) a direct test of the idea that imaginal disc tissue can undergo intercalation in response to a positional information discontinuity. Although imaginal disc fragments cannot yet be grafted together with controlled orientation, they can be mixed together with tungsten needles until they form a coherent mass which can be transplanted into adult hosts. Genetic markers can be used to identify which structures in the final pattern arose from each of the starting components. Fragments from opposite ends of the wing disc were first shown to undergo only duplication when cultured intact or mixed with identical fragments before being cultured in adults for the standard seven-day period. But when these fragments were mixed with each other prior to culture, they were both stimulated to regenerate other parts of the pattern. In fact, both components were capable of producing, at some frequency, all of the pattern elements scored, and we interpret this as resulting from intercalation between the two components in the mixture. Fragments from each side of the wing disc were also stimulated to regenerate when combined together, but this occurred less frequently than in the first series, possibly because of the barrier posed by the anterior-posterior compartment boundary (Garcia-Bellido et al., 1973).

The pairs of fragments used in the above experiments differed from each other predominantly in the circumferential, rather than radial parameter, and presumably most of the intercalation observed had occurred in the circumferential direction. More recently, attempts have been made to induce intercalation in the radial, or proximal-distal direction. Presumptive proximal and distal fragments of the leg (Strub, 1978) or wing (Haynie and Schubiger, 1978) disc were mixed together and cultured. The proximal fragment in each case produced abundant distal structures, but whether these resulted from intercalation or distal transformation cannot be determined. However, the distal fragments in each of these experiments failed to regenerate proximal structures as might have been expected if intercalation had occurred in the distal to proximal direction. A possible explanation for this result is that the proximal-distal axis is divided into a series of repeated segments as has been demonstrated for cockroach legs (Bohn, 1970; Bulliere, 1971). If that were the case, then confrontations between homologous levels of different segments would lead to no intercalation, and confrontations between non-homologous levels would lead to intercalation only within the segment. The latter result would probably not have been detected in the experiments so far done.

VII. BILATERALLY SYMMETRICAL FIELDS

The polar coordinate model, or more generally the idea that pattern regulation occurs principally by intercalation between different positions confronted during wound healing, makes some specific predictions regarding the behavior of bilaterally symmetrical fields such as that of the genital disc. Cuts which are made parallel to the line of symmetry should produce fragments which will regenerate or duplicate, just as in the case of asymmetrical discs. However, when cuts are made so as to produce bilaterally symmetrical fragments, then these fragments can heal in a way which brings together only identical positions from right and left sides. Such a situation would not be expected to lead to growth, and no regeneration or duplication should occur. These predictions have been tested and confirmed using the male genital disc (Bryant and Hsei, 1977) where asymmetrical fragments regenerated or duplicated depending on their size, but four different symmetrical fragments, in general, failed to show any regulation. Only one of the four types of symmetrical fragments tested showed a low frequency of regeneration during culture.

A bilaterally symmetrical derivative of the wing disc is produced by the mutation *wingless* in *Drosophila* (Sharma and Chopra, 1976). We have tested bilaterally symmetrical fragments cut from these discs and shown that they fail to either regenerate or duplicate during culture (Bryant and A. James, unpublished).

If the theory of intercalation is correct, it should be possible to stimulate regeneration of bilaterally symmetrical fragments by mixing them, prior to culture, with bilaterally symmetrical fragments from other parts of the imaginal disc. This prediction has also been confirmed with the male genital disc (Littlefield and Bryant, unpublished). Fragment *02*, consisting primarily of presumptive genitalia, was mixed with genetically marked fragment *35*, which is largely presumptive anal plates. After culture and metamorphosis, each fragment had regenerated, at high frequency, structures normally derived from the other component in the combination. We assume that this regeneration occurred by intercalation between the two components of the mixture, a conclusion which is supported by the fact that many of the structures produced contained cells derived from both of the original components in the combination (Fig. 7).

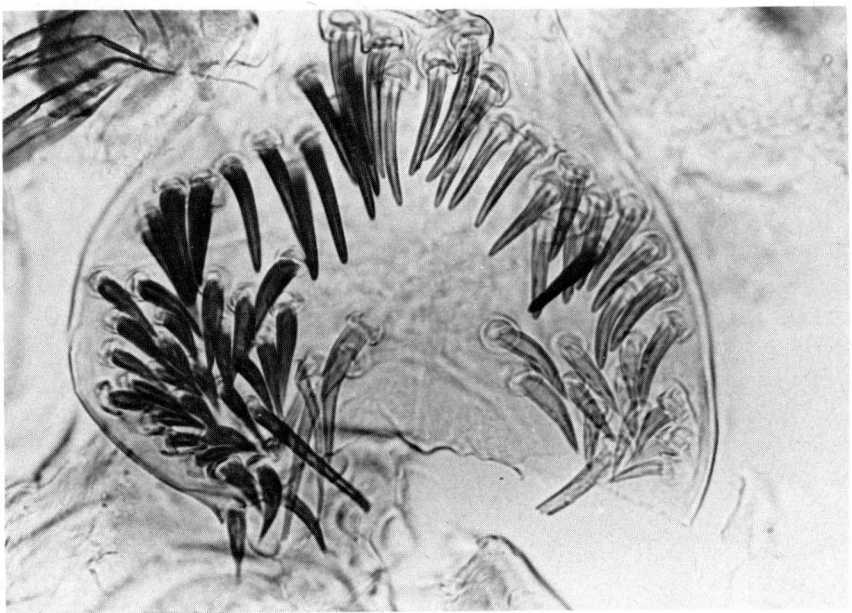

Fig. 7. Mosaic claspers, parts of the male genitalia produced by interaction between fragments of *yellow* (light bristles) and wild-type (dark bristles) male genital discs. The dark bristles were produced by regeneration from a *35* fragment, stimulated by mixing prior to adult culture with a *yellow 02* fragment. Littlefield and Bryant (unpublished).

We have recently retested these predictions using the female genital disc, with essentially similar results (Littlefield and Bryant, 1979). Bilaterally symmetrical fragments usually show a pattern of wound-healing suggestive of contacts being established between identical positions on right and left sides (Fig. 8). Such fragments, show, after

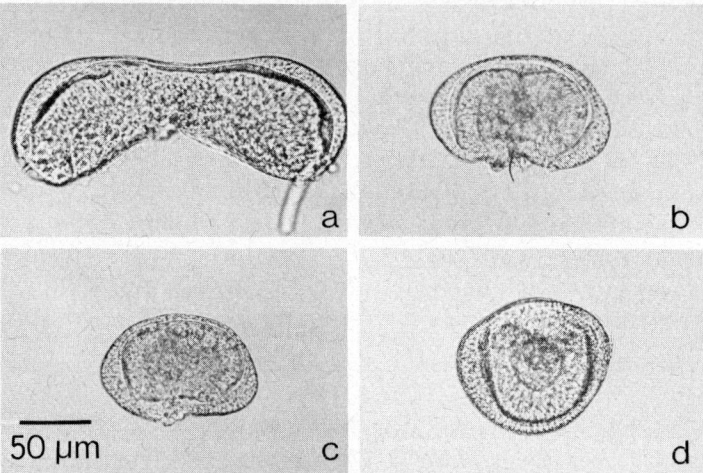

Fig. 8. Photographs of a bilaterally symmetrical (24) fragment of the female genital disc (a) immediately after cutting and the same fragment after (b) one, (c) two and (d) seven days of adult culture. From Littlefield and Bryant (1979).

metamorphosis, fusion of groups of pattern elements from right and left sides, and cultured fragments show a similar picture, with in most cases no regeneration or duplication (Fig. 9). However, some of these fragments did show a low frequency (a maximum of about 20%) of regeneration of missing parts, as with one of the bilaterally symmetrical fragments of the male genital disc. The frequency of regeneration in these pieces could be markedly increased, and regeneration could be induced in normally non-regenerating fragments, by mixing them with their complementary bilaterally symmetrical fragments. This, of course, we interpret as being due to intercalation, as with the male genital disc.

VIII. INTERPRETATION

Although the lack of regulation seen in cultured bilaterally symmetrical fragments could be accounted for by their symmetrical mode of healing, the occasional regeneration seen in some of these fragments call for an explanation. We have recently proposed (Littlefield and Bryant, 1979) an interpretation of these results which is based on the

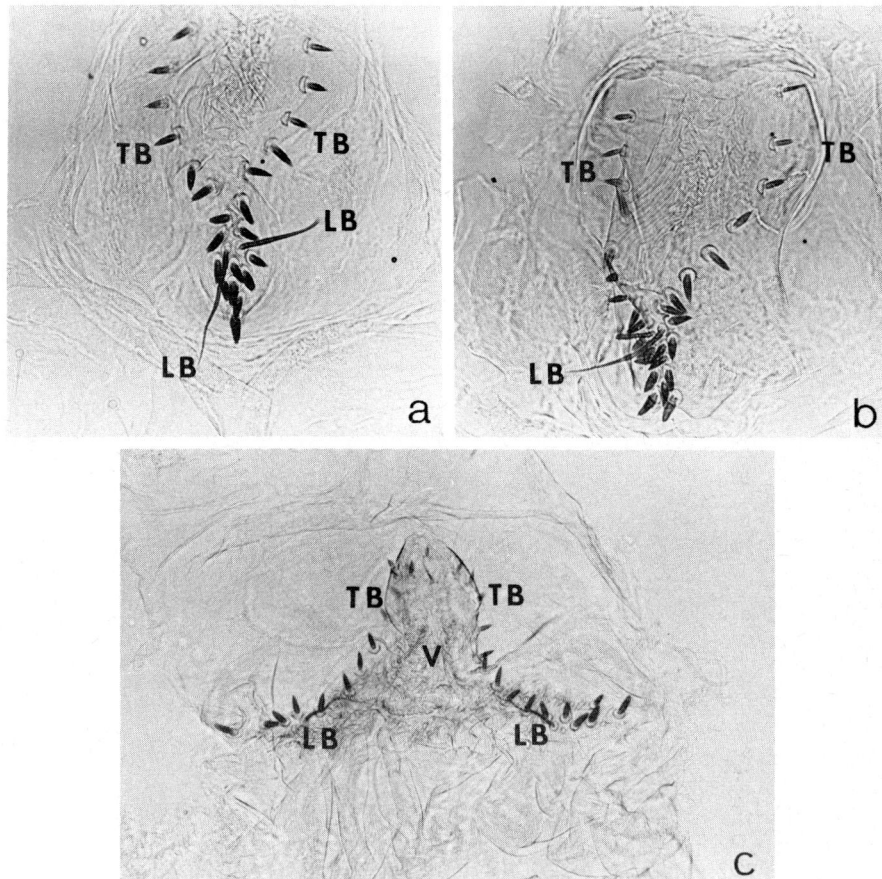

Fig. 9. Female genital structures from (a) immediately metamorphosed and (b) seven-day cultured bilaterally symmetrical (24) fragments, and (c) an immediately metamorphosed whole female genital disc. In (a) note the left-right fusion of the parts of the pattern marked by the long bristles (L.B.). In (b) a similar picture is seen, indicating that no regeneration or duplication has occurred during the culture period. T.B., thorn bristles of the vaginal plates. From Littlefield and Bryant (1979).

finding that each genital disc is not organized into a thickened disc epithelium and a thin peripodial membrane, as are other imaginal discs, but rather consists of a sac of epithelium which is thickened throughout most of the structure. Fate-mapping studies (Ehrensperger and Nöthiger, personal communication; Emmert, 1972; Littlefield and Bryant, 1979) and considerations of the mode of eversion and cell lineage of the discs suggest that both dorsal and ventral parts of the epithelium give rise to adult structures (Fig. 10), and therefore presumably both

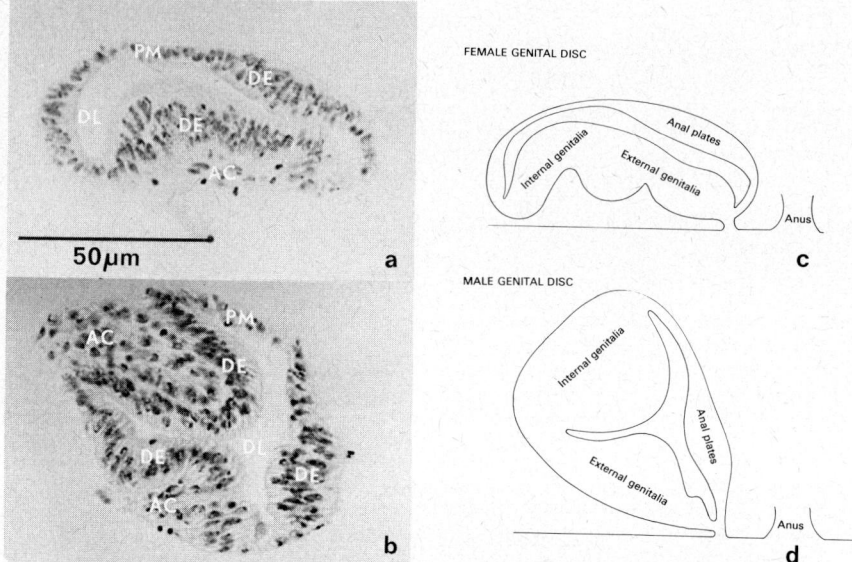

Fig. 10. (a) and (b). Parasaggital sections of the female (a) and male (b) genital discs, stained with Feulgen and Fast Green. A.C., adepithelial cells; D.E., disc epithelium; D.L., disc lumen; P.M., a small region of the disc where the cells are less columnar than elsewhere — the only region that might be considered to be peripodial membrane, (c) and (d). The approximate positions of the primordia for adult structures in the genital discs. From Littlefield and Bryant (1979); (d) adapted from Ehrensperger and Nöthiger (pers. comm.).

carry positional information. This, then, raises the possibility that the occasional cases of regeneration in bilaterally symmetrical fragments might be due to an alternative mode of wound healing in which the dorsal and ventral cut edges fuse to one another, rather than the right-to-left healing considered so far. Dorsal-ventral healing would provide positional value discontinuities which would, by intercalation, lead to either regeneration or duplication of the pattern (Fig. 11). Although most of the bilaterally symmetrical fragments we studied (Fig. 8) seemed to show left-to-right healing, after wound contraction the two kinds of wound healing proposed here may not be readily distinguishable in such simple observations of tissue morphology.

The three-dimensional model for the genital disc proposed in Fig. 11 accounts not only for the results on bilaterally symmetrical fragments, but also for many of the other results so far obtained. Thus, intercalation between different bilaterally symmetrical fragments in a cultured mixture is expected, and when the disc is cut parallel to the line of symmetry, the resulting asymmetrical fragments will either regenerate (if larger than half the disc) or duplicate (if smaller than half the disc)

PATTERN SPECIFICATION IN IMAGINAL DISCS 311

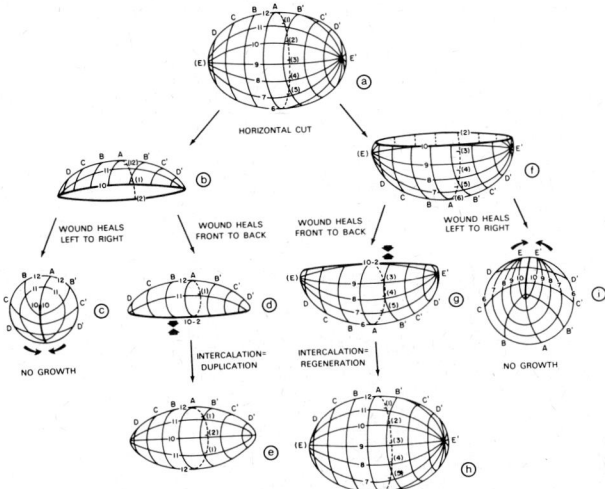

Fig. 11. Polar coordinate model specifying position in a bilaterally symmetrical field. (a) Following a cut in the horizontal direction, transecting the axis of symmetry, the mode of regulation is dependent on the mode of wound healing in the fragment. Wound healing from left to right leads to the confrontation of cells having identical positional values and no stimulation of growth (b-c and f-i). Wound healing from back to front (dorsal to ventral in the genital disc) leads to intercalation between the confronted positions. According to the shortest intercalation rule (French *et al.*, 1976) fragments containing less than half of the positional values (d-e) will duplicate, whereas fragments containing more than half of the positional values (g-h) will regenerate. From Littlefield and Bryant (1979).

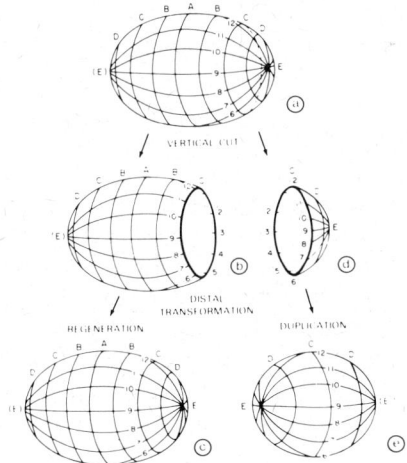

Fig. 12. Polar coordinate model specifying position in a bilaterally symmetrical field. (a) Following a cut in the vertical direction, parallel to the axis of symmetry, regulation proceeds according to the complete circle rule for distal transformation (French *et al.*, 1976) with the formation of all of the more distal positional values from the proximal wound surface. A cut at level C' leads to regeneration of D'E' in the larger fragment (b-c) and duplication of D'E' in the smaller fragment (d-e).

according to the complete circle rule for distal transformation (Fig. 12).

Although the peripodial membrane is not thought to give rise to adult structures in the leg, wing and other asymmetrical discs, this has never been proven. Furthermore, even if it does not give adult derivatives, it is conceivable that it carries positional information which can interact with that in the disc epithelium. If so, it would be worthwhile to consider an asymmetrical version of the three-dimensional model in Figure 11 to apply to these discs. In fact, such a model makes, for the most part, predictions similar to those derived from the two-dimensional model. It does, however, predict that for certain fragments, the mode of regulation would depend on the mode of healing, a prediction which we are at present attempting to test.

IX. DISC INTERACTIONS AND THE UNIVERSALITY OF POSITIONAL INFORMATION

The fact that much of the regulative behavior of several imaginal discs is consistent with a single formal model supports in a general way the idea proposed by Wolpert (1971) that different developing fields in the organism might make use of the same array of positional information. According to this idea, cells in different fields would interpret positional information differently, by reference to their different determined states established earlier in development.

The discovery of intercalation in imaginal discs (Haynie and Bryant, 1976) made it possible to test the postulate that the same positional signals are used in different imaginal discs. If this idea is correct, then it should be possible to detect intercalation between parts of different imaginal discs which are mixed together before culture. Strub (1977) found that the developmental behavior of dissociated leg disc cells could be profoundly modified by mixing them with wing disc cells, a result which was interpreted as due to interaction between the two types of cells. A specific test for interaction between a wing-disc fragment and other imaginal discs has been carried out by Wilcox and Smith (1977) using as a test fragment the duplicated presumptive notum produced in place of a wing disc in the *wingless* mutant, and we (Bryant *et al.*, 1978) have repeated and extended this experiment using the presumptive notum from normal wing discs as a test piece. These test pieces are always restricted to producing notum when cultured intact, and with very rare exceptions, they produce only notum when mixed with identical presumptive notum fragments before culture. However, when mixed with other parts of the wing disc or whole wing discs, during subsequent culture they regenerate other parts of the wing pattern at high frequencies, presumably as a result of intercalation between the two components of the mixture (Haynie and Bryant, 1976; Bryant, et al., 1978). In other experiments, it was shown (Adler and Bryant, 1977; Wilcox and Smith, 1977) that heavily irradiated wing disc tissue would also elicit this response from the notum fragment, whereas similarly irradiated presumptive notum would not do so.

The crucial experiment, then, was to examine the behavior of similar test fragments when mixed with imaginal discs other than wing (heterotypic discs) prior to culture. In most of our experiments, we mixed unirradiated test pieces with heavily irradiated heterotypic discs in order to prevent growth and transdetermination in the latter which would have confused the results.

In both Wilcox and Smith's experiments and those of our group, the haltere disc and leg disc were very effective in eliciting regeneration of the notum test piece. From this, it seems reasonable to conclude that the cells in these different discs can and perhaps do use the same signals for positional information. This, of course, opens the possibility of using different test pieces mixed with different disc fragments in order to determine the homological relationships between positional values in different discs.

The non-thoracic imaginal discs (eye, antenna, genital) were found by Wilcox and Smith to be incapable of eliciting regeneration by the *wingless* notum, and we have indeed found them to be much less effective than the thoracic discs. In our experiments, there were reproducible quantitative differences between these discs as well as between thoracic discs with the order of effectiveness being wing > haltere > 1st leg > 2nd leg > 3rd leg > antenna > male genitalia > eye > female genitalia.

A possible explanation for the quantitative differences in effectiveness of different discs in causing regeneration of the notum piece was that different discs use different subsets of the positional values of a universal field. However, in our experiments, there were no clear differences in the range of pattern elements regenerated by the test piece in response to different heterotypic disc types. We therefore suggest that the positional signals may, in fact, be common to all imaginal discs, but that cells from different disc types might differ in their ability to adhere to or form junctions with wing disc cells, an idea which is well established in the imaginal disc literature (see, for example, Garcia-Bellido, 1966a,b).

X. PROSPECTS

The model we have proposed for pattern formation and regulation in imaginal discs as well as other systems focuses attention on local cell interactions rather than global control systems in controlling both pattern and growth. The main problem we now face is to devise ways to analyze the nature of the subtle differences between cells in different parts of a pattern and the nature of the cell interactions which establish those differences.

One approach to this problem is to establish a method where pattern regulation can occur *in vitro*, so that molecular probes can be used to track cell surface components, and the cell interactions can be analysed by interfering with them chemically. Fortunately, imaginal disc tissue can now be grown successfully *in vitro* (Davis and Shearn, 1977) so that these approaches may be feasible.

An alternative strategy which can be contemplated with *Drosophila* is a mutational analysis of the process of pattern formation and growth control. Such a strategy was pursued by Curt Stern and his coworkers for many years (see Stern, 1968) during which time many pattern mutations were studied using genetic mosaics. Most of them seem to have been mutations affecting the response of cells to positional information, rather than affecting the specification of positional information itself (see Bryant, 1978). A possible reason for this is that any mutation affecting positional signalling would be likely to be lethal if, as seems to be the case, cells in different fields of the same organism use the same positional signals. One of our present approaches to the problem then, focuses on lethal mutations affecting pattern, in particular mutations such as *lethal (2) giant discs* (Bryant and Schubiger, 1971) and *lethal (3)c43^{hs1}* (Martin et al., 1977) in which not only pattern formation, but growth control in imaginal discs is abnormal. By studying the cellular basis for such defects we hope to gain an understanding of the cell functions and interactions in the normal organism which control the spatial organization and growth of developing tissues.

ACKNOWLEDGEMENTS

The author's investigations are supported by grant HD06082 from the National Institutes of Health. It is a pleasure to thank the colleagues mentioned in the text for the use of their unpublished results.

REFERENCES

Adler, P.N. and Bryant, P.J. (1977). *Develop. Biol.* **60**, 298-304.
Auerbach, C. (1936). *Trans. Roy. Soc. Edinb.* **58**, 787-815.
Bohn, H. (1970). *Wilhelm Roux' Arch.* **165**, 303-341.

Bryant, P.J. (1970). *Develop. Biol.* **22**, 389-411.
Bryant, P.J. (1971). *Develop. Biol.* **26**, 637-651.
Bryant, P.J. (1975a). *J. Exp. Zool.* **193**, 49-77.
Bryant, P.J. (1975b). *Ciba Found. Symp.* **29**, 71-93.
Bryant, P.J. (1978). *In* "Genetics and Biology of Drosophila" (T.R.F. Wright and M. Ashburner, eds.). In press.
Bryant, P.J., Adler, P.N., Duranceau, C., Fain, M.J., Glenn, S., Hsei, B., James, A.A., Littlefield, C.L., Reinhardt, C.A., Strub, S. and Schneiderman, H.A. (1978). *Science* **201**, 928-930.
Bryant, P.J. and Hsei, B. W. (1977). *Amer. Zool.* **17**, 595-611.
Bryant, P.J. and Schneiderman, H.A. (1969). *Develop. Biol.* **20**, 263-290.
Bryant, P.J. and Schubiger, G. (1971). *Develop. Biol.* **24**, 233-263.
Bullière, D. (1971). *Develop. Biol.* **25**, 672-709.
Crick, F. (1970). *Nature* **225**, 420-422.
Davis, K. and Shearn, A. (1977). *Science* **196**, 438-440.
Duranceau, C. (1977). Ph. D. Thesis, University of California, Irvine.
Emmert. W. (1972). *Wilhelm Roux' Arch.* **171**, 109-120.
French, V., Bryant, P.J. and Bryant, S.V. (1976). *Science* **193**, 969-981.
French, V. and Bullière, D. (1975a). *C. R. Acad. Sci. Paris* **280**, 53-56.
French, V. and Bullière, D. (1975b). *C. R. Acad. Sci. Paris* **280**. 295-298.
Fristrom, D. and Fristrom, J.W. (1975). *Develop. Biol.* **43**, 1-23.
Garcia-Bellido, A. (1966a). *Develop. Biol.* **14**, 278-306.
Garcia-Bellido, A. (1966b). *Exp. Cell Res.* **44**, 382-392.
Garcia-Bellido, A., Ripoll, P. and Morata, G. (1973). *Nature New Biol.* **245**, 251-253.
Gateff, E.A. and Schneiderman, H.A. (1975). *Wilhelm Roux' Arch.* **176**, 171-189.
Gehring, W. (1966). *J. Embryol. Exp. Morphol.* **15**, 77-111.
Ginter, E.K. and Kusin, B.A. (1969), *Drosoph. Inform. Serv.* **44**, 74.
Ginter, E.K. and Kuzin, B.A. (1970). *Ontogenese* **1**, 492-500.
Goodwin, B. and Cohen, M.H. (1969)., *J. Theor. Biol.* **25**, 49-107.
Hadorn, E., Bertani, G. and Gallera, J. (1949). *Wilhelm Roux' Arch.* **144**, 31-70.
Haynie, J.L. and Bryant, P.J. (1976). *Nature* **259**, 659-662.
Haynie, J. and Schubiger, G. (1978). *Develop. Biol.* In press.
Lawrence, P.A., Crick, F.H.C. and Munro, M. (1972). *J. Cell. Sci.* **22**, 815-853.
Littlefield, C.L. and Bryant, P.J. (1979). *Develop. Biol.* In press.
Loosli, R. (1959). *Develop. Biol.* **1**, 24-64.
Lüönd, H. (1961). *Develop. Biol.* **3**, 615-656.
Madhavan, M.M. and Schneiderman, H.A. (1977) *Wilhelm Roux' Arch.* **183**, 269-305.
Martin, P., Martin, A. and Shearn, A. (1977). *Develop. Biol.* **55**, 213-232.
McMahon, D. (1973). *Proc. Nat. Acad. Sci. U.S.* **70**, 2396-2400.
Meinhardt, H. and Gierer, A. (1974). *J. Cell Sci.* **15**, 321-346.
Nöthiger, R. (1972). *In* "Biology of Imaginal Discs" (H. Ursprung and R. Nöthiger, eds.), pp. 1-34. Springer, New York.
Poodry, C.A. and Schneiderman, H.A. (1970). *Wilhelm Roux' Arch.* **166**, 1-44.
Reinhardt, C.A. (1978). Submitted to *Develop. Biol.*
Reinhardt, C., Hodgkin, N. and Bryant, P.J. (1977). *Develop. Biol.* **60**, 238-257.
Rose, S.M. (1952). *Amer. Nat.* **86**, 337-354.
Roseland, C. (1976). Ph.D. Thesis, University of California, Irvine.
Schläpfer, T. (1963). *Wilhelm Roux' Arch.* **154**, 378-404.
Schubiger, G. (1968). *Wilhelm Roux' Arch.* **160**, 9-40.

Schubiger, G. (1971). *Develop. Biol.* **26,** 277-295.
Schubiger, G. (1974). *Wilhelm Roux' Arch.* **174,** 303-311.
Sharma, R.P. and Chopra, V.L. (1976). *Develop. Biol.* **48,** 461-465.
Sprey, Th. E. and Oldenhave, M. (1974). *Neth. J. Zool.* **24,** 291-310.
Stern, C. (1968). "Genetic Mosaics and Other Essays." Harvard University Press, Cambridge.
Strub, S. (1977). *Nature* **269,** 688-691.
Strub, S. (1978). Submitted to *Develop. Biol.*
Stumpf, H.F. (1966). *Nature* **212,** 430-431.
Tobler, H. (1966). *J. Embryol. Exp. Morphol.* **16,** 609-633.
Ursprung, H. (1959). *Wilhelm Roux' Arch.* **151,** 504-558.
Ursprung, H. (1962). *Develop. Biol.* **4,** 22-39.
Ursprung, H. (1972). *In* "Biology of Imaginal Discs" (H. Ursprung and R. Nöthiger, eds.), pp. 93-107. Springer, New York.
Van der Meer, J.M. and Ouweneel, W.J. (1974). *Wilhelm Roux' Arch.* **174,** 361-373.
Vogt, M. (1946). *Biol. Zbl.* **65,** 223-238.
Webster, G. (1971). *Biol. Rev.* **46,** 1-46.
Wilcox, M. and Smith, R. J. (1977). *Develop. Biol.* **60,** 287-297.
Wildermuth, H. (1968). *Wilhelm Roux' Arch.* **160,** 41-75.
Wolpert, L. (1968). *In* "Towards a Theoretical Biology. I. Prolegomena" (C.H. Waddington, ed.), pp. 125-133. Edinburgh University Press.
Wolpert, L. (1971). *Curr. Topics Develop. Biol.* **6,** 183-224.

Pattern Formation and Compartments in the Tarsus of *Drosophila*

P. A. Lawrence
*MRC Laboratory of Molecular Biology
Hills Road, Cambridge
CB2 2QH, England*

G. Morata
*Centro de Biologia Molecular
C.S.I.C.
Facultad de Ciencias, C-X
Universidad Autonoma de Madrid
Madrid-34, Spain*

I. Introduction 317
II. Normal Mesothoracic Tarsus 319
III. The Effect of Homoeotic Mutants on Tarsal Pattern 319
IV. Antennal-Tarsal Homology 322
References 323

I. INTRODUCTION

Progress in our understanding of patterns in animal development depends on model systems. The epidermis of *Drosophila* is a good model partly because the patterns themselves are often simple and contain discrete elements (such as bristles), and partly because there are numerous mutations which affect the pattern in defined ways. With *Drosophila* we can therefore look forward to an understanding of pattern formation in terms of the role of specific genes.

Last year at these meetings (Morata and Lawrence, 1978a) we reviewed the compartment hypothesis in detail. Here we only summarise some recent observations and stress the relationship between compartments and pattern. The relevant tenets of the compartment hypothesis are:

(i) Small groups of founder cells are set aside according to position; each group generates a polyclone of cells which forms a compartment in the adult or larva (Garcia-Bellido *et al.*, 1973, 1976; Crick and Lawrence, 1975).

(ii) Specific 'selector' genes control the developmental pathway followed by a polyclone (Garcia-Bellido, 1975) — for example when the selector gene *engrailed* is active the posterior rather than the anterior pattern is constructed (Garcia-Bellido and Santamaria, 1972; Morata and Lawrence, 1975; Lawrence and Morata, 1976). Likewise when *bithorax* is active the metathoracic rather than the mesothoracic pattern is formed (Lewis, 1963).

(iii) Polyclones may be autonomous units in pattern formation; the borders of gradients of positional information may coincide with compartment boundaries, and growth may be locally controlled within polyclones (Crick and Lawrence, 1975).

(iv) Different polyclones, with different selector genes active, interpret the same positional information — a prepattern — in different ways. There is therefore an underlying homology (Stern, 1954).

(v) Different polyclones may interact in the formation of local pattern (Santamaria and Garcia-Bellido, 1975).

We picture development of *Drosophila* as a series of discrete steps: segmentation separates the embryo into a number of polyclones which are probably identical in size (Lawrence *et al.*, 1978a; Wieschaus, personal communication). Typically, these are each subdivided into an anterior and posterior daughter polyclone with the *engrailed* gene. Subdivisions occur to establish separate polyclones for the germ layers and to generate larval and adult cells; it is simplest to imagine that these subdivisions are also binary and all depend on common mechanisms. Pattern is established by positional information which is reiterated in each growing polyclone, the differential response being determined by the binary code of active and inactive selector genes (Garcia-Bellido, 1975). The code in any particular polyclone will depend on its history — the series of genetic decisions it has undergone in earlier development.

We intend to illustrate some of these points by reference to one example only — the bristle pattern in the tarsus of the mesothoracic leg. This tarsal pattern can be formed in at least three places — the mesothorax as in wildtype, the metathorax as in flies mutant for *bithorax* and *postbithorax* and the head in *spineless-aristapedia* flies.

II. NORMAL MESOTHORACIC TARSUS

The bristle pattern of a typical mesothoracic tarsus is shown in Fig. 1; it is almost perfectly mirror-symmetric. From clonal analysis (Steiner, 1976; Lawrence and Morata, 1977) we know that the mesothoracic polyclone is subdivided into anterior and posterior groups of founder cells at, or soon after, the blastoderm period. Detailed analysis of the tarsus (Lawrence *et al.*, 1978b) has shown that the compartment boundary demarcating the interface between these two polyclones does not coincide with the mirror plane, but runs down the bristle rows most immediately posterior to that plane in both dorsal and ventral faces. For example, the posterior ventral row contains bristles and bracts from both anterior and posterior polyclones, the provenance of specific bristles and bracts varying from individual to individual. Apparently the bristle-forming cells are recruited from a band of epidermal cells which span the interface between the two polyclones, and they *move* into a straight line thereafter (Lawrence *et al.*, 1978b). This is a clear example of cooperation between polyclones; the bristle row seems to be set up without regard to the origin of cells forming it.

III. THE EFFECT OF HOMOEOTIC MUTANTS ON TARSAL PATTERN

Two observations tell us that the mirror symmetry of the tarsus is more apparent than real. First, the compartment boundary is asymmetrically placed. Second, only the posterior half is affected by the *engrailed* mutation, which adds extra bristles there but not in the anterior half (Lawrence *et al.*, 1978b). Third, in the metathorax, which is homologous to the mesothorax, the tarsal bristle pattern is itself asymmetric. Thus, although they make the same pattern in the mesothoracic tarsus, the anterior and posterior polyclones are different — as is clearly shown in the more proximal leg segments where the arrangement and structure of the bristles is quite different in the anterior and posterior compartments (Steiner, 1976). We have suggested (Lawrence and Morata, 1976; Morata and Lawrence, 1977) that the difference depends critically on the selector gene *engrailed* which, when active, selects the posterior developmental pathway. Other selector genes are important in the development of tarsal pattern; for example, the *bithorax* and *postbithorax* mutations transform the anterior and

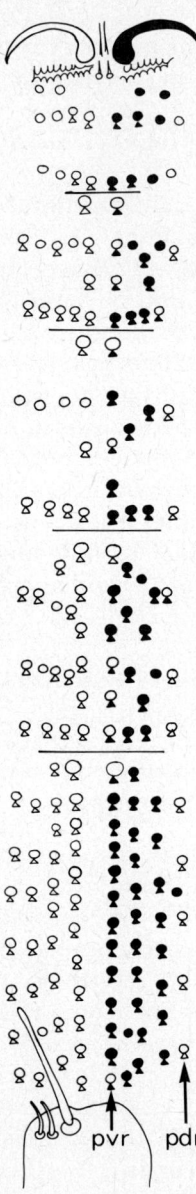

Fig. 1. The bristle pattern in a normal mesothoracic tarsus. The mirror symmetry of the pattern is apparent, but note that bristles typically part of the posterior compartment (shown in black) are less than 40% of the total. The bristle pattern is that of a single leg but, because most clones do not completely fill the compartment, the posterior bristles are labelled after the study of several large clones. The provenance of bristles and bracts in both the posterior ventral row (*pvr*) and the posterior dorsal row (*pdr*) varies. Data from Lawrence *et al.*, 1978b.

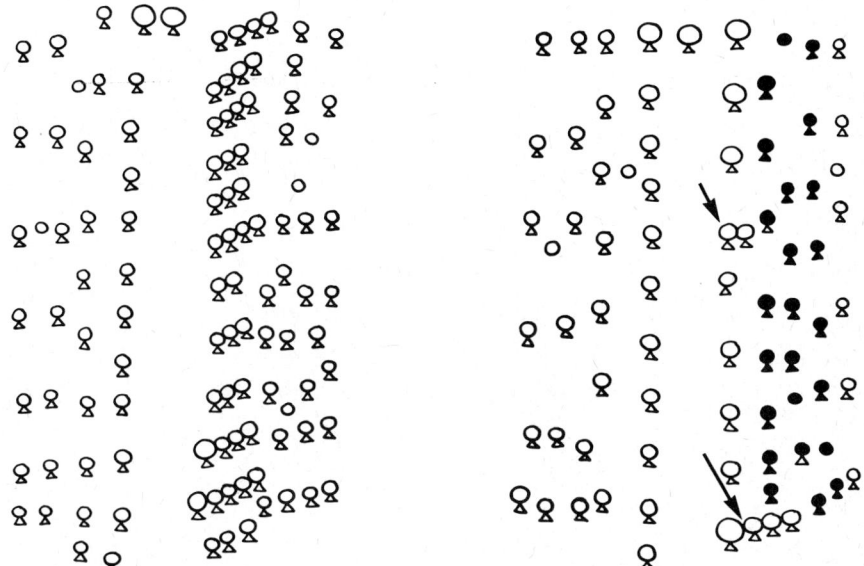

Fig. 2a. Left, the normal bristle pattern of the basitarsus of the metathorax. Note the transverse rows in the posterior compartment.

Fig. 2b. Right, a metathoracic basitarsus bearing a posterior clone of marked cells mutant for *postbithorax*. Marked bristles and bracts are shown in black. Note that transverse rows are almost completely eliminated but some pbx^+ cells form partial transverse rows (arrows).

posterior metathoracic compartments into mesothoracic ones (reviewed in Morata and Lawrence, 1977). These two genes apparently work separately in the two polyclones, determining the repertoire of bristles and the way they are arranged (Lawrence *et al.*, 1978b).

Underlying the different patterns of the three legs there are common fields of positional information. This is most clearly illustrated in legs mosaic for meta - and mesothoracic cells which form single integrated patterns, each cell autonomously forming bristle elements appropriate for its segmental type. Moreover, when clones of cells mutant for *bithorax* or *postbithorax* are made in the third leg, they autonomously express a bristle pattern appropriate to the position of that clone. For example, if a *postbithorax* clone is formed in the posterior compartment (where there are transverse rows of bristles in the third, but not in the second leg) the transverse rows are eliminated (Fig. 2).

The pattern of bristles in the basitarsus of mosaic legs is particularly interesting. In wildtype second legs there is a posterior ventral row of

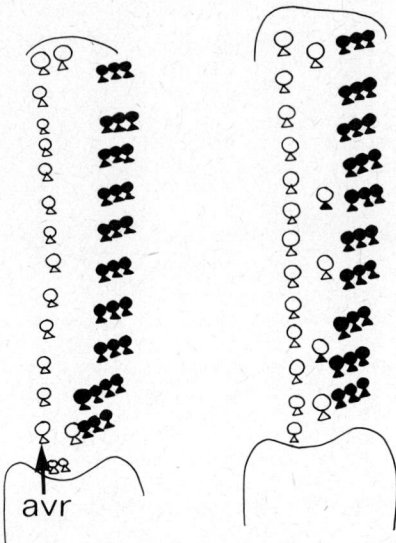

Fig. 3a. Left, ventral face of a typical basitarsus of the metathorax. Note the gap between the anterior ventral row (avr) and the transverse rows. Bristles and bracts marked by posterior clones are shown in black.

Fig. 3b. Right, ventral face of a basitarsus from a metathoracic leg bearing an anterior clone of cells mutant for *bithorax*. This clone makes a pattern typical of anterior mesothorax. Note the development of bristles in the gap. These bristles may be a partial posterior ventral row. Data from Lawrence *et al.*, 1978.

bristles to which the anterior polyclone sometimes contributes. This row of bristles is not found in third legs. When the anterior polyclone is mesothoracic, and the posterior metathoracic, occasional bristles are formed along the line where the posterior ventral row would be if the leg were entirely mesothoracic (Fig. 3). These observations illustrate how patterns are the result of an interaction between the two components; (i) underlying positional information which is pervasive and common to different compartments; (ii) the determined state of the cells which is cell autonomous and, because it depends on the local state of selector genes, is unique to each compartment.

IV. ANTENNAL—TARSAL HOMOLOGY

The mutation *aristapedia* produces a tarsus in place of the distal antenna. This tarsus has a pattern similar to that of a mesothoracic leg and clonal analysis has shown that it develops in the same way. For example, clones in the anterior half of the tarsus can cross the mirror plane and mark bristles in the posterior ventral row, just as in the normal tarsus. However, in the antenna, separation of anterior and posterior polyclones occurs in larval life — this is much later than in the normal tarsus (Morata and Lawrence, 1978b). It is thus possible for the descendents of a cell marked in the first stage larva to fill the whole homoeotic tarsus, while clones in the normal tarsus are always restricted to either anterior or posterior compartments. In spite of this difference there is an underlying homology between the mesothorax and the head segment, so that in *aristapedia* anterior and posterior antenna are transformed into anterior and posterior tarsus respectively. The anterior compartment in the head includes most of the head capsule and all of the eye, while the posterior includes the remaining part of the antenna and head (Morata and Lawrence, 1978b). This example again emphasises the underlying systems that different patterns have in common, a homology that is revealed by mutations in single genes.

REFERENCES

Crick, F.H.C. and Lawrence, P.A. (1975). *Science* **189**, 340-347.
Garcia-Bellido, A. (1975). *In* "Cell Patterning". *Ciba Foundation Symp.* **29**, 161-182.
Garcia-Bellido, A., Ripoll, P. and Morata, G. (1973). *Nature New Biol.* **245**, 251-253.
Garcia-Bellido, A., Ripoll, P. and Morata, G. (1976). *Develop. Biol.* **48**, 132-137.
Garcia-Bellido, A. and Santamaria, P. (1972). *Genetics* **72**, 87-104.
Lawrence, P.A., Green, S.M. and Johnston, P. (1978a). *J. Embryol. Exp. Morphol.* **43**, 223-245.
Lawrence, P.A. and Morata, G. (1976). *Develop. Biol.* **50**, 321-337.
Lawrence, P.A. and Morata, G. (1977). *Develop. Biol.* **56**, 40-51.
Lawrence, P.A., Struhl, G. and Morata, G. (1978b). *J. Embryol. Exp. Morph.* submitted.
Lewis, E.B. (1963). *Amer. Zool.* **3**, 33-56.
Morata, G. and Lawrence, P.A. (1975). *Nature* **255**, 614-617.
Morata, G. and Lawrence, P.A. (1977). *Nature* **265**, 211-216.
Morata, G. and Lawrence, P.A. (1978a). 36th *Symp. Soc. Develop. Biol.* in press.
Morata, G. and Lawrence, P.A. (1978b). *Nature* **274**, 473-474
Santamaria, P. and Garcia-Bellido, A (1975). *Wilhelm Roux' Arch.* **178**, 233-245.
Steiner, E. (1976). *Wilhelm Roux' Arch* **180**, 9-30.
Stern, C. (1954). *Amer. Sci.* **42**, 213-247.

Subject Index

A

Abdomens, double
  anterior determinant and, 103-107
  bicaudal mutation and, 193-198
  insects, in, 101-103
  photoreversal of uv induction of, 105ff
  uv induction of, 103ff
Acetylcholinesterase
  ascidian embryos and, 35-37, 41-44
  actinomycin D and ascidian, 43, 48
Actinomycin D
  ascidian acetylcholinesterase and, 43, 48
  ascidian alkaline phosphatase and, 43-48
  ascidian tyrosinase and, 43, 48
Alkaline phosphatase
  ascidian embyros and, 36-38
  actinomycin D and ascidian, 43-48
  maternal ascidian mRNA determinant for, 43-46
Anabaena
  intercellular interactions and pattern formation in, 247-266
Antennal-tarsal homology
  Drosophila, in, 322-323
Anterior determinant, in *Smittia*
  analysis of, 103-117
  cellular localization of, 108-110
  embryo as a test system for, 103-107
  gradient hypothesis for, 118-120
  molecular characterization of, 110-115
  RNA and, 113-117
  RNase and, 113-115
  topographical localization of, 107-108

Antero-posterior gradient
  bicaudal mutation phenotypes and, in *Drosophila*, 198
Antero-posterior pattern determination
  bicaudal mutation of *Drosophila* and, 191-200, 209-210
  cytoplasmic determinants and, 101, 108-110, 120-122
Apical tuft determinants
  cleavage plane and, 68ff
  localization during cleavage of, 57-70
Ascidiacea
  cytoplasmic determinants and, 29-51
Ascidian
  developmental mosaicism and, 30ff
  morphogenetic determinants and, 30ff
Ascidian embryo
  Conklin-Ortolani fate map and, 35
  restricted developmental potential of blastomeres and, 34-37
  visible cytoplasmic segregations in, 32-33
Ascidian morphogenetic determinants
  cell lineage and, 33-34
  cytoplasmic segregation altered and, 41-42
  determinate cleavage and, 33-34
  differentiation without cleavage and, 37-41
  restricted developmental potential of blastomeres and, 34-37
Asters
  localization of determinants in *Cerebratulus* and, 61ff
Autoradiography
  RNA synthesis in ascidian development, demonstration by, 46
  sulfating sites, demonstration by, 82-83, 89-91, 93

Axis
  fixation of polar, 80-87
Axolotl
  o maternal effect mutation of, 167-183

B

Bacteria
  vegetal pole and, 17-18
Behavior
  epithelial hydra, 289
Bicaudal mutation
  anterior determinant as model for, 197-198
  antero-posterior pattern of *Drosophila* and, 191-200
  dorso-ventral discontinuities, of *Drosophila*, 197
  gradient as model for, 198
  left-right discontinuities, *Drosophila* and, 197
  Meinhardt model for, in *Drosophila*, 120, 198-200
  models, testing of, for, 197-200
  posterior embryo effects of, in *Drosophila*, 196
  phenotypes of, in *Drosophila*, 191-197
  somatic tissue effect, in *Drosophila*, 193-196
Bicaudal phenotypes
  bimodal distribution of, in *Drosophila*, 193
Bilaterally symmetrical fields
  imaginal disc, in *Drosophila*, 306-308
Bilaterally symmetrical fragments
  model for interpretation of regeneration of, 311
  regeneration of, interpretation of, 308-312
Biochemistry
  differentiation and intercellular interactions in cyanobacteria, of, 252-254
Bithynia
  polar lobes and, 6ff
Bristle pattern
  Drosophila, 317-323
  homoeotic mutant effect on tarsal, 319-323

C

Caenorhabditis elegans
  description of, 151-152
  early development and mutants of, 149-165
  maternal effect, temperature-sensitive mutants, 149-165
Cell
  pole, 138-144
Cell differentiation
  epithelial hydra, in, 288
Cell division
  developmental potential localization and, 53-76
Cell interactions
  Drosophila imaginal discs and, 295-316
Cell movement
  epithelial hydra, in, 287
Cell-surface patterning
  Tetrahymena thermophila, in, 215-246
Cellular determination
  germ plasm and, 128, 144-145
Cellular localization
  anterior determinants, of, 108-110
Centrifugation
  polar lobes, large, 15-16
  production of double abdomens, 103
  vegetal body, 10
Cephalons
  double, in insects, 101-103
Cerebratulus
  cleavage and localization of determinants in, 57ff
Chimeric hydra, 282-286
  interstitial cell elimination and, 274
Chironomidae
  morphogenetic determinant in, 97-126
Ciliary meridians
  longitude control of, in *Tetrahymena*, 230-233
  Tetrahymena thermophila, in, 222-225
Ciliary structures
  Tetrahymena, as reference points in positioning in, 235-238
Ciona
  enzyme development in cleavage arrested embryos of, 37-41
  enzyme development and actinomycin D in embryos of, 43-46, 48
  larval structure of, 31
  maternal mRNA determinant for alkaline phosphatase in embryos of, 43-46
  mRNA synthesis in embryos of, 46-47
  nuclear transplantation in enucleated eggs of, 42

Cleavage
  determinant localization and, 57-74
  zygote defective temperature sensitive mutants, of, 160-163
Cleavage clock phenomenon, 65-68
Cleavage plane
  Cerebratulus apical tuft determinants and, 68ff
  Cerebratulus gut determinant and, 68ff
  ctenophore oral-aboral axis and, 71
  ctenophore symmetry and, 69ff
  determinant localization and, 68-74
  symmetry and, 68ff
Cleavage-stage embryos
  ontogeny of determinant localization and, 57-60
Colchicine
  interstitial cell elimination from hydra and, 272-274
Compartments
  Drosophila, in the tarsus of, 317-323
  homoeotic mutations and, 122-123, 319-323
  pattern formation and, 122-123, 318ff
Complete circle rule
  polar coordinate model, of, 304
Conklin-Ortolani fate map
  ascidian embryo and, 35
Contractile vacuole pore
  latitude control of, 228-230
  longitude control of, 230-232
  positioning of, 223-225
Cortical clearing
  polar axis fixation and, 89-90
Ctenophore
  cleavage and localization of combplate cilia determinants in, 62ff
  cleavage and localization of photocyte determinants in, 62ff
Ctenophore symmetry
  cleavage plane and, 69ff
Culture in vitro
  pole cells in, 141-144
Cyanobacteria
  intercellular interactions and, 247-266
  pattern formation and, 247-266
Cylindrospermum
  intercellular interactions and, 247-266
  pattern formation and, 247-266
Cytochalasin B
  cleavage and cytoplasmic localizations, influence on, 21, 37-41, 63ff
  polar axis fixation and, 80-82
  prevention of fucoidin localization by, 82-84
Cytoplasm
  polar lobe, 6-16
Cytoplasmic localization
  electrical current in, of *Fucus*, 94-95
  Fucus, mechanism of, 94-95
  Fucus, microfilaments in, 94
  germ plasm, 127
  polar lobe cortex and, 16-22
Cytoplasmic segregations
  ascidian embryos and visible, 32-33
  ascidian morphogenetic determinants and altered, 41-42

D

Determinant localization
  ascidian development and, 32-46
  cleavage plane in relation to symmetry and, 68-74
  ontogeny of, during cleavage, 57-60
  polar lobes and, 6-16
Dorsal mutation phenotypes
  dominant, 206-207
  Drosophila, 200-209
  interpretation of, 208-209
  recessive, 200-205
  temperature sensitive, in *Drosophila*, 206
Dorso-ventral discontinuities
  bicaudal mutation of *Drosophila* and, 197
Dorso-ventral gradient
  dorsal mutation phenotypes and, in *Drosophila*, 209
Dorso-ventral pattern formation
  Drosophila maternal effect mutants and, 200-210
Double abdomens
  anterior determinant and, 103-117
  bicaudal mutation and, 193-198
  insects, in, 101-103
  photoreversal of uv induction of, 105ff
Double cephalons
  insects, in, 101-103
Double-membrane vesicles
  large polar lobes and, 12-15

Drosophila
  antennal-tarsal homology in, 322-323
  bicaudal mutation in, 191-198
  bristle pattern of, 317-323
  cell interactions and, imaginal discs, 295-316
  compartments in the tarsus of, 317-323
  dorsal mutation in, 200-209
  embryo, fate map of, 188-189
  germ plasm of, 127-146
  growth control and, imaginal discs, 295-316
  imaginal discs and cell interactions, 295-316
  imaginal disc, description of, 296-298
  pattern formation and, imaginal discs, 295-316
  pattern formation in the tarsus of, 317-323
  pole cells of, 127-146
  pole cell-specific nuclear body of, 140-141
Drosophila melanogaster
  embryogenesis of, 187-189
  grandchildless mutations of, 137-138
  spatial coordinates of embryo of, 185-211
Drosophila subobscura
  grandchildless mutations of, 137

E

Electrical current
  cytoplasmic localization of *Fucus*, 94-95
  polar axis fixation and, 84-85
Electrophysiology
  epithelial hydra, and behavior of, 289
Embryogenesis
  Drosophila melanogaster, 187-189
  insect, 97-98, 104
Embryonic pattern formation in insects
  genetic approach to, 187-191
  gradients vs. localized determinants in, 98, 100-101, 118-120, 189-190, 197-200
  models of, 98, 100-101, 118-122, 189-191, 208-210
Epithelial hydra
  behavior of, 289
  cell differentiation in, 288
  cell movement in, 287
  development of, 267-293
  electrophysiology and behavior of, 289
  histological structure of, 288-289
  structure of, 275-277
  tissue determination and, 280-282
  tissue movement in, 287

F

Fate map
  ascidian embryo, of, 35
  Drosophila embryo, of, 188-189
Fucan
  ricin as localization probe for, 91-93
Fucoidin
  sulfation dependence for transport of, 91-93
  transport of, 87-91
Fucoidin localization
  cytochalasin B prevention of, 82-84
Fucus
  polar deposition of a sulfated polysaccharide in, 77-96

G

Gamma irradiation
  hydra interstitial cell, as eliminator of, 274
Gastrula arrest
  o maternal effect mutation, 168-169
  o maternal effect mutation, correction of, 169-170
Gastrulation
  bicaudal mutations and, in *Drosophila*, 194-196
  dorsal mutations and, in *Drosophila*, 201-203
Genetic control
  early development, of, 167-183
Germ cell determination
  polar granules and, 133-137
Germ plasm
  Drosophila, of, 127-146
  properties of, 128-132
Grandchildless mutations, 137-138
Growth control
  Drosophila, imaginal discs and, 295-316
Gut determinant
  cleavage plane and, 68ff
Gut
  developmental determinants and *Cerebratulus*, 57-60

H

Heterocysts
  biochemical studies of differentiation of, 252-264

Heterocysts (cont'd)
  enzymes retained in isolated, 255-258
  first generation of isolated, 252-255
  formation, control of, 260-264
  immature, 262-264
  intercellular interactions and formation of, 249-251, 260-264
  isolated, first generation of, 252-255
  isolated, metabolically active and intact, 258-260
  isolated, second generation of, 255-260
  isolated, with retained enzymes, 255-258
  mature, 260-261
  pattern formation and, in cyanobacteria, 249-252
  vegetative cells and immature, 262-264
  vegetative cells and mature, 260-261
Histological structure
  epithelial hydra, of, 288-289
Homoeotic genes
  compartments and, 122-123, 318-323
Homoeotic mutations
  antero-posterior decision and, 122-123
  bristle pattern, effect on tarsal, 319-323
  compartments and, 122-123, 319-323
H-test
  maternal effect zygote defective mutants, for, 153-154
Hydra
  development of epithelial, 267-293
  growth of, 268-269
  interstitial cells, selective elimination of, 272-282
  interstitial cell theory for, 270-271
  nerve cell role in development of, 271-272
  pattern control in, 289-290
  tissue structure of, 268-269
Hydra chimeras, 282-286
Hydra development, 268-272

I

Imaginal discs of *Drosophila*
  cell interactions and, 295-316
  description of, 296-298
  growth control and, 295-316
  pattern formation and, 295-316
  intercalation in, 305-306

  pattern regulation in, 298-301
  polar coordinate model of regulation of, 303-305
  positional information and universality of, 312-313
  wound healing in, 301-303
In vitro culture
  pole cells in, 141-144
Induction
  ascidian neural ectoderm and, 36-37
  epithelial hydra and, 279-280
Insect embryogenesis
  gradient vs. mosaic model of, 98, 100-101, 118-120, 189-191, 197-200, 208-210
Intercalation
  bilaterally symmetrical fields and, 306-308
  imaginal discs, in *Drosophila*, 305-306
  polar coordinate model, of, 303-304
Intercellular interactions
  biochemical studies of, in cyanobacteria, 252-254
Interspecific transplantation
  polar plasm, of, 132
Interstitial cells
  colchicine as eliminator of, from hydra, 272-274
  gamma irradiation as eliminator of, from hydra, 274
  hydra, selective elimination of, 272-282
  hydra without, development of, 267-293
Interstitial cell theory
  hydra, for, 270-271
Ion accumulation
  polar axis fixation and, 84-85

L

Latitude control
  contractile vacuole pore, of *Tetrahymena*, 228-230
  long-range positioning and, 226-230
Left-right discontinuities
  bicaudal mutation of *Drosophila* and, 197
Long-range positioning
  ciliary meridians, of, 232-233
  contractile vacuole pore and, 228-232
  latitude control and, 226-230
  longitude control and, in *Tetrahymena*, 230-233

Long-range positioning (cont'd)
  measuring system and, in *Tetrahymena* 238-242
  relational systems of, in *Tetrahymena*, 225-233
Longitude control
  ciliary meridians, of, 232-233
  contractile vacuole pore, of, 230-232
  long-range positioning and, in *Tetrahymena*, 230-233
  oral primordium, of, 230

M

Maternal effect mutations
  bicaudal, in *Drosophila*, 191-198
  dorsal mutation in *Drosophila*, 200-210
  Drosophila melanogaster embryo spatial coordinates and, 185-211
  gradients and *Drosophila* embryo and, 190-191
  grandchildless, 137-138
  o, of the Mexican axolotl, 167-183
  screening for zygote-defective, 153-155
  temperature sensitive, in *Caenorhabditis elegans*, 149-165
Measuring system
  long-range positioning, in *Tetrahymena*, 238-242
Meinhardt model
  pattern formation, of, 120, 198-200
Mesothoracic tarsus
  normal, bristle pattern of, 319
Metamerization
  antero-posterior decision in embryonic pattern formation, and, 120-122
Mexican axolotl
  see axolotl
Microfilaments
  Fucus cytoplasmic localization and, 94
Microvilli
  polar lobe and, 20
Midgut
  pole cells and, 139
Model
  embryonic pattern formation, for, 118-122, 189-191, 197-200
  polar axis fixation, for, 85-87
Molecular characterization
  anterior determinants of, 110-115
  polar granules, of, 133-137
  RNA synthesis in *Ciona* embryos, of, 47

Mollusc development
  role of polar lobes in, 3-27
Mosaicism
  ascidians and developmental, 30ff
mRNA
  alkaline phosphatase and maternal ascidian, 43-46
  ascidian development and synthesis of, 46-47
  masked, as anterior determinant in *Smittia*, 115-117
  polar lobe and, 23-25
  ribonucleoprotein particles containing, in insect eggs, 117-118
Multisheet vesicles
  polar lobe, large, and, 14-15
Muscle determinant
  ascidian embryo and mitochondrial-associated, 42-43
Mutations
  grandchildless, 137-138
  homoeotic, 122-123, 319-323
  maternal effect, of *Drosophila melanogaster*, 185-211
  o maternal effect, 167-183
  temperature sensitive maternal effect, in *Caenorhabditis elegans*, 149-165

N

Nematodes
  temperature sensitive maternal effect mutants of, 149-165
Nerve cells
  hydra development, role in, 271-272
  hydra without, development of, 267-293
Nuclear activation
  stability of, and nuclear transplantation, 173-177
  stability of, and *o*+ substance, 173-177
Nuclear body
  pole cell-specific, 140-141
Nuclear transplantation
  in ascidian eggs, 42
  nuclear activation stability and, 173-177
  o+ substance and, 173-177
Nucleus
  o+ substance interactions, retention of capacity for, 178

O

o gene
  o+ substance synthesis, 170-171
o maternal effect mutation, 167-183
  biochemical characterization of mutant eggs, 171-173
  cytological characterization of mutant eggs, 171-173
  description of, 168-173
  gastrular arrest, correction of, 169-170
o mutant phenotype
  characterization of correction of, 177
o+ substance
  axolotl and, 167-183
  characteristics of, 170-171
  mode of action of, 178-181
  nucleus interaction with, retention of capacity for, 178
  stability of its nuclear activation, 173-177
Ontogeny
  determinant localization during cleavage, of, 57-60
  Drosophila polar plasm, of, 130-132
Oral primordium
  latitude control of, in *Tetrahymena*, 226-230
  long-range positioning of, 230
  longitude control of, in *Tetrahymena*, 230
  positional control of, in *Tetrahymena*, 226-230
Oral-aboral axis
  ctenophore cleavage and establishment of, 71
Ova-deficient maternal effect mutation, 167-183
  also see *o* maternal effect mutation

P

Pattern control
  hydra, 289-290
Pattern formation
  compartments and, 318ff
  Drosophila, in the tarsus of, 317-323
  Drosophila imaginal discs and, 295-316
  intercellular interactions and, in cyanobacteria, 247-266
  physiology of heterocyst, in cyanobacteria, 249-252
  physiology of spore, in cyanobacteria, 251-252
  polyclones and, 318ff
Pattern regulation
  imaginal disc, *Drosophila*, 298-301
  Paramecium, in, 217
  Stentor, in, 216
Patterning
  cell-surface, 215-246
Photoreversal
  uv induction of double abdomens and, 105ff
Polar axis fixation
  cytochalasin B sensitive process for, 80-82
  in *Fucus*, 80-87
  ion accumulation and, 84-85
  model for, 85-87
Polar coordinate model
  bilaterally symmetrical fields and, 306-308
  imaginal disc, of regulation of, 303-305
Polar deposition
  control of, of a sulfated polysaccharide, 77-96
Polar granules
  Drosophila, properties of, 132-137
  purification of, 133-134
  synthesis of proteins of, 136
Polar lobes
  Bithynia and, 6ff
  centrifugation of large, 15-16
  consequences of removal of, 10-11
  cytoplasm of, 6-16
  cytoplasmic determinants, vegetal body of, 6-11
  cytoplasmic localization mechanisms and cortex of, 16-22
  double-membrane vesicles of large, 12-15
  microvilli and, 20
  mollusc development and, 3-27
  multisheet vesicles and large, 14-15
  RNA and, 22-25
  special structures in small, 6-9
  surface differentiations of, 16-22
  transcription and, 22-25
  translation and, 22-25
  ultrastructure, 3-27
  u,trastructure of large, 11-15
  vegetal body in, 6-11, 23
  vesicular aggregation and, 8
Polar plasm
  autonomy of, and transplantation, 129-130
  interspecific transplantation of, 132
  ontogeny of, 130-132
  properties of, 128-132

Pole cells
    general properties of, 138-140
    Drosophila, of, 127-146
    in double abdomen phenotypes, 99, 101, 117-119, 196
    in vivo and in vitro culture, 141-144
Pole cell-specific nuclear body, 140-141
Polyclones
    pattern formation and, 318
Polysaccharide
    polar deposition of sulfated, 77-96
Positional information
    Drosophila bristle pattern, 318ff
    imaginal disc interactions and universality of, 312-313
Positioning
    contractile vacuole pore, of, 223-225
    duality of systems of, in *Tetrahymena*, 233-235
    latitude control and long-range, 226-230
    long-range, and longitude control, 230-233
    long-range, of ciliary meridians, 232-233
    long-range, of contractile vacuole pore, 230-232
    long-range, relational systems of, 225-233
    measuring system and, 238-242
    short-range, of ciliary meridians, 222-225
    underlying mechanisms of, in *Tetrahymena*, 233-242
Proteins
    temperature sensitive for synthesis, 158-160
Protein composition
    polar granules, of, 134-137

R

Regeneration
    bilaterally symmetrical fragments, interpretations of, 308-312
    bilaterally symmetrical fragments, model of, 311
Rhizoid site
    fucoidin localization at, 87-91
Rhizoid-specific products
    localization of, 87-93
Ribonucleoprotein particles
    mRNA-containing, in insect eggs, 117-118
Ricin
    fucan localization probe, as a, 91-93

RNA
    anterior determinants in *Smittia* and, 112-117
    polar lobe morphogenetic factors and, 22-25
RNase
    anterior determinants and, in *Smittia*, 113-115
R-test
    maternal effect zygote defective mutants, for, 153-155

S

Sagittal plane
    ctenophore and, 71-72
Selector genes, 123, 318ff
Short-range positioning
    ciliary meridians, of, 222-225
Size determinants
    hydra chimera and, 282-284
Smittia
    morphogenetic determinant in, 97-126
Spatial coordinates
    Drosophila melanogaster, of embryo of, 185-211
Spores
    intercellular interactions and formation of, 251-252
    pattern formation physiology and cyanobacteria, 251-252
S-test
    maternal effect zygote defective mutants, for, 153-155
Sulfating sites
    autoradiographic demonstration of, 82-83, 89-91, 93
Sulfation
    fucoidin transport's dependence on, 91-93
Symmetrical fields
    bilaterally, imaginal discs and, 306-308
Symmetry
    cleavage plane and, 68-74
    time of establishment, 68-74

T

Tarsus
    Drosophila, bristle pattern of normal mesothoracic, 319
    effect of homoeotic mutants on pattern of, 319-323

Temperature sensitive mutations
    dorsal mutation, in *Drosophila*, 206
    isolation and characterization of zygote-defective, 152-160
    maternal effect, in *Caenorhabditis elegans*, 149-165
Temperature sensitive zygote defective mutants
    morphology of, 160-163
Temperature sensitivity
    critical times of, 155-160
Temperature shifts
    zygote-defective mutants, effects on, 156-158
Tentacle formation
    hydra chimera and, 285
Tentacular plane
    axis of symmetry in ctenophore and, 71-72
Tetrahymena thermophila
    cell-surface patterning in, 215-246
    description of, 218-221
    relational systems of long-range positioning in, 225-233
Tissue determination
    epithelial hydra and, 280-282
Tissue movement
    epithelial hydra, in, 287-288
Topographical localization
    anterior determinants of *Smittia*, for, 107-108
Transplantation
    polar plasm and its autonomy, 129-130
Tyrosinase
    ascidian embryos and, 38-41, 43, 45

U

Ultrastructure
    polar lobe, large, 11-15
Uv irradiation
    induction of double abdomens, 103ff
    action spectrum for, 110-112
    photoreversal of, 105ff

W

Wound healing
    imaginal disc, in *Drosophila*, 301-303

X

X-irradiation
    interstitial cell theory for hydra and, 271

Y

Yolk nuclei
    pole cells and, 138-139

Z

Zygote
    Fucus, 77-96
Zygote-defective mutants
    isolation and characterization of, 152-160
    morphology of, 160-163
    critical temperature sensitivity times and, 155-160

V

Vegetal body, 6-11
    centrifugation of, 10
    morphogenetic significance of, 10-11
    vesicular aggregation and, 8
Vegetal pole
    bacteria and, 17-18
Vegetative cells
    heterocysts and, interactions between immature, 262-264
    heterocysts and, interactions between mature, 260-261
Vesicular aggregation
    vegetal body and, 7